BIOCHEMISTRY AND MOLECULAR BIOLOGY
IN THE POST GENOMIC ERA

THE FUTURE OF METABOLIC ENGINEERING

BIOCHEMISTRY AND MOLECULAR BIOLOGY IN THE POST GENOMIC ERA

Additional books and e-books in this series can be found on Nova's website under the Series tab.

BIOCHEMISTRY AND MOLECULAR BIOLOGY
IN THE POST GENOMIC ERA

THE FUTURE OF METABOLIC ENGINEERING

ABHISHEK SHARMA
AND
DHRUTI AMIN
EDITORS

nova
science publishers
New York

DOI: https://doi.org/10.52305/VEAH4499

NOTICE TO THE READER

Library of Congress Cataloging-in-Publication Data

ISBN: 978-1-68507-362-6

Published by Nova Science Publishers, Inc. † New York

Contents

FOREWORD

Pragya Tiwari
Department of Biotechnology, Yeungnam University,
Gyeongsan, Gyeongbuk, Republic of Korea

The last two decades have witnessed increasing data on metabolic networks in biological species, projecting a need to elucidate cellular metabolism via scientific interventions. The existing and forthcoming developments in genetic manipulation of biological systems have empowered the field of metabolic engineering to explore/attempt alterations in metabolic pathways and entire pathway reconstitution in biological systems, putting forth the concept of "system metabolic engineering". The ever-expanding toolbox of "metabolic engineering" has been benefited by the strategies to insert/delete or overexpress targeted genes and regulation/modulation of metabolic networks, defining new paradigms in genetic manipulation of biological systems. The field of metabolic engineering has made huge strides in creating "designer organisms" with desired modifications, highlighting prospective avenues in multi-faceted applications, including agriculture, environment, and health care. The advent of metabolic engineering in industrial biotechnology has been a recent emerging concept that aims to boost industrial/large-scale production of value-added metabolites, foodstuffs, and biofuels on a commercial scale. With its initial implementation, the science of metabolic engineering has come a long way and witnessed new developments, while addressing the associated challenges to improve and redirect metabolic flux towards targeted production of the desired metabolite, however, much remains to be addressed for realistic application of these goals. Scientific interventions towards targeted and enhanced production of desired metabolites may be attempted through various strategies: limiting metabolic flux towards undesired by-products and improving precursor supply, promoting the uptake of substrates, pathway reconstitution, and optimization of constituent enzymes and extracellular production of targeted metabolite, respectively. In addition, the increased understanding of the metabolic networks in biological species has aided in altering the copy number of genes and promoter regulation and thereby changes in tuning gene expression, ensuring the optimal supply of co-factors and enhancing the function of pathway enzymes.

The field of metabolic engineering has witnessed significant success in genetic manipulation of microbial/plant systems, towards the production of plant-based vaccines, value-added metabolites of therapeutic/industrial importance, improving trait of biological system/yield of targeted metabolites, environmental applications, biofuels/smart crops in agriculture and a number of diverse applications in present time. As "biosynthetic platforms" for targeted metabolite production, plant systems hold an edge over microbial systems demonstrating an evolved system with organized genetic assembly, gene expression, and categorization of metabolites. New advances in metabolic engineering approaches such as virus-based expression systems, CRISPR-Cas for genome editing, gene-silencing approaches, among others were employed for the production of value-added metabolites in homo/heterologous systems. The capacity to manipulate/re-design biological systems has improved the biosynthetic property of the microbial/plant systems, projecting novel avenues in the era of metabolic engineering. Furthermore, metabolic engineering of algal systems has also been explored with some success, as exemplified by the production of specialized metabolites in *Synechococcus* sp. PCC 7942, carotenoid production in *Chlamydomonas reinhardtii,* and optimization of microalgae bio-refinery in *Phaeodactylum tricornutum* among other key examples, and needs to be further optimized and studied. Metabolic engineering of microbial species has been extensively studied, with distinct species genetically manipulated for the production of value-added metabolites, however, challenges in terms of low metabolic flux, inactivity/low activity of enzymes, and limited information of metabolic pathways in microbial system limits the complete applications. In contrast, plant systems represent a "self-reliant" system for sustainable production of value-added metabolites, with consistent and stable production, as exhibited by key examples on antimicrobial production, edible vaccines, plant defense-related metabolites, CRISPR-Cas mediated multiplex gene editing in *Glycine max,* and others. Metabolic engineering of microbes/plant systems offers an attractive platform for biotechnological interventions towards the production of value-added metabolites.

The book entitled "The Future of Metabolic Engineering" by Nova Publishers, USA (edited by Dr. Abhishek Sharma and Dr. Dhruti Amin, C. G. Bhakta Institute of Biotechnology, Uka Tarsadia University, Surat, Gujarat, India), provides a comprehensive insight into several aspects of metabolic engineering of biological systems and their socio-economic importance. The book will be of particular interest to the scientific community and researchers working in areas of genetic engineering and academicians for understanding various concepts/research in genetic engineering areas, serving as the basis for academics/research in the area of biotechnological interventions.

I acknowledge and extend my best wishes for the successful endeavor.

PREFACE

The last two decades of scientific research have approached the "Metabolic Engineering" of host viz. animal, plant, and bacterial cells to produce the desired products by modifying metabolic pathways at the biochemical and molecular levels. Conceptually, metabolic engineering takes into account the identification of major blocks or control points in a metabolic pathway at the molecular level followed by removal of these limitations with the help of various cellular engineering interventions. Understanding the metabolic pathways requires appropriate experiment setup, molecular biology and biochemistry methods, computational modelling, data analysis, and interpretation to allow the researchers to manipulate them as per their needs.

This book comprises a total of 12 chapters from multiple contributors around the world, including Brazil, Egypt, India, Saudi Arabia, and Turkey. This book provides deep insight into the past, present, and future of metabolic engineering in the animal, microbial, and plant system, communicating interdisciplinary research and relevant results in biochemistry, molecular biology, applied microbiology, cellular physiology, cellular nutrition in health and disease, and biochemical engineering.

Chapter 1 – In this chapter, the authors represent advances in metabolic engineering and CRISPR applications.

Chapter 2, 3 – In these chapters, the authors reported the application of metabolic engineering and nano-therapeutic in cancer biology especially focused on breast cancer cells.

Chapter 4 – deals with invertebrate and vertebrate cell cultures in bio-therapeutics.

Chapter 5 – represents insight into the trends timeline of metabolic engineering in microbes and its reflection on the future scope.

Chapter 6 – provides deep insight into the metabolic engineering approach for optimizing microalgae biorefinery using *Phaeodactylum tricornutum* as a model organism.

Chapter 7 – Authors reported deciphering the metabolic adjustments of engineered plants using GC-MS.

Chapter 8, 9, 10, 11, 12 – These chapters represent the deep insight into recent development and prospects of plant-based vaccines, biosynthesis of secondary metabolites, and regulatory processes of important industrial medicinal and aromatic plants, respectively, using the metabolic engineering approaches in promotion of human health and environment.

This book will help postgraduate students, research scholars, postdoctoral fellows, and teachers from different disciplines interested in fortifying, expanding their knowledge, and knowing the research gaps about metabolic engineering and its application in the microbiology and biotechnology fields.

Abhishek Sharma
Dhruti Amin

In: The Future of Metabolic Engineering ISBN: 978-1-68507-362-6
Editors: A. Sharma and D. Amin

Chapter 1

ADVANCES IN METABOLIC ENGINEERING: APPLICATIONS OF CRISPR AND BEYOND

***Enuh Blaise Manga*[1], *Pınar Aytar Çelik*[1,2,*], *Belma Nural Yaman*[1,3] and *Ahmet Çabuk*[1,4]**

[1]Department of Biotechnology and Biosafety,
Graduate School of Natural and Applied Sciences,
Eskişehir Osmangazi University, Eskişehir, Turkey
[2]Environmental Protection and Control Program,
Eskişehir Osmangazi University, Eskişehir, Turkey
[3]Department of Biomedical Engineering,
Faculty of Engineering and Architecture,
Eskişehir Osmangazi University, Eskişehir, Turkey
[4]Department of Biology, Faculty of Science and Letters,
Eskişehir Osmangazi University, Eskişehir, Turkey

ABSTRACT

Metabolic engineering has rapidly evolved thanks to the continuous development of technologies to sequence and manipulate the genome of

* Corresponding Author's E-mail: pinaraytar@gmail.com.

living organisms. Gene editing with transcription activator-like effector nucleases (TALENS) and zinc finger nucleases (ZFNs) to induce double-stranded breaks (DSBs) was revolutionary but the discovery of the Clustered Regularly Interspaced Short Palindromic Repeats (CRISPR) Cas system in bacteria and the possibility for its reprogramming to target the desired locus in genes offered a more accurate, faster, and cheaper method for editing genomes. Over the years, the technology has been adapted towards various engineering goals by using variants of the Cas protein (CRISPR associated protein) and/or guide RNA (gRNA) to target single and multiple sites simultaneously. Reasons for these wide applications are due to the relative ease to customize the technology for either editing or regulation of gene expression, its ability to generate markerless mutants, and the scalability of CRISPR tools to other non-conventional cell factories. As such, challenges previously encountered in metabolic engineering are increasingly being overcome, bringing forth new hope for the viability of next-generation industrial biotechnology. Though new challenges are still encountered, an increase in the efficiency of computational tools, better machine learning algorithms, and the availability of sequencing data on conventional and non-conventional cell factories with continuous exploration of the microbial dark matter will continue to expand discovery and innovation possibilities using CRISPR technologies. This chapter aims to discuss the contributions of CRISPR tools in metabolic engineering and their future potential.

Keywords: metabolic engineering, CRISPR, gene editing

INTRODUCTION

Microbial metabolic engineering encapsulates cell manipulation approaches aimed at improving yields of bio-based products from microbes in the industry with an overall aim to increase production sustainability and better human lives. Genetic engineering has been a key contributor in this field with various approaches to improve strains for heterologous production of proteins, increasing the range of substrates, development of novel pathways for new products and the degradation of xenobiotics, improving the physiology of cells, and reducing the production of byproducts to increase the yield of desirable products (Nielsen 2001). Other approaches involve the production of high-demand products from simple, widely available, and cheap starting products which are more sustainable compared to the traditional non-renewable and limited products (Keasling 2010). This has been done through pathway rerouting (Hugenholtz 1999), improving excretion from the cell through the engineering of efflux systems for overexpression in the case of intracellular products such as amino acids (Jones et al. 2015), and the application of high-throughput engineering (Rothstein et al. 2020).

Considering previous gene-editing technologies such as ZFNs and TALENS, CRISPR has become the state of the art in genetic engineering due to its many advantages. Components of the CRISPR/Cas system such as the Cas protein, gRNA, donor DNA have been modified in many ways to successfully achieve different engineering goals in microbes. These engineering approaches make use of one or more gene editing, gene activation, and/or interference in conventional and non-conventional cells (Ding et al. 2020). Highly accurate and markerless engineering has been reported with a single base, gene and gene cluster engineering, simplex and multiplex engineering, or static and dynamic self-automated intelligent microbial systems. These tools have helped in the engineering of several previously difficult to engineer species (Peters et al. 2015; Si et al. 2015; Kwon et al. 2020; Ng et al. 2020). Multiplex and cluster editing has played a great role in systems and synthetic biology through providing tools to gain comprehensive knowledge on the physiology and metabolism of cells, the design of cell chassis for next-generation cell factories, and self-regulating intelligent systems (Dai et al. 2015; Kelwick et al. 2020; Liu et al. 2020; Liu et al. 2020).

A search using the keywords CRISPR and metabolic engineering in the Pubmed Central database returned 9431 results this date which shows a continuous increase in CRISPR-based tools developed, publications, and patents over the last few years. Many successes have been reported for species across bacteria and yeasts (Hsu et al. 2014; Sander and Joung 2014; Mougiakos et al. 2018; Phithakrotchanakoon et al. 2018). A search on the Espacenet database with the keywords CRISPR and metabolic engineering showed 688 patent files at this time. However, with the increase in wealth of knowledge and application scope in metabolic engineering also comes new challenges to overcome. The new technologies developed need more data than ever for validation, more computational tools and power for analysis as well as more researchers to understand and apply these technologies. Key technical challenges such as off-target effects through optimal gRNA design, and toxicity are continuously being addressed (Long et al. 2015; Zhao et al. 2020). This chapter will discuss the background of genome editing, the recent applications of CRISPR technologies in metabolic engineering, challenges, and future perspectives.

BACKGROUND: GENE EDITING TALENS, ZFNS, CRISPR

Gene editing is a scientific field that arose many years ago. Since the discovery of the structure of the DNA by Watson and Crick, scientists have been intrigued by its functioning and how it could be used to improve the lives of humans. As DNA polymerization and sequencing technologies evolved alongside technologies for the manipulation of genetic material to study their structures, function, and regulation, more relationships between genotypes and phenotypes began to be revealed. Genes associated with certain phenotypes that were causes of "limitations" in human systems were revealed including those that could as well be improved to reveal desirable phenotypes and functions. This led to the development of gene editing as an approach to modify molecules to improve human lives. Successful gene editing, therefore, required tools that could "do away" with certain genes associated with unwanted phenotypes and "insert" those that bring forth beneficial phenotypes (Adli 2018). Understanding the structural characteristics of DNA and how it interacts with other molecules such as proteins was vital towards developing tools to achieve this aim. Hence, since the discovery that DNA could interact in various ways with proteins, scientists have since researched different ways to form chimeras that can achieve targeted strand breaks on the DNA molecule (Gupta et al. 2014; Adli 2018).

ZFNs and TALENs are the first meganuclease chimera systems that were synthesized and used in genetic engineering. These are artificial structures obtained from fusing nuclease domains from restriction enzymes with DNA recognition domains of other proteins. The first endeavor to do this was the combination of ZFNs domain proteins with nucleases to cut DNA. The effectiveness was quite good but limitations such as complexity of design arouse which limited its application. Though their DNA binding specificities could be modified to recognize different DNA sequences, the limitation of this technology was the difficulty and complexity involved in the building of the meganucleases (Sander et al. 2014; Adli 2018). With the discovery of more nucleases and DNA binding protein domains, the TALENs were developed with much more specificity and ease but customization and editing efficiency were still quite low.

Not until 2012, the CRISPR system of bacteria which had been revealed as a bacterial defense mechanism against bacteriophages was shown to be reprogrammable for the targeted editing of genetic material. The components of this system were Cas9 (spCas9) protein which has a nuclease function, a gRNA composed of a CRISPR RNA (crRNA), and a scaffold RNA (tracrRNA) from *Streptococcus pyogenes* (Jinek et al. 2012). The crRNA is synthesized from foreign DNA previously incorporated in the bacteria CRISPR arrays. It also bears a protospacer region that can recognize its complementary sequence in the same source when there is a new encounter with the cell. To achieve editing, a gRNA/Cas9 complex is formed that moves to and binds on the recognition site causing double-stranded breaks which can be repaired either through a homology-directed repair pathway (HDR) or non-homologous end-joining pathways (NHEJ) (Figure 1). The NHEJ pathway is absent or weak in most prokaryotes, therefore, introducing a challenge in repairing double-strand breaks and performing clean edits. However, precise edits can be achieved with the HDR pathway through mutation, insertion of a single or double-stranded DNA donor (Sander et al. 2014). By convention, this forms the basis of the type II CRISPR system according to classification. This is further classified into 6 types (type I-VI) and 33 subtypes (Mougiakos et al. 2016; Makarova et al. 2020). A classification of Cas proteins and their mechanisms of action is discussed elsewhere (Mougiakos et al. 2016). The success of CRISPR tools is due to the relative ease of use, low cost, and easy customization. This has led to a drastic increase in various CRISPR-based tools applicable in many scientific areas to achieve a wide range of research goals.

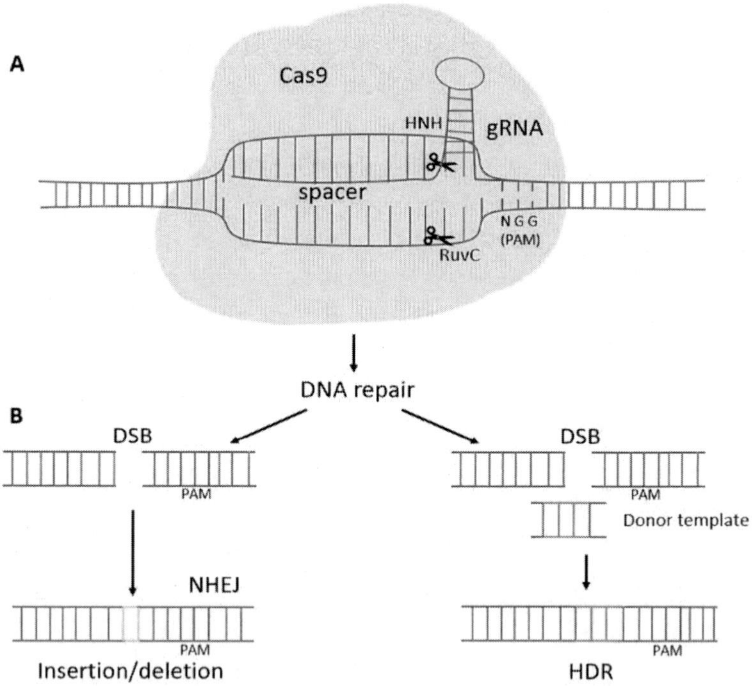

Figure 1. Mechanism of DSB by the CRISPR/Cas system. A) Structure of the Cas9 protein with the HNH and RuvC domains and the cutting site near the NGG PAM recognition sequence. B) HDR and NHEJ mediated DSB repair mechanisms (Lee et al. 2020).

CRISPR tools are highly efficient in achieving markerless single and multiple edits in a genome simultaneously. Through a detailed understanding of the mechanism and structures in this system, researchers have been able to design many CRISPR-based approaches applicable in various areas in metabolic engineering. This has helped in the development of frameworks for rational engineering through the improvement of the catalytic function of Cas endonucleases, gRNA specificity, PAM (protospacer adjacent motif) requirements, and reducing off-target effects (Jiang et al. 2017). This is also aided by the application of deep learning convolutional neural network methodologies to predict gRNA activity in prokaryotic systems (Wang et al. 2019). The development of a CRISPR technology requires certain considerations such as choosing a proper architecture for the gRNA using a good platform for design, choosing suitable target sites and range for gRNA, appropriate method of delivery of the components, and controlling off-target effects (Sander et al. 2014). This information can be easily obtained today on available platforms with semi-automated functionality and easy user interfaces. The mitigation of off-target effects is a key challenge that is usually of high interest to researchers.

CRISPR technologies are now an important component for the development of tools for next-generation metabolic engineering. These approaches are geared towards making industrial bioprocesses that produce biobased chemicals that are more competitive. This includes metabolic engineering for the use of low-cost substrates, designing energy-saving processes to increase commercial competitiveness against synthetic products (Yu et al. 2019). General approaches in metabolic engineering that have leveraged CRISPR technologies include gene and genome editing, and gene expression regulation. These have been applied in strain development, degradation of xenobiotics, improvement of microbial physiology, increasing microbial product yields, etc. (Yu et al. 2019).

CRISPR/Cas System Development and Delivery

The design of the CRISPR components is generally done in silico (Mougiakos et al. 2016). This can be done on publicly available platforms or outsourced to third-party companies. It is however important to note during design that prediction of tracrRNA, necessary for the functioning of the type II CRISPR system is usually challenging because of their differences in structures and non-canonical base pairs formed due to its interactions with the CRISPR RNA, Cas9, or itself. Due to its central role, getting its structure right is therefore important to design an effective tool. In this light, computational models to predict tracrRNA more accurately are continuously being developed (Chyou and Brown 2019). It could also take a series of iterative wet-lab experiments to try different combinations to obtain what works best.

After the components have been designed, synthesized, and assembled, vehicles such as replicating plasmids (Jinek et al. 2012; Ungerer and Pakrasi 2016) and suicide/integrating plasmids (Wang et al. 2019) can be used to introduce and maintain the CRISPR system in the host cell. Possible methods that could be exploited for the delivery of vehicles include exploring natural mechanisms of the microbes such as competence, conjugation, and through physical manipulation by electroporation or chemical induction. To facilitate CRISPR component uptake, host cell components such as restriction enzymes, native CRISPR Cas systems, and integration (through homologous recombineering and using transposons) which constitute its defense mechanism against foreign DNA must be considered (Freed et al. 2018).

Using CRISPR for Gene Editing

Strand breaks in the genome are caused by the activity of the nuclease domain of the Cas protein. For gene-editing experiments, the commonly used nuclease is spCas9 because it is well characterized and requires just a gRNA (Jinek et al. 2012; Freed et al. 2018), and because it works in a broad spectrum of hosts (Freed et al., 2018). Other systems using Cpf1 (Ungerer and Pakrasi 2016), native Cas proteins (used to avoid heterologous Cas protein toxicity) (Zhang et al. 2018), CasX and CasY (with highly compact systems uncovered from uncultivated microbes through metagenomics) (Burstein et al. 2017) with potential in metabolic engineering are still to be fully characterized (Freed et al. 2018). The discovery and characterization of new nucleases will continue revealing different structures and mechanisms of action which may lead to the discovery of new methods to achieve gene editing more efficiently.

Strain development has become very useful in improving the substrate range of microorganisms, increasing the production yield, or using heterologous hosts for the production of high-value bioproducts. Genome engineering was traditionally done through homologous recombination of genetic material but was limited. With the development of CRISPR, a wide range of options is now possible from genome-wide editing approaches to single base editing with relatively high accuracy and precision (Song et al. 2015). Despite slow adoption in bacteria systems as compared to eukaryotic systems, applications in strain development involving gene editing and gene expression regulation continue to be improved. CRISPR-based techniques have promising industrial applications in microbial strain typing and tracking, virome identification, the development of bacteria vaccines, and advanced antimicrobials. Patent applications and acquisition has also increased significantly over the years (Choi and Lee 2016). Most CRISPR applications make use of heterotrophs but have also been successfully adapted for autotrophs such as cyanobacteria (Carroll et al. 2018).

Using CRISPR/Cas for Gene Regulation

Gene regulation in cells is primarily associated with the activity of activator and repressor proteins on RNA polymerase's interactions with DNA that affects its ability to transcribe it. Gene repression can be achieved when a repressor blocks the binding of RNA polymerase on a promoter or blocks the movement of RNA polymerase on a DNA strand to be transcribed. On the other hand, gene activation occurs through the inhibition of a native repressor or by introducing a stronger promoter that enhances the recruitment of RNA polymerase.

The HNH and RuvC domains of the Cas9 protein carry out the catalytic function of the protein. Each one of them cleaves a strand resulting in DSBs (DNA double-strand breaks) near the PAM sequence of the target (Figure 1). Mutations can be induced in the sequences coding for these domains to modify their catalytic activity (Adli 2018). A modification of the Cas9 nuclease protein to catalytically inactive dCas9 (nuclease-dead Cas9) results in a complete loss of cleavage activity. When combined with gRNA, it can still be directed to a target site but it only interferes with initiation and elongation during DNA transcription. They can act as repressors or activators of DNA transcription with sequence-specific activity. These mutant versions have found applications in strain development and metabolic engineering with some successes despite its recent emergence (Schultenkämper et al. 2020). In CRISPR interference (CRISPRi), transcription is halted through the blockage of RNA polymerase while for CRISPR activation (CRISPRa) dCas9 binds upstream on promoters recruiting further binding activation components thereby activating transcription (Mougiakos et al. 2016; Yao et al. 2018). These mechanisms are summarized in Figure 2. dCas9-mediated gene regulation is a more effective approach than traditionally used RNA interference (RNAi) and small RNAs which were relatively fewer tools and often created only binary changes at the level of gene expression. In addition, issues related to an overload of messenger RNA observed due to small RNA use are resolved alongside the benefits of more concise mechanisms (Wu et al. 2020).

CRISPRa is promising for strain optimization in metabolic engineering demonstrated transcription activation. Protein subunits from RNA polymerase can be fused with dCas9 to obtain activator-like effects on promoters of bacteria. However, activation activity is highly dependent on the strength of the promoter it binds to (Bikard et al. 2013). Challenges such as sequence limitations, below optimal PAM site positions, and inherent promoter characteristics still need to be overcome to improve the efficiency of this method. Hopefully, predictive models can help researchers identify the most effective target sites and genes prone to activation to overcome these challenges (Fontana et al. 2020).

On the other hand, CRISPRi has also been used for various engineering goals. Applications of the modified endonuclease protein have inspired diverse metabolic engineering approaches to produce biobased products (Mougiakos et al. 2018). The general idea is usually to completely silence competing pathways or redirect the flow of precursors towards the synthesis of a product of interest. For example, metabolomics data was obtained and explored to identify bottlenecks in the pathway for the production of L-homoserine in *Escherichia coli* W3110. After this, systematic metabolic engineering and analysis using CRISPRi were carried out leading to an increase in yields up to 37.57 g/l. Using this approach, rational design strategies in other microbial cell factories can be developed for the production of high-value products (Liu et al. 2020). Other applications of pathway engineering to obtain higher yields include 5- Aminolevulinic acid (5-ALA) in *E. coli* from structurally unrelated carbon sources (Miscevic et al. 2020), repression in *Bacillus licheniformis* through simultaneous multiple gene repressions with high efficiency of up to 94.00% to produce L-valine (Zhan et al. 2020) and *Corynebacterium glutamicum* to enhance the production of lysine in a multiplex fashion (Park et al. 2018).

Even for traditionally difficult to engineer species like cyanobacteria, other Cas proteins such as Cas12 have been used to obtain clean point mutations through gene insertions and deletions with high efficiencies (Behler et al. 2018). More specifically, CRISPR/Cpf1 has been used to generate markerless point mutations in model cyanobacteria *Synechococcus*, *Synechocystis*, and *Anabaena* (Ungerer et al. 2016). Cyanobacteria are autotrophic and unlike their heterotrophic counterparts, had not been extensively exploited for metabolic engineering because the tools often used were cumbersome.

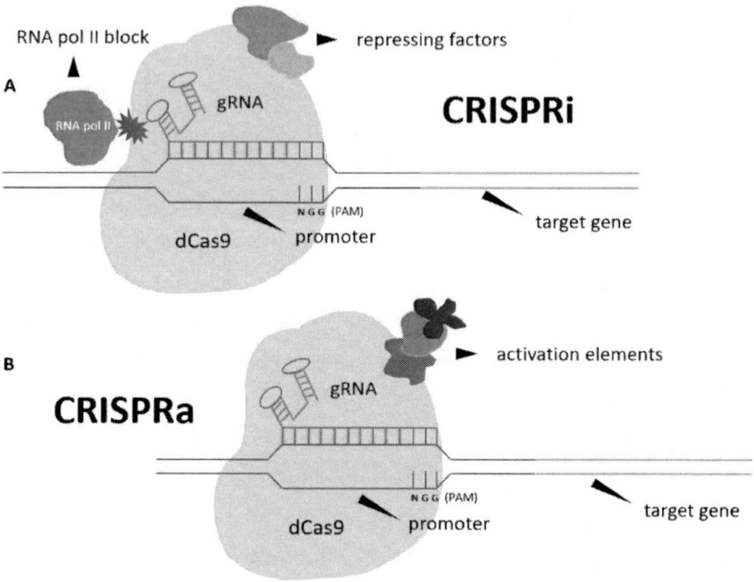

Figure 2. A) The CRISPR/Cas system mechanism by CRISPRi. dCas9 acquires repression factors and blocks RNA polymerase. B) CRISPRa showing activation by recruiting activation elements. The cleaving domains of Cas9 beside the PAM recognition site are inactive (Lee et al. 2020).

From a more general perspective, CRISPR technologies have had applications beyond static metabolic engineering approaches into more advanced dynamic approaches. In static metabolic engineering, the intrinsic metabolism of the mutant is allowed to account for changes made through gene editing. This, however, does have limitations such as possible metabolic imbalance, accumulation of intermediates, and growth disruption. The more advanced autonomous regulation-based cell factories seem to provide a fix to this issue by implementing control switches (Wu et al. 2020). Due to the ever-changing conditions that exist in bioreactors during fermentations, autonomously regulated gene expression in cell factories could become very useful. Autonomous dual control systems using CRISPRi and biosensor metabolites have been reported and demonstrated to increase nutraceutical N-acetylglucosamine (GlcNAc) production in *Bacillus subtilis* up to 131.6 g/l (Wu et al. 2020).

Other previously reported systems include temperature and light-sensitive mutants (Richter et al. 2016), quorum sensing-based systems in *E. coli* with multiple control points and modules that can be independently tuned with predictable capabilities (Dinh et al. 2019), and large genetic circuits engineered using dCas9 (Nielsen et al. 2014; Gander et al. 2017; Zhang et al. 2018). Challenges in designing circuits such as the leakage of components in transcription and absence of digital response, arbitrary interconnectedness and other general variations that are organism dependent are still to be overcome (Gander et al. 2017). There is also a need to research and develop more controllers to facilitate automation and tuning of pathways in cell factories. The research and development of cell factories that are sensitive to other key conditions in bioreactors could lead to the development of more responsive and efficient producer cells.

Anti-CRISPR

As CRISPR is the immune system of bacteria, phages have also simultaneously developed methods to evade the destructive effects of the Cas9 nuclease. The system is countered with small proteins of 50-150 amino acids called anti-CRISPR proteins (Acrs) with no similarities to known proteins. They prevent viral DNA destruction through many mechanisms like interfering with recruitment and DNA binding of Cas, mimicking some inhibitors and occupying PAM sites, binding to catalytic domains of Cas9, etc. (Yao et al. 2018). These anti-CRISPR mechanisms are numerous and are still being uncovered. Their applications in metabolic engineering center around mitigating the off-target effects of gRNA (Yao et al. 2018).

Multiple gene mutations are often required to understand and characterize pathways for high-value products in cell factories. The generation of markerless and clean mutants is often challenging even in model organisms due to the lack of efficient genome editing tools. In some of such cases, anti-CRISPR proteins can play a role in the control and regulation of Cas nuclease expression. This was applied by Wasels and colleagues where CRISPR/Cas9 system was improved with a gene encoding an anti CRISPR protein. They succeeded to obtain clean deletions of *bdhA*, *bdhB*, and *bdhC* which are butanol dehydrogenase genes to impair solvent production in *Clostridium acetobutylicum* to understand the role of alcohol dehydrogenase in solventogenesis. It showed fast, clean, accurate, and iterative modifications with potential for applicability in other bacterial species (Wasels et al. 2020). So far, Acrs are used to control the expression of CRISPR nucleases and reduce off-target effects which are important for efficiency and specificity in genetic engineering. It is expected that as more Acr proteins are discovered and studied novel mechanisms and applications will be revealed that can improve gene editing in metabolic engineering (Liu et al. 2020). As a complement, since Acrs have previously "unseen" structures, machine learning algorithms to predict the structures of new Acrs based on sequence information only are already being developed including web servers to rapidly test and validate new Acrs (Eitzinger et al. 2020). This will contribute to rapidly speed up the development of efficient gene editing tools for metabolic engineering.

Synthetic Morphology

Another approach for improving the efficiency of microbial cell factories is the optimization of cell morphology (programming morphogenesis) and the spatial organization of enzymes. Coined in 2008, the expression was explained to imply that cells could be reprogrammed to take specific desired structural and tissue arrangements (Gramelsberger 2020). Though the structural traits of bacteria are highly evolutionarily conserved and regulated by complex systems, they still offer an opportunity for exploitation to design new cell factory types through metabolic engineering. Imparting bacteria with novel spatial properties hold great promise for metabolic engineering in industrial biotechnology. The applications are not just limited to single cells, but they can be further extended to multicellular systems as well (Volke and Nikel 2018). The outcome of these approaches is the optimization of growth rates, increase cell volumes, and a simplification of the downstream processes. This becomes very useful for the enhancement of synthetic capacity in the production of intracellular products such as polyhydroxyalkanoates (PHAs) (Huo et al. 2020) and enhancing cell to cell communications through engineering multi-strain interaction patterns and population interactions such as in biofilms (Ding et al. 2020). CRISPR is currently suitable for the engineering of larger genomes and many genes simultaneously, therefore it offers the possibility to develop cell factories based on synthetic morphology engineering with relative ease compared to other methods.

Volke and Nikel, using *E. coli* and *Halomonas bluephagenesis* successfully engineered mutants using CRISPR/Cas9 to accumulate more polyhydroxybutyrate granules than wild types leading to the generation of more mutants with similar morphological changes (Volke and Nikel 2018). Morphology engineering can also require direct editing on genes related to cell morphology, cytoskeleton, membrane-bound, and membrane-associated proteins. In *E. coli* for example, CRISPRi was used to repress *ftsZ* and *mreB* essential genes in morphology which led to longer and larger cells that increased the accumulations of polyhydroxybutyrate (PHB) from 45% to 68%. The trade-off however was a reduction in the overall cell growth (Elhadi et al. 2016). In another study, CRISPRi was also used to repress the expression of *ftsZ*, *prpC*, and *gltA* genes to increase PHB accumulation in *Halomonas* sp. TD01 through increasing cell length and redirecting substrates towards poly(3-hydroxybutyrate-co-3-hydroxyvalerate) (PHBV) and PHB synthesis respectively (Tao et al. 2017).

Despite the increase in cell size and metabolite production, it is still not yet clearly established how other physiological functions such as cell division, growth rates, and growth in fermentation broth are affected. Research on the applications of more structural genes, the effects of these approaches on other systems, and the application of proven approaches on other species is recommended to understand the implications in the development of efficient cell factories.

The Use of Conventional and Non-Conventional Cell Factories

Before the discovery of CRISPR tools for metabolic engineering, the available tools for genome engineering were quite cumbersome and difficult to use limiting their application to a few commonly used bacteria cell factories such as *E. coli* and *Saccharomyces cerevisiae*. This limited the use of microorganisms that have natural abilities to produce high-value products, resist industrial conditions, or have useful metabolic pathways. For example, methylotrophic yeasts with high protein secretion and glycosylation are useful for the production of proteins while thermotolerant yeasts could be engineered for efficient production in high-temperature processes (Cai et al. 2019). Another option could be to engineer these phenotypes into common cell factories. With other tools, however, this was more difficult added to the complexity involved in the engineering of *S. cerevisiae* to include desired phenotypes such as thermotolerance or to increase the range of usable substrates. Therefore, engineering hosts with these natural characteristics to increase product yield, titers, or rates of production tend to be a more efficient option (Löbs et al. 2017).

A CRISPR tool to engineer the methylotrophic yeast *Pichia pastoris* known to have a relatively poor intrinsic recombinant machinery compared to *S. cerevisiae* and lacked tools for gene disruption and cassette integration was developed. This was achieved through optimizing CRISPR/Cas9 expression showing up to 100% targeting rates and multiplexing capacity (Weninger et al. 2016). Also, the limitation of targeted gene mutagenesis in the thermophilic yeast *Ogataea thermomethanolica* TBRC656 was overcome by the CRISPR/Cas9 plasmid system for integrative and episomal expression with up to 97% rate of disruption in some genes (Phithakrotchanakoon et al. 2018). With the continuous exploration of these new organisms, future CRISPR tools will heavily leverage emerging insights from increasing sequencing data which will increase our understanding of their metabolism. Tools such as Genome-wide CRISPR/Cas9 loss of function screens for functional genomics are expected to greatly help improve engineering research in non-conventional yeasts (Löbs et al. 2017). Other genera such as *Acetobacterium, Clostridium, and Glucunobacterium*, etc. are natural producers of high amounts of acetic acid using a variety of substrates. Therefore, pathway engineering in these non-conventional factories could lead to the production of cost-effective amounts of acetic acid (Li et al. 2020).

CRISPR/Cas tools have been reported to have a toxic unwanted effect in some organisms which can limit their applications in a wide range of species. As a solution to this, approaches to make use of endogenous systems are researched. This was the case when the CRISPR/Cas system of *Clostridium tyrobutyricum* was used to engineer the same host for butanol production by redirecting carbon flow to butanol production after replacing the *cat1* gene with *adhE1/adhE2* obtaining 26.6 g/l of the product. Highly efficient single and multiplex editing was obtained with efficiencies of up to 100% (Zhang et al. 2018). The versatility of this tool holds great promise for novel cell factory development with increasingly exciting approaches being tested with impressive results.

Microbial cells vary in their ability to efficiently deliver in industrial processes. Cells termed conventional cell factories are suitable for industrial biotechnology because they have been extensively studied and thrive well in industrial conditions. With the CRISPR technology, many tools have been developed that make possible the engineering of unconventional cells that have desired properties to make them more industrially suitable (Raschmanová et al. 2018).

Using Microbes as Chassis for Synthetic Biology

Rational design and construction of robust bacterial chassis for metabolic engineering present key examples of bacterial species that have secured a place in industrial bioproduction. The emergence of novel bacterial chassis is also considered in the light of the unique properties of their physiology and metabolism, and the practical applications in which they are expected to outperform other microbial platforms (Calero et al. 2019). However, scientific capabilities have advanced from simple gene mutations to the insertion of rationally designed, complex synthetic circuits and the creation of entirely synthetic genomes. It is time to develop the next generation of synthetic biology chassis; robust organisms that can provide the advanced physiology novel synthetic circuits will require to move synthetic biology from the laboratory into fieldable technologies. This can be accomplished by developing chassis-specific genetic toolkits that are as extensive as those for *E. coli*. However, the holy grail of synthetic biology would be the development of a universal toolkit that can be ported into any chassis. Toolboxes include gene expression regulation and genome-wide editing which have been developed due to improved understanding of cell processes (Liu et al. 2019). CRISPR/Cas technologies have enabled the development of many other tools such as plasmid-based tools for the so-called "build-your-own" cell factories (Nora et al. 2019). Key model organisms are *E. coli*, *B. subtilis*, *Corynebacterium glutamicum*, and *S. cerevisiae* which are well-studied hosts for synthetic biology (Liu et al. 2020).

Synthetic biology leverages the power of CRISPR for genome-wide editing or downregulation of several genes simultaneously that can affect several pathways at the same time. The result is structural changes such as enlarged cells for storage of more products, improvement in flux towards the desired product, and improvement in titer during downstream purification (Zhang et al. 2020).

CRISPR Enabled Trackable Genome Engineering (CREATE)

This is also another rapid strategy to develop variants for the production of high-value products in a fast way. Hundreds and even thousands of mutations can be tested at the same time at various steps of complex metabolic networks followed by deep screening for mutants producing the highest titers (Liang et al. 2017; Bassalo et al. 2018). Studies that have shown success with this concept include the investigation of 16,000 mutations in the lysine metabolism in *E. coli* which led to the identification of multiple routes affecting the synthesis of Lysine (Bassalo et al., 2018) and increasing the titers of isopropanol in *E. coli* through a design-build-test-learn cycle with 1000 strains (Liang et al. 2017). This clearly shows that this methodology speeds up traditional engineering workflows and can lead to a better understanding of complex metabolic pathways as well as the discovery of new ones.

Multiplexed Automated Genome Engineering (MAGE)

Conventional genome editing usually requires time to generate many variants and subsequently screen those with the desired mutation in the unedited background. This represents a bottleneck in the engineering of efficient mutants to be used as cell factories. With the precise editing ability of the CRISPR system and higher efficiencies in the engineering of multiple genes in a single round, CRISPR tools have improved the efficiency of these methods helping to broaden the scope of genome manipulations beyond many bacterial hosts using MAGE. The diversity of genetic mutants obtained by mismatches, insertions, or deletions can be rapidly and continuously obtained (Wang et al. 2009; Bao et al. 2016). This will enable researchers to test multiple hypotheses at the same time.

Success with MAGE is currently limited to cells with well-known physiologies such as *E. coli.* As more cells are being characterized, the extension of this technology to other cell factories is expected to lead to more efficient and cheaper production of many more high-value products

(Bao et al. 2016). In bacteria, genome-scale insertion of T7 promoters with co-selection MAGE (CoS-MAGE) for the optimization of aromatic amino acids biosynthesis has been demonstrated (Wang et al. 2012). Due to difference in requirements for each cell, a similar approach to MAGE applied in yeasts known as yeast oligo-mediated genome engineering (YOGE) provides better efficiency suitable for yeast cell engineering though functions at lower efficiency than its equivalent in bacteria (Horwitz et al. 2015; Jakočiūnas et al. 2015). Future approaches such as synthetic sequence pre-placing, with the benefits of an increase in computational power, high-throughput sequencing, and automation will bring game-changing innovations that will improve strain development (Malcı et al. 2020).

Engineering Biosynthetic Gene Clusters

Some beneficial natural products in microbes are coded by genes in clusters named biosynthetic gene clusters. Biosynthetic gene clusters can be recovered from microbial (meta) genomes from extensive sequencing and bioinformatic data (Li et al. 2017). Many of these clusters are not expressed in their hosts when cultured in laboratory conditions as a result, using heterologous expression in well-characterized hosts could provide an avenue for the discovery of the products that they code for (Lin et al. 2020). Large gene clusters are often required in metabolic engineering but synthesizing these genes is often expensive and difficult even with polymerase chain reaction and restriction enzyme methods. These gene clusters can be recombineered in heterologous systems through knockout and integration for the production of high-value products with a clean and markerless CRISPR-based method (Choi et al. 2018). With CRISPR/Cas9 technology, it was possible to obtain clones from targeted clusters of up to 100kb in a single step (Jiang et al. 2015). More recently, ExoCET (Exonuclease Combined with RecET recombination) was developed to clone a 106kb gene cluster for salinomycin into operational plasmids which can be transformed into host cells. In another study, gene positioning into expression vectors with nucleotide precision was achieved with Cas9 nucleases without further subcloning (Wang et al. 2018).

Conclusion

The CRISPR/Cas system is highly effective in genome manipulations and scalable to other engineering methods and hosts. Key issues such as decreasing off-targets through effective gRNA design, limited ability to recognize new PAM sequences need to be continuously addressed. Controlling the expression of the CRISPR nucleases in cell hosts can be further researched as a means to reduce off-target effects during engineering and developing more algorithms to increase the predictive power of gRNA off-target effects. An increase in the development of Next-generation sequencing platforms and other high throughput genome sequencing and mining technologies in combination with developments in machine learning and synthetic biology is expected to improve the design-build-test-learn cycle for the development of more efficient microbial cell factories of the future. The growing applications that range from static engineering through dynamic engineering, systems, and synthetic biology, will increasingly give researchers more efficient tools with increased versatility in hypothesis testing for effective rational strain design. The cost of engineering and industrial-scale production is also expected to keep reducing such that bioproducts can become increasingly competitive on the market and promote transitions to more sustainable bioeconomies.

References

Adli, Mazhar. (2018). "The CRISPR Tool Kit for Genome Editing and Beyond." *Nature Communications*, *9*, 1911, DOI: 10.1038/s41467-018-04252-2.

Bao, Zehua, Cobb, Ryan E. & Zhao, Huimin. (2016). "Accelerated Genome Engineering through Multiplexing." *WIREs Systems Biology and Medicine*, *8*, 5–21. DOI: 10.1002/wsbm.1319.

Bassalo, Marcelo C., Garst, Andrew D., Choudhury, Alaksh, Grau, William C., Oh, Eun J., Spindler, Eileen, Lipscomb, Tanya. & Gill, Ryan T. (2018). "Deep Scanning Lysine Metabolism in *Escherichia coli*." *Molecular Systems Biology*, *14*(11), e8371. DOI: 10.15252/msb.20188371.

Behler, Juliane, Vijay, Dhanya, Hess, Wolfgang R. & Akhtar, M. Kalim. (2018). "CRISPR-Based Technologies for Metabolic Engineering in Cyanobacteria." *Trends in Biotechnology*, *36*(10), 996–1010. DOI: 10.1016/j.tibtech.2018.05.011.

Bikard, David, Jiang, Wenyan, Samai, Poulami, Hochschild, Ann, Zhang, Feng. & Marraffini, Luciano A. (2013). "Programmable Repression and Activation of Bacterial Gene Expression Using an Engineered CRISPR-Cas System." *Nucleic Acids Research*, *41*(15), 7429–37. DOI: 10.1093/nar/gkt520.

Cai, Peng, Gao, Jiaoqi. & Zhou, Yongjin. (2019). "CRISPR-Mediated Genome Editing in Non-Conventional Yeasts for Biotechnological Applications." *Microbial Cell Factories*, *18*, 63. DOI: 10.1186/s12934-019-1112-2.

Calero, Patricia. & Nikel, Pablo I. (2019). "Chasing Bacterial Chassis for Metabolic Engineering: A Perspective Review from Classical to Non-Traditional Microorganisms." *Microbial Biotechnology*, *12*(1), 98–124. DOI: 10.1111/1751-7915.13292.

Choi, Kyeong Rok, Cho, Jae Sung, Cho, In Jin, Park, Dahyeon. & Lee, Sang Yup Lee. (2018). "Markerless Gene Knockout and Integration to Express Heterologous Biosynthetic Gene Clusters in *Pseudomonas putida*." *Metabolic Engineering*, *47*, 463–74. DOI: 10.1016/j.ymben.2018.05.003.

Chu, Luan Long, Montecillo, Jake Adolf V. & Bae, Hanhong. (2020). "Recent Advances in the Metabolic Engineering of Yeasts for Ginsenoside Biosynthesis." *Frontiers in Bioengineering and Biotechnology*, *8*, 139. DOI: 10.3389/fbioe.2020.00139.

Dai, Zongjie. & Nielsen, Jens. (2015). "Advancing Metabolic Engineering through Systems Biology of Industrial Microorganisms." *Current Opinion in Biotechnology*, *36*, 8–15. DOI: 10.1016/j.copbio.2015.08.006.

Ding, Qiang, Diao, Wenwen, Gao, Cong, Chen, Xiulai. & Liu, Liming. (2020). "Microbial Cell Engineering to Improve Cellular Synthetic Capacity." *Biotechnology Advances*, *45*, 107649. DOI: 10.1016/j.biotechadv.2020.107649.

Ding, Wentao, Zhang, Yang. & Shi, Shuobo. (2020). "Development and Application of CRISPR/Cas in Microbial Biotechnology." *Frontiers in Bioengineering and Biotechnology*, *8*, 711. DOI: 10.3389/fbioe.2020.00711.

Dinh, Christina V. & Prather, Kristala L. J. (2019). "Development of an Autonomous and Bifunctional Quorum-Sensing Circuit for Metabolic Flux Control in Engineered *Escherichia coli*." *Proceedings of the National Academy of Sciences of the United States of America*, *116* (51), 25562-25568. DOI: 10.1073/pnas.1911144116.

Eitzinger, Simon, Asif, Amina, Watters, Kyle. E., Iavarone, Anthony T., Knott, Gavin J., Doudna, Jennifer A. & Minhas, Fayyaz ul Amir Afsar. (2020). "Machine Learning Predicts New Anti-CRISPR Proteins." *Nucleic Acids Research*, *48*(9), 4698–4708. DOI: 10.1093/nar/gkaa219.

Elhadi, Dina, Lv, Li, Jiang, Xiao-Ran, Wu, Hong. & Chen, Guo-Qiang. (2016). "CRISPRi Engineering *E. coli* for Morphology Diversification." *Metabolic Engineering*, *38*, 358–69. DOI: 10.1016/j.ymben.2016.09.001.

Fontana, Jason, Sparkman-Yager, David, Zalatan, Jesse G. & Carothers, James M. (2020). "Challenges and Opportunities with CRISPR Activation in Bacteria for Data-Driven Metabolic Engineering." *Current Opinion in Biotechnology*, *64*, 190–98, DOI: 10.1016/j.copbio.2020.04.005.

Gander, Miles W., Vrana, Justin D., Voje, William E., Carothers, James M. & Klavins, Eric. (2017). "Digital Logic Circuits in Yeast with

CRISPR-DCas9 NOR Gates." *Nature Communications*, *8*(1), 15459. DOI: 10.1038/ncomms15459.

Gramelsberger, Gabriele. (2020). "Synthetic Morphology: A Vision of Engineering Biological Form." *Journal of the History of Biology*, *53*, 295-309. DOI: 10.1007/s10739-020-09601-w.

Gupta, Rajat M. & Musunuru, Kiran. (2014). "Expanding the Genetic Editing Tool Kit: ZFNs, TALENs, and CRISPR-Cas9." *The Journal of Clinical Investigation*, *124*(10), 4154–61. DOI: 10.1172/JCI72992.

Horwitz, Andrew A., Walter, Jessica M., Schubert, Max G., Kung, Stephanie H., Hawkins, Kristy, Platt, Darren M., Hernday, Aaron D., et al. (2015). "Efficient Multiplexed Integration of Synergistic Alleles and Metabolic Pathways in Yeasts via CRISPR-Cas." *Cell Systems*, *1*(1), 88–96. DOI: 10.1016/j.cels.2015.02.001.

Hugenholtz, Jeroen. & Kleerebezem, Michiel. (1999). "Metabolic Engineering of Lactic Acid Bacteria: Overview of the Approaches and Results of Pathway Rerouting Involved in Food Fermentations." *Current Opinion in Biotechnology*, *10*(5), 492–97. DOI: 10.1016/S0958-1669(99)00016-6.

Huo, Kaiyue, Zhao, Fengjie, Zhang, Fang, Liu, Ruihua & Yang, Chao. (2020). "Morphology Engineering: A New Strategy to Construct Microbial Cell Factories." *World Journal of Microbiology and Biotechnology*, *36*, 127. DOI: 10.1007/s11274-020-02903-5.

Jakočiūnas, Tadas, Bonde, Ida, Herrgård, Markus, Harrison, Scott J., Kristensen, Mette, Pedersen, Lasse E., Jensen, Michael K. & Keasling, Jay D. (2015). "Multiplex Metabolic Pathway Engineering Using CRISPR/Cas9 in *Saccharomyces cerevisiae*." *Metabolic Engineering*, *28*, 213–22. DOI: 10.1016/j.ymben.2015.01.008.

Jiang, Wenjun, Zhao, Xuejin, Gabrieli, Tslil, Lou, Chunbo, Ebenstein, Yuval. & Zhu, Ting F. Cas9-Assisted Targeting of Chromosome Segments CATCH Enables One-Step Targeted Cloning of Large Gene Clusters, *Nature Communications*, *6*, 8101. DOI: 10.1038/ncomms9101.

Jiang, Fuguo. & Doudna, Jennifer A. (2017). "CRISPR-Cas9 Structures and Mechanisms." *Annual Review of Biophysics*, *46*, 505–29, 2017. DOI: 10.1146/annurev-biophys-062215-010822.

Jinek, Martin, Chylinski, Krzysztof, Fonfara, Ines, Hauer, Michael, Doudna, Jennifer A. & Charpentier, Emmanuelle. (2012). "A Programmable Dual-RNA-Guided DNA Endonuclease in Adaptive Bacterial Immunity." *Science*, *337*(6096), 816–21. DOI: 10.1126/science.1225829.

Jones, Christpher M., Hernández Lozada, Nestor J. & Pfleger, Brian F. (2015). "Efflux Systems in Bacteria and Their Metabolic Engineering Applications." *Applied Microbiology and Biotechnology*, *99*(22), 9381–93. DOI: 10.1007/s00253-015-6963-9.

Keasling, Jay D. (2010). "Manufacturing Molecules Through Metabolic Engineering." *Science*, *330*(6009), 1355–58. DOI: 10.1126/ science.1193990.

Kelwick, Richard J. R., Webb, Alexander J. & Freemont, Paul S. (2020). "Biological Materials: The Next Frontier for Cell-Free Synthetic Biology." *Frontiers in Bioengineering and Biotechnology*, *8*, 399. DOI: 10.3389/fbioe.2020.00399.

Kwon, Seong W., Paari, Kuppusamy A., Malaviya, Alok. & Jang, Yu-Sin. (2020). "Synthetic Biology Tools for Genome and Transcriptome Engineering of Solventogenic Clostridium." *Frontiers in Bioengineering and Biotechnology*, *8*, 282. DOI: 10.3389/fbioe.2020.00282.

Lee, Sang Yup, Kim, Hye Mi. & Cheon, Seungwoo. (2015). "Metabolic Engineering for the Production of Hydrocarbon Fuels." *Current Opinion in Biotechnology*, *33*, 15–22. DOI: 10.1016/ j.copbio.2014.09.008.

Lee Hyunbae., Subburaj Saminathan., Tu Luhua., Lee Ka-Yeon., Park Gwangsu. & Lee Geung-Joo. (2020). "Overview of CRISPR/Cas9: a chronicle of the CRISPR system and application to ornamental crops. " *Korean Journal of Agricultural Science*, *47*, 903-920. https://doi. org/10.7744/kjoas.20200075.

Li, Lei, Jiang, Weihong. & Lu, Yinhua. (2017). "New Strategies and Approaches for Engineering Biosynthetic Gene Clusters of Microbial Natural Products." *Biotechnology Advances*, *35*(8), 936–949. DOI: 10.1016/j.biotechadv.2017.03.007.

Li, Jian, Rong, Lanxin, Zhao, Yu, Li, Shenglong, Zhang, Cuiying, Xiao, Dongguang, Foo, Jee Loon. & Yu, Aiqun. (2020). "Next-Generation Metabolic Engineering of Non-Conventional Microbial Cell Factories for Carboxylic Acid Platform Chemicals." *Biotechnology Advances*, *43*, 107605. DOI: 10.1016/j.biotechadv.2020.107605.

Liang, Liya, Liu, Rongming, Garst, Andrew D., Lee, Thomas, Nogué, Violeta, Sanchez, Beckham, Gregg T. & Gill, Ryan T. (2017). "CRISPR Enabled Trackable Genome Engineering for Isopropanol Production in *Escherichia coli*." *Metabolic Engineering*, *41*, 1–10.DOI: 10.1016/j.ymben.2017.02.009.

Lin, Zhenquan, Nielsen, Jens. & Liu, Zihe. (2020). "Bioprospecting Through Cloning of Whole Natural Product Biosynthetic Gene Clusters." *Frontiers in Bioengineering and Biotechnology*, *8*, 526. DOI: 10.3389/fbioe.2020.00526.

Liu, Yanfeng, Shin, Hyun-dong, Li, Jianghua. & Liu, Long. (2015). "Toward Metabolic Engineering in the Context of System Biology and Synthetic Biology: Advances and Prospects." *Applied Microbiology and Biotechnology*, *99*(3), 1109–18. DOI: 10.1007/s00253-014-6298-y.

Liu, Yanfeng, Liu, Long, Li, Jianghua, Du, Guocheng. & Chen, Jian. (2019). "Synthetic Biology Toolbox and Chassis Development in *Bacillus subtilis*." *Trends in Biotechnology*, *37*(5), 548–62. DOI: 10.1016/j.tibtech.2018.10.005.

Liu, Peng, Zhang, Bo, Yao, Zhen-Hao, Liu, Zhi-Qiang. & Zheng, Yu-Guo. "Multiplex Design of the Metabolic Network for Production of L-Homoserine in *Escherichia coli*." *Applied and Environmental Microbiology*, *86*(20), e01477-20. DOI: 10.1128/AEM.01477-20.

Liu, Qiong, Zhang, Hongxia. & Huang, Xiaotian. (2020). "Anti-CRISPR Proteins Targeting the CRISPR-Cas System Enrich the Toolkit for

Genetic Engineering." *The FEBS Journal*, *287*(4), 626–44. DOI: 10.1111/febs.15139.

Liu, Yanfeng, Su, Aanqi, Li, Jianghua, Ledesma-Amaro, Rodrigo, Xu, Peng, Du, Guocheng. & Liu, Long. (2020). "Towards Next-Generation Model Microorganism Chassis for Biomanufacturing." *Applied Microbiology and Biotechnology*, *104*(21), 9095–9108. DOI: 10.1007/s00253-020-10902-7.

Löbs, Ann-Kathrin, Schwartz, Cory. & Wheeldon, Ian. (2017). "Genome and Metabolic Engineering in Non-Conventional Yeasts: Current Advances and Applications." *Synthetic and Systems Biotechnology*, *2*(3), 198–207. DOI: 10.1016/j.synbio.2017.08.002.

Long, Matthew R., Ong, Wai Kit. & Reed, Jennifer L. (2015). "Computational Methods in Metabolic Engineering for Strain Design." *Current Opinion in Biotechnology*, *34*, 135–41. DOI: 10.1016/j.copbio.2014.12.019.

Makarova, K. S., Wolf, Y. I., Iranzo, J., Shmakov, S. A., Alkhnbashi, O. S., Brouns, S. J. J., Charpentier, E., et al. (2020). "Evolutionary Classification of CRISPR–Cas Systems: A Burst of Class 2 and Derived Variants." *Nature Reviews Microbiology*, *18*(2), 67–83. DOI: 10.1038/s41579-019-0299-x.

Malcı, Koray, Walls, Laura E. & Rios-Solis, Leonardo. (2020). "Multiplex Genome Engineering Methods for Yeast Cell Factory Development." *Frontiers in Bioengineering and Biotechnology*, *8*, 589468. DOI: 10.3389/fbioe.2020.589468.

Miscevic, Dragan, Mao, Ju-Yi, Kefale, Teshager, Abedi, Daryoush, Moo-Young, Murray. & Perry Chou, C. (2020). "Strain Engineering for High-Level 5-Aminolevulinic Acid Production in *Escherichia coli*." *Biotechnology and Bioengineering*, DOI: 10.1002/bit.27547.

Mougiakos, Ioannis, Bosma, Elleke F., Ganguly, Joyshree, Oost, John van der. & Kranenburg, Richard van. (2018). "Hijacking CRISPR-Cas for High-Throughput Bacterial Metabolic Engineering: Advances and Prospects." *Current Opinion in Biotechnology*, *50*, 146–57. DOI: 10.1016/j.copbio.2018.01.002.

Mougiakos, Ioannis, Bosma, Elleke F., Vos, Willem M. de, Kranenburg, Richard van. & Oost, John van der. (2016). "Next Generation Prokaryotic Engineering: The CRISPR-Cas Toolkit." *Trends in Biotechnology*, *34*(7), 575–87. DOI: 10.1016/j.tibtech.2016.02.004.

Ng, I-Son, Keskin, Batuhan B. & Tan, Shih-I. (2020). "Critical Review of Genome Editing and Synthetic Biology Applications in Metabolic Engineering of Microalgae and Cyanobacteria." *Biotechnology Journal*, *15*(8), 1900228. DOI: 10.1002/biot.201900228.

Nielsen, Jens. (2001). "Metabolic Engineering." *Applied Microbiology and Biotechnology*, *55*(3), 63–83. DOI: 10.1007/s002530000511.

Nielsen, Alec A. K. & Voigt, Christopher A. (2014). "Multi-Input CRISPR/Cas Genetic Circuits That Interface Host Regulatory Networks." *Molecular Systems Biology*, *10*(11), 763. DOI: 10.15252/msb.20145735.

Nora, Luisa Czamanski, Westmann, Caua Antunes, Guazzaroni, Maria-Eugenia, Siddaiah, Chandranayaka, Gupta, Vijai Kumar. & Silva-Rocha, Rafael. (2019). "Recent Advances in Plasmid-Based Tools for Establishing Novel Microbial Chassis." *Biotechnology Advances*, *37*(8), 107433. DOI: 10.1016/j.biotechadv.2019.107433.

Park, Jaehyun, Shin, Hyojung, Lee, Sun-Mi, Um, Youngsoon. & Woo, Han Min. (2018). "RNA-Guided Single/Double Gene Repressions in *Corynebacterium glutamicum* Using an Efficient CRISPR Interference and Its Application to Industrial Strain." *Microbial Cell Factories*, *17*(1), 4. DOI: 10.1186/s12934-017-0843-1.

Peters, Gert, Coussement, Pieter, Maertens, Jo, Lammertyn, Jeroen. & De Mey, Marjan. (2015). "Putting RNA to Work: Translating RNA Fundamentals into Biotechnological Engineering Practice." *Biotechnology Advances*, *33*(8), 1829–44. DOI: 10.1016/j.biotechadv.2015.10.011.

Phithakrotchanakoon, Chitwadee, Puseenam, Aekkachai, Wongwisansri, Sriwan., Eurwilaichitr, Lily, Ingsriswang, Supawadee, Tanapongpipat, Sutipa. & Roongsawang, Niran. "CRISPR-Cas9 Enabled Targeted Mutagenesis in the Thermotolerant Methylotrophic

Yeast *Ogataea thermomethanolica*." *FEMS Microbiology Letters*, *365*(11). DOI: 10.1093/femsle/fny105.

Qi, Hao, Li, Bing-Zhi, Zhang, Wen-Qian, Liu, Duo. & Yuan, Ying-Jin. (2015). "Modularization of Genetic Elements Promotes Synthetic Metabolic Engineering." *Biotechnology Advances*, *33*(7), 1412–19. DOI: 10.1016/j.biotechadv.2015.04.002.

Raschmanová, Hana, Weninger, Astrid, Glieder, Anton, Kovar, Karin. & Vogl, Thomas. (2018). "Implementing CRISPR-Cas Technologies in Conventional and Non-Conventional Yeasts: Current State and Future Prospects." *Biotechnology Advances*, *36*(3), 641–65. DOI: 10.1016/j.biotechadv.2018.01.006.

Richter, Florian, Fonfara, Ines, Bouazza, Boris, Schumacher, Charlotte Helene, Bratovič, Majda, Charpentier, Emmanuelle. & Möglich, Andreas. (2016). "Engineering of Temperature- and Light-Switchable Cas9 Variants." *Nucleic Acids Research*, *44*(20), 10003-10004. DOI: 10.1093/nar/gkw930.

Rothstein, Samuel M., Sen, Swastik. & Mansell, Thomas J. (2020). "Towards High-Throughput Genome Engineering in Lactic Acid Bacteria." *Current Opinion in Biotechnology*, *61*, 181–88. DOI: 10.1016/j.copbio.2019.12.015.

Sander, Jeffrey D. & Joung, J. Keith. (2014). "CRISPR-Cas Systems for Editing, Regulating and Targeting Genomes." *Nature Biotechnology*, *32*(4), 347–55, DOI: 10.1038/nbt.2842.

Schultenkämper, Kerstin, Brito, Luciana F. & Wendisch, Volker F. (2020). "Impact of CRISPR Interference on Strain Development in Biotechnology." *Biotechnology and Applied Biochemistry*, *67*(1), 7-21. DOI: 10.1002/bab.1901.

Si, Tong, HamediRad, Mohamad. & Zhao, Huimin. (2015). "Regulatory RNA-Assisted Genome Engineering in Microorganisms." *Current Opinion in Biotechnology*, *36*, 85–90. DOI: 10.1016/j.copbio.2015.08.003.

Tao, Wei, Lv, Li. & Chen, Guo-Qiang. (2017). "Engineering *Halomonas* Species TD01 for Enhanced Polyhydroxyalkanoates Synthesis via

CRISPRi." *Microbial Cell Factories*, *16*(1), 48. DOI: 10.1186/s12934-017-0655-3.

Ungerer, Justin. & Pakrasi, Himadri B. (2016). "Cpf1 Is A Versatile Tool for CRISPR Genome Editing Across Diverse Species of Cyanobacteria." *Scientific Reports*, *6*, 39681. DOI: 10.1038/srep39681.

Volke, Daniel C. & Nikel, Pablo I. (2018). "Getting Bacteria in Shape: Synthetic Morphology Approaches for the Design of Efficient Microbial Cell Factories." *Advanced Biosystems*, *2*(11), 1800111. DOI: 10.1002/adbi.201800111.

Wang, Harris H., Isaacs, Farren J., Carr, Peter A., Sun, Zachary Z., Xu, George, Forest, Craig R. & Church, George M. (2009). "Programming Cells by Multiplex Genome Engineering and Accelerated Evolution." *Nature*, *460*(7257), 894–98. DOI: 10.1038/nature08187.

Wang, H. H., Kim, H., Cong, L. & Jeong, J. (2012). "Bang, D. and Church, G. M., Genome-Scale Promoter Engineering by Coselection MAGE." *Nature Methods*, *9*(6), 591–93. DOI: 10.1038/nmeth.1971.

Wang, Hailong, Li, Zhen, Jia, Ruonan, Yin, Jia, Li, Aiying, Xia, Liqiu, Yin, Yulong., et al. (2018). "ExoCET: Exonuclease *in Vitro* Assembly Combined with RecET Recombination for Highly Efficient Direct DNA Cloning from Complex Genomes." *Nucleic Acids Research*, *46*(5), e28. DOI: 10.1093/nar/gkx1249.

Wang, Lei. & Zhang, Juhua. (2019). "Prediction of SgRNA On-Target Activity in Bacteria by Deep Learning." *BMC Bioinformatics*, *20*(1), 517. DOI: 10.1186/s12859-019-3151-4.

Wasels, François, Chartier, Gwladys, Hocq, Remi, Ferreira. & Nicolas Lopes. (2020). "A CRISPR/Anti-CRISPR Genome Editing Approach Underlines the Synergy of Butanol Dehydrogenases in *Clostridium acetobutylicum* DSM 792." *Applied and Environmental Microbiology*, *86*(13), e00408-20. DOI: 10.1128/AEM.00408-20.

Weninger, Astrid, Hatzl, Anna-Maria, Schmid, Christian, Vogl, Thomas. & Glieder, Anton. (2016). "Combinatorial Optimization of CRISPR/Cas9 Expression Enables Precision Genome Engineering in the Methylotrophic Yeast *Pichia pastoris*." *Journal of Biotechnology*, *235*, 139–49. DOI: 10.1016/j.jbiotec.2016.03.027.

Wu, Yaokang, Liu, Yangeng, Lv, Xueqin, Li, Jianghua, Du, Guocheng. & Liu, Long. "Applications of CRISPR in a Microbial Cell Factory: From Genome Reconstruction to Metabolic Network Reprogramming." *ACS Synthetic Biology*, *9*(9), 2228–38. DOI: 10.1021/acssynbio.0c00349.

Yao, Ruilian, Liu, Di, Jia, Xiao, Zheng, Yuan, Liu, Wei. & Xiao, Yi. (2018). "CRISPR-Cas9/Cas12a Biotechnology and Application in Bacteria." *Synthetic and Systems Biotechnology*, *3*(3), 135–49. DOI: 10.1016/j.synbio.2018.09.004.

Zhan, Yangyang, Xu, Yong, Zheng, Pengling, He, Min, Sun, Shanhu, Wang, Dong, Cai, Dongbo, Ma, Xin. & Chen, Shouwen. (2020). "Establishment and Application of Multiplexed CRISPR Interference System in *Bacillus licheniformis*." *Applied Microbiology and Biotechnology*, *104*(1), 391–403. DOI: 10.1007/s00253-019-10230-5.

Zhang, Jie, Zong, Wenming Hong, Wei, Zhang, Zhong-Tian. & Wang, Yi. (2018). "Exploiting Endogenous CRISPR-Cas System for Multiplex Genome Editing in *Clostridium tyrobutyricum* and Engineer the Strain for High-Level Butanol Production." *Metabolic Engineering*, *47*, 49–59. DOI: 10.1016/j.ymben.2018.03.007.

Zhang, Shuyi. & Voigt, Christopher A. (2018). "Engineered dCas9 with Reduced Toxicity in Bacteria: Implications for Genetic Circuit Design." *Nucleic Acids Research*, *46*(20), 11115–25. DOI: 10.1093/nar/gky884.

Zhang, Xu, Lin, Yina, Wu, Qiong, Wang, Ying. & Chen, Guo-Qiang. (2020). "Synthetic Biology and Genome-Editing Tools for Improving PHA Metabolic Engineering." *Trends in Biotechnology*, *38*(7), 689–700. DOI: 10.1016/j.tibtech.2019.10.006.

Zhao, Jing, Fang, Huan. & Zhang, Dawei. (2020). "Expanding Application of CRISPR-Cas9 System in Microorganisms." *Synthetic and Systems Biotechnology*, *5*(4), 269–76. DOI: 10.1016/j.synbio.2020.08.001.

In: The Future of Metabolic Engineering ISBN: 978-1-68507-362-6
Editors: A. Sharma and D. Amin

Chapter 2

SIGNIFICANT ROLE OF METABOLIC ENGINEERING IN THE UNDERSTANDING OF CANCER

Aarti Gayakwad, Neha P. Shukla and Giftson J. Senapathy*

C. G. Bhakta Institute of Biotechnology,
Uka Tarsadia University, Bardoli, Gujarat, India

ABSTRACT

Metabolic engineering is the modification of genetic and regulatory processes to develop products of interest. It developed many techniques and tools for a better explanation of metabolism in living cells. The modification of biosynthetic pathwaysthrough metabolic engineering could help in treating diseases like cancer. In order to develop or modify a cancer treatment, it is necessary to have a clear understanding on its metabolic pathways. In this chapter, many metabolic engineering tools were discussed such as metabolic flux analysis (MFA), metabolomics, biomarkers and thermodynamic control analysis in the context of cancer. MFA is an experimental fluxomics technique in that ^{13}C MFA is precisely a powerful tool for quantifying intracellular

*Corresponding Author's E-mail: meetgiftson@gmail.com.

metabolic fluxes in cancer cells. Metabolomics, a study of numerous metabolites, revolves around the possible ways of understanding cancer metabolism using specific metabolic engineering tools. Fluorodeoxyglucose-positron emission tomography is one of the earliest metabolic markers used for cancer detection. To date, different biomarkers were discovered to check the significant role of amino acids, sugars and lipids in the proliferation of cancer cells. In fact, metabolic engineering paved the way for the discovery of many cancer drugs which were in practice now. We have also discussed on the discovery of certain plant-based drugs that could possibly minimize the treatment induced side effects. Overall, this chapter revolves around the possible ways to study and treat cancer with the help of metabolic engineering.

Keywords: metabolic engineering, cancer, natural drugs, metabolomics, biomarkers

1. Introduction

Globally, cancer is the one of the leading causes of the death (Boyd et al., 2021). According to World Health Organisation (WHO), cancer is a condition where normal cells become indefinitely proliferating abnormal cells. These abnormal cells divide uncontrollably and destroy other body tissues. It is a multistage process which generally progresses from a precancerous lesion to a malignant tumour caused by many different factors. These changes arise from the interaction of a person's genetic factors along with many external agents (World Health Organization, 2020). Cancer metabolism is one of the oldest areas of research, which covers the alterations of metabolic activities in cancer cells. Some common altered metabolic features are noted in many types of cancer cells and this reprogrammed metabolism is considered to be a hallmark of cancer (DeBerardinis and Chandel, 2016). For example, researchers reported the correlation of high-fat diet intake, high fecal and low serum cholesterol bile with colorectal cancer.

Also, it was reported that many unsaturated fatty acids increase the risk of prostate cancer and breast cancer (Schmidt and Patel, 2021).

Metabolic engineering in general is a science about alteration of metabolic pathways with a deliberate modification of cell metabolism to produce desired compounds. It uses variety of databases, component libraries and conditions to generate the maximum output rate of a desired chemical compound. It also prevents inhibitors and conditions that influence the growth rate and other active functions in the organism to reach the goals. Metabolic engineering is gaining commercial applications for the production of functional metabolites. This requires up-regulation or down-regulation of defined proteins in a metabolic pathway, such that the cell forms a new product (Yadav and Tanveer et al., 2018). Using metabolic engineering, metabolic alterations in chronic diseases like cancer can be studied in a better way to develop novel drugs.

In this chapter, the use of different metabolic engineering tools for the study and design of cancer drugs were discussed.

2. Engineers' Approaches to Study Cancer Metabolism

To improve damaged cellular properties, metabolic engineers combine different pathways and systemic analysis with specific molecular biological techniques. Nearly a century ago, Otto Warburg first described the contrasting metabolic states of tumors and related tissues. Through animal models he proposed "Warburg Effect", (Zuand Guppy, 2004) which acts as a tool to understand the differencing behaviours of many plants based anticancer drugs *in vivo* and *in vitro*. Engineers include methodologies, reviews and applied researches that focus the applicability of metabolic engineering in the area of cancer metabolism.

Dai and Locasale have given an idea about the theory of flux analysis and described its current applications to explain the contribution of reprogrammed metabolism in tumor growth. Nilsson and Nielsen later discussed about quantitative prognosis of cancer metabolism through genome-scale metabolic modelling. Dong then explained isotope tracing and network approaches for reprogrammed cancer metabolic pathways (Metallo and Deberardinis, 2017).

With the help of network analysis and stable-isotope tracers, metabolic reduction of glutamine, serine and glycine (Metallo and Gameiro et al, 2011), altered glycolysis, (Vander and Locasale, 2010) and one-carbon metabolism (Locasale, 2013) were studied in cancer cells. These pathways are accountable to take out cellular building blocks and energy and use them for cellular growth in cancer cell. To date, many other researches have been done in cancer metabolism to explore how the engineering approaches can be applied in metabolic reframe of cancerous cell. Based on these studies, different tools were developed for the study and the treatment of the cancer (Figure 1).

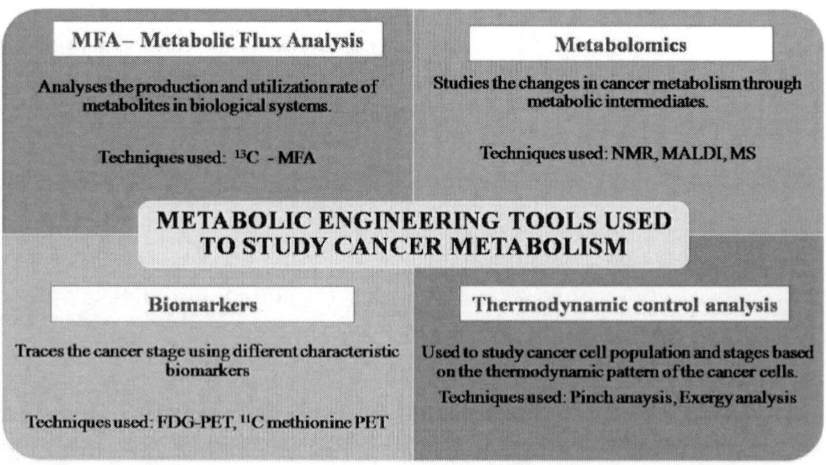

Figure 1. Important metabolic engineering tools to study cancer. Where, MFA- Metabolic Flux Analysis, NMR- Nuclear Magnetic Resonance, MALDI – Matrix Assisted Laser Desorption and/or Ionization, MS- Mass Spectroscopy, FDG PET- Fluorodeoxyglucose (FDG)-Positron Emission Tomography.

3. Metabolic Engineering Tools used in Cancer Metabolism Study

3.1. Metabolic Flux Analysis

Metabolic flux analysis (MFA) is a useful tool for quantitative characterization of the metabolic phenotype in mammalian and microbial cell line (Metallo and Deberardinis, 2017; Wiechert, 2001). It is an experimental flux omics technique mostly used to research the production and utilization rates of metabolites in biological system. MFA may use ^{13}C, NMR and mass spectrometry techniques to determine pathway fluxes (Hiller et al., 2013; Dong et al., 2017). Still ^{13}C-MFA is not widely used by many cancer biologists due to certain challenges.

In MFA, stoichiometric model was used to calculate intracellular fluxes for the major intracellular reactions and mass balances are applied around the intracellular metabolites. The final outcomes will be given in the form of a diagram which shows the biochemical reaction of flux calculations. This information obtained can be extensively assisted in metabolic engineering which would help to identify acute branch points in pathways (Xiao and Liang, 2012).

^{13}C metabolic flux analysis (^{13}C-MFA) is a powerful path for quantifying intracellular metabolic fluxes in cancer cells. For the calculation of the ^{13}C-MFA, three inputs are necessary: 1) external rates; 2) isotopic labeling and (3) metabolic model (Figure 2). By new advanced ^{13}C-MFA, varieties of techniques have been recently developed to delineate the secondary metabolism, to improve the validity and resolution of the flux, and examine the metabolism for different organisms. With the help of ^{13}C-MFA, cell factories will be more rationally planned and applied to improve the biochemical production (Antoniewicz, 2018).

Many drugs were developed with the help of metabolic engineering for cancer treatment. Different drugs will target different metabolic pathway for the treatment of cancer.

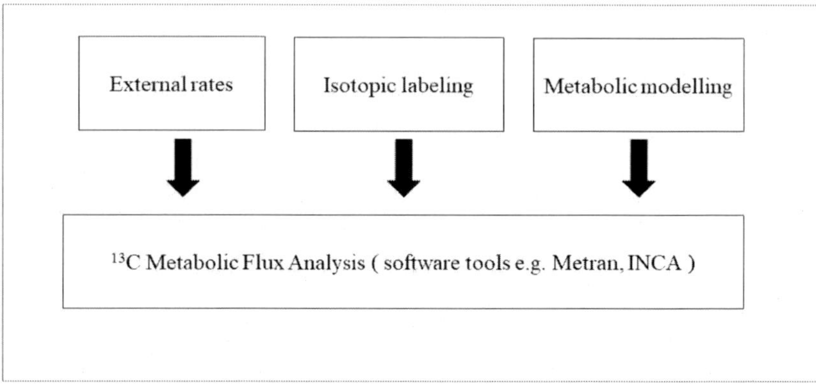

Figure 2. ^{13}C Metabolic Flux Analysis (image modified from Antoniewicz, 2018). METRAN and INCA (Isotopomer Network Compartmental Analysis) are specific software packages used for metabolic flux analysis and other tracer experiments.

Table 2. Cancer Drugs That Target Metabolic Pathways (modified from Schmidt and Patel et al. (2021)

Metabolic Pathway	**Target**	**Drug**
Enzymes mutated in cancer	Mutant isocitrate dehydrogenase 1	AG-120; BAY1436032; LY3410738; FT-2102
Nucleic acid synthesis	Thymidylate synthase (TS); Dihydro folate reductase (DHFR); Glycinamide ribonucleotide formyl transferase (GARFT); Dihydroorotate dehydrogenase(DHODH)	5-Fluorouracil; capecitabine; pemetrexed; Raltitrexed; Methotrexate; Pemetrexed; Brequinar; Leflunomide
Glutamine metabolism	Glutaminase1(GLS1)	CB-839; IPN60090
Glycolysis	GLUT1; Pyruvate KinaseM2(PKM2); TEPP-46	WZB117; BAY-876
Lipid metabolism	ATP-citrate Lyase (ACLY)	SB-204990

3.2. Metabolomics

In cancer, the changes occur at the genomic level which affects the expression of transcription process, protein expression, and its function. The metabolome gives operational readout of these changes. In sequential, it affects the protein activity with alteration of RNA transcription and DNA replication. Metabolomics is the study of thousands of metabolites, small intermediate and substance and outcome of the cell metabolism. In advance research, the field of metabolomics has advanced instrumentation (Schmidt and Patel et al., 2021). Many major techniques like mass spectrometry (MS) and nuclear magnetic resonance (NMR), matrix-assisted laser desorption/ionisation (MALDI), PET, computed tomography and radionucleide scans were used for the detection of cancer. Several limitations of imaging includes limited availability and short half-life of some radiotracers, impotence to find smaller tumors, poor image resolution and inability to prominent tumors from non-malignant through hyper metabolic processes (Schmidt and Patel et al., 2021).

3.3. Biomarkers

Fluorodeoxyglucose-positron emission tomography (FDG-PET) is one of the earliest metabolic markers used for the cancer detection. It is widely used for tumor imaging due to increased glucose metabolism in most of the tumors. Further, ^{18}F-labeled sugars, such as D-mannose, D-fructose, D-lactose and D-galactose, have been made to study as PET tracers in preclinical settings and also allow the findings of some cancer cells that use sugar and other molecules as an energy source. Utilization of amino acid has been checked by [^{11}C]-tyrosine PET in soft tissue sarcomas and pituitary adenomas. Also, [^{11}C] methionine PET has been assessed in brain tumors. To understand the role of glutaminolysis in cancer, a series of glutamine-dependant PET tracers, like [^{18}F]-(2S,

4R)-4-fluorolutamine, L-[5-^{11}C]-glutamine and [^{18}F]-(2S, 4S)-4-(3-fluoropropyl) glutamine, have been synthesized.

Like amino acids and sugars, lipids play an important role in the proliferation of cancer cells. ^{11}C-fatty acids have been used as PET radio tracers to study β-oxidation. Radio labeled forms of the phospholipid head group choline, [^{18}F]-fluoro ethyl choline and [^{18}F]-fluoro methyl choline, were used in the clinic for the detection of the cancer. In contrast biofluids, such as blood, serum, saliva, urine and sweat, are the most convenient source of material for the detection of biomarkers. If some other diseases were present along with cancer, then the biofluids will not able to find the differences of specific metabolic alterations due to cancer. Hormone biomarkers and serum antigens were used as diagnostic markers for prostate adenocarcinoma, ovarian cancer, thyroid cancer, nonseminomatous germ-cell tumors, hepatocellular carcinoma and others (Schmidt and Patel et al., 2021).

3.4. Thermodynamic Control Analysis

In bioenergetics, the thermodynamic model shows Warburg effect. In cancer cells, the entropy production rate, temperature and heat production is observed to be higher than the metabolic efficiency rate as compared to the non-cancerous cell. This model also gives an idea on the long-term control of cancer cell populations and cancer prevention. This model anticipates that there is a sufficient amount of nutrients present in circulating blood, similar to the situation of excessive energy seen in cancer cells. Mainly there are two thermodynamic solutions to control the cancer cell population, i.e., (i) decreasing rate of entropy production in cancerous cells and (ii) increasing entropy production rate for the normal cells. Recent anti-cancer therapies could have worked on this basis (Luo et al., 2006).

Luo et al., 2006 stated that, if entropy production of non-cancerous cells increases when compared to cancer cells, the entropy flows from normal to cancer cells eliminating the growth of the tumor. They also explained that the square wave electricpulses helps to increase the normal cells entropyrate. Similarly, when nutrients are rich, any conditions that higher the entropy rate of normal cells credibly have an anti-cancer effect.

Under the limited amount of the nutrient availability, this model anticipates that cancerous cells with a high production rate of entropy will have a lesser chance of survival in comparison to non-cancerous cells. Supporting this statement, several studies have shown that glucose or glutamine expropriation triggers the cell death in cancerous cells but not in normal cells. Parallely other studies have also reported that restriction in calories decreased the risk of cancer in different mammals.

4. Metabolic Engineering of Plant Derived Anticancer Agents

Plants contain higher number of bioactive compounds, many of which have been used since pre-history for their therapeutic properties to treat a range of illnesses. Recently, these metabolites have focused on their anti-neoplastic activities against a variety of cancers (Senapathy et al., 2020). It was proved that glycosides were used as anti-cancer agents, namely Podophyllotoxin (PPT) (aryl-tetralinlignanor alkaloids). The indole alkaloids (vincristine and vinblastine from *Catharanthus roseus*), di-terpenoid alkaloids (taxol and its analogues from *Taxus* and *Corylus* species) and quinoline alkaloids (camptothecin from *Camptotheca acuminata*) are the three main forms of alkaloids (Antoniewicz et al., 2018).

Podo Phyllo Toxin (PPT), an aryl tetra linlignan, found naturally in roots and rhizomes of many herbs such as *P. hexandrum, Podophyllum peltatum* and *L. narbonense.* This compound shows many vigorous antimitotic and antiviral activities. The primary mode of action of PPT is the inhibition of Topo-II and protein tyrosin kinase. Transcriptomic analyses of *P. hexandrum* Royle (having anticancer lignan PPT) were used to investigate the expression of key genes involved in the production of this compound (Antoniewicz, 2018). The higher expression of secoisolariciresinol dehydrogenase (SD), chorismate mutase (CM), p-coumarate 3-hydroxylase (PCH), prephenate dehydrogenase (PD), caffeic acid 3-O- methyltransferase (CMT), cinnamoyl-Co-A reductase (CCR) cinnamyl alcohol dehydrogenase (CAD), arogenate dehydratase (ADH) and cinnamate 4-hydroxylase (C4H) correlated with a higher yield of PPT in the root and rhizomes of *P. hexandrum* (Antoniewicz, 2018).

Isolated from the bark of the Pacific yew tree (*Taxusbrevifolia*), Taxol is an expensive anti-cancer drug due to its low availability from a natural source and a high extraction taxol. Nineteen enzyme-catalyzed steps were involved in the biosynthesis of taxol (Changxing and Galani et al., 2020; Iqbal and Abbasi, 2017). It has been studied extensively against ovarian, breast, and lung cancers (Iqbal and Abbasi, 2017).

Taxanes stabilize the assembly of microtubule to prevent de-assembling. Accordingly, it blocks the progression of mitosis and prolonged activation of checkpoint. It leads to apoptosis or reversion of the cancerous cell to the resting phase (G0 phase) in cell cycle. It is also involved in the stimulation of apoptotic cascades by down regulating Bcl2 and upregulating Fas/FasL and Bax (Ionkova and Sasheva, 2013). The important challenge to use of taxol was identified to be its restricted accessibility and insufficient quantity.

The discovery of natural therapeutics is anticipated to revolutionize cancer treatment in the upcoming years. Biotechnological tools like cell culture and *in vitro* propagation are cost-effective for producing anticancer compounds at an industrial level. A plethora of literature has been published on the optimization of biomass growth using these techniques to increase the production of podophyllotoxin (from Linum, Podophyllum), taxanes (from Taxus), camptothecin (from *C. acuminate*), and terpeneindole alkaloids (from *C.roseus*). Studies have revealed the importance of different elicitors both abiotic and biotic, are equally important for the activation of genes involved in the metabolic pathways, which could enhance the production of anticancer compounds. Many plant derived products were used to treat cancer with the help of metabolic engineering (Iqbal and Abbasi, 2017; Ionkova and Sasheva, 2013).

Future Aspects and Conclusion

Metabolic engineering was extensively studied in past decade to treat many chronic diseases. As discussed in this chapter, different types of tools were used to study and treat different types of cancers. ^{13}C metabolic fluxanalysis (^{13}c-mfa) is a powerful path for quantifying in tricellular metabolic fluxes in cancer cells. Metabolomics is the study of thousands of metabolites, small intermediate and substance and outcome of the cell metabolism. Specific biomarkers are used for the specific type of cancer detection. With the help of thermodynamic model we can analyze the entropy level of normal cells and cancerous cells. Till date, cancer cannot be completely cured but through the emergence of different therapies, it is treatable. Also these therapies have many side effects on cancer patients. To minimize the side effects, many plant-based drugs were employed to treat cancer. Many plant materials show anti-cancerous activity with no side effects. Besides these advantages, many metabolic engineering tools have certain limitations. In future research works, these limitations can be worked out and by using different metabolic engineering tools diagnosis and treatment of cancer can be made effective for the well-being of the society.

References

Antoniewicz, M.R. (2018). A guide to ^{13}C metabolic flux analysis for the cancer biologist. *Experimental & molecular medicine*, *50*(4), 1-13.

Boyd, A. D., Song, X., & Furgal, C. M. (2021). A systematic literature review of cancer communication with indigenous populations in Canada and the United States. *Journal of Cancer Education*, *36*(2), 310-324.

Changxing, L., Galani, S., Hassan, F.U., Rashid, Z., Naveed, M., Fang, D., & Jianhua, L. (2020). Biotechnological approaches to the production of plant-derived promising anticancer agents: An update and overview. *Biomedicine & Pharmacotherapy*, *132*, 110918.

De Berardinis, R.J., and Chandel, N.S. (2016). Fundamentals of cancer metabolism. *Science advances*, 2(5), e1600200.

Dong, W., Keibler, M.A., & Stephanopoulos, G. (2017). Review of metabolic pathways activated in cancer cells as determined through isotopic labeling and network analysis. *Metabolic engineering*, *43*, 113-124.

Ehrlich, A., Booher, S., Becerra, Y., Borris, D.L., Figg, W.D., Turner, M.L., &Blauvelt, A. (2004). Micellar paclitaxel improves severe psoriasis in a prospective phase II pilot study. *Journal of the American Academy of Dermatology*, *50*(4), 533-540.

García-Granados, R., Lerma-Escalera, J.A., &Morones-Ramírez, J.R. (2019). Metabolic engineering and synthetic biology: synergies, future, and challenges. *Frontiers in bioengineering and biotechnology*, *7*, 36.

Hiller, K., Wegner, A., Weindl, D., Cordes, T., Metallo, C.M., Kelleher, J.K., &Stephanopoulos, G. (2013). NTFD—a stand-alone application for the non-targeted detection of stable isotope-labeled compounds in GC/MSdata. *Bioinformatics*, *29*(9), 1226-1228.

Ionkova, I., Sasheva, P., Ionkov, T., &Momekov, G. (2013). Linumnarbonense: A new valuable tool for biotechnological production of a potent anticancer lignan Justicidine B. *Pharmacognosymagazine*, *9*(33), 39.

Iqbal, J., Abbasi, B.A., Mahmood, T., Kanwal, S., Ali, B., Shah, S.A., & Khalil, A.T. (2017). Plant-derived anticancer agents: A green anticancer approach. *Asian Pacific Journal of Tropical Biomedicine*, *7*(12), 1129-1150.

Luo, L., Molnar, J., Ding, H., Lv, X., &Spengler, G. (2006). Physicochemical attack against solid tumors based on the reversal of direction of entropy flow: an attempt to introduce thermodynamics in anticancer therapy. *Diagnostic Pathology*, *1*(1), 1-7.

Leonard, E., Fowler, Z.L., &Koffas, M. (2007). Metabolic engineering. In *Systems biology* (pp. 301-359). Springer, Dordrecht.

Locasale, J.W. (2013). Serine, glycineandone-carbon units: cancer metabolism in full circle. *Nature Reviews Cancer*, *13*(8), 572-583.

Metallo, C.M., & Deberardinis, R.J. (2017). Engineering approaches to study cancer metabolism. *Metabolic engineering*, *43*, 93.

Metallo, C.M., Gameiro, P.A., Bell, E.L., Mattaini, K.R., Yang, J., Hiller, K., &Stephanopoulos, G. (2012). Reductive glutamine metabolism by IDH1 mediates lipogenes is under hypoxia. *Nature*, *481*(7381), 380-384.

Sabzehzari, M., Zeinali, M., &Naghavi, M.R. (2020). Alternative sources and metabolic engineering of Taxol: Advances and future perspectives. *Biotechnology Advances*, *43*, 107569.

Senapathy, G.J., George, B.P., &Abrahamse, H. (2020). Exploring the Role of Phytochemicals as Potent Natural Photosensitizers in Photo dynamic Therapy. *Anti-Cancer Agents in Medicinal Chemistry (Formerly Current Medicinal Chemistry-Anti-Cancer Agents)*, *20*(15), 1831-1844.

Schmidt, D.R., Patel, R., Kirsch, D.G., Lewis, C.A., VanderHeiden, M.G., & Locasale, J.W. (2021). Metabolomics in cancer research and emerging applications in clinical oncology. *CA: a cancer journal for clinicians*.

Vander Heiden, M.G., Locasale, J.W., Swanson, K.D., Sharfi, H., Heffron, G.J., Amador-Noguez, D., & Cantley, L.C. (2010). Evidence for an alternative glycolytic pathway in rapidly proliferating cells. *Science*, *329*(5998), 1492-1499.

Wiechert, W. (2001). ^{13}C metabolic flux analysis. *Metabolic engineering*, 3(3), 195-206.

Woolston, B.M., Edgar, S., &Stephanopoulos, G. (2013). Metabolic engineering: past and future. *Annual review of chemical and biomolecular engineering*, *4*, 259-288.

World Health Organization (2020). *Global Cancer Profile 2020.* https://gco.iarc.fr/today/data/factsheets/cancers/20-Breast-fact-sheet.pdf

Xiao, J., Liang, D., Liu, Y., Zhang, H., Liu, Y., Zhao, H., &Chen, Y.H. (2012). Taxol, a microtubule stabilizer, improves cardiac functional recovery during post ischemic reperfusion in rat in vitro. *Cardiovascular therapeutics*, *30*(1), 12-30.

Yadav, D., Tanveer, A., Malviya, N., and Yadav, S. (2018). Expanding the concepts and tools of metabolic engineering to elucidate cancer metabolism. *Biotechnology progress*, 28(6), 1409-1418.

Yadav, D., Tanveer, A., Malviya, N., and Yadav, S. (2018). Overview and principles of bioengineering: the drivers of omics technologies. In *Omics Technologies and Bio-Engineering* (pp.3-23). Academic Press.

Zamboni, N., Fendt, S.M., Rühl, M., &Sauer, U. (2009).^{13}C-based metabolic flux analysis. *Nature protocols*, *4*(6),878-892.

Zu, X.L., &Guppy, M. (2004). Cancer metabolism: facts, fantasy, and fiction. *Biochemical and biophysical research communications*, *313*(3), 459-465.

In: The Future of Metabolic Engineering ISBN: 978-1-68507-362-6
Editors: A. Sharma and D. Amin

Chapter 3

GENOME EDITING IN METABOLICALLY ENGINEERED BREAST CANCER CELLS WITH A SPECIAL FOCUS ON NANO-THERAPEUTIC APPLICATIONS

P. Parizad, R. Krishnamurthy, PhD and Giftson J. Senapathy*, PhD
C. G. Bhakta Institute of Biotechnology,
Uka Tarsadia University, Bardoli, Gujarat, India

ABSTRACT

Globally Breast cancer accounts for 1 in 4 cancers in women. Breast cancer is diagnosed mostly in later stage in India due to which metastases occur and the treatment also fails after sometime. So, the treatment modality has to be personalized focusing on genetic - epigenetic aberrations of breast cancer. The genome editing technology is opening up the possibility to modify as well as target the effected genes directly, hence extends our ability to contribute to the diseases genetically. Recent advancements in programming cells with genome editing will be discussed here like zinc finger nucleases, transcription

* Corresponding Author's E-mail: meetgiftson@gmail.com.

activator-like effector nucleases and clustered regularly interspaced short palindromic repeat- Cas- associated nucleases and the possible hybrid nanocarriers for delivery of the components. Nanoparticles display theranostics approach by delivering the nucleases. This chapter revolves around possible ways to treat breast cancer using programmed cells with genetic modifications attached to the nanocarriers which could possibly serve as a promising direction for personalized medicine in breast cancer.

Keywords: breast cancer, nanomedicine, phytotherapy, genome editing

1. Introduction

According to report one million new cases of cancers listed in the world. Breast cancer is most common among females and it constitutes the 18% of all other cancers of women. The prediction of breast cancer incidence may increase to 85 per 1 lakh women by 2021 [1]. According to WHO Global Cancer profile (2020) the estimated past and future of total cases of breast cancer per year is 1,655,589 in 2012, 2,069,792 in 2018 and 2,778,850 in 2040 [2]. In India, there has been a steady increase in breast cancer in almost all the PBCRs. This increase leads to a bigger health challenge of women. To tackle this health situation different health and other agenises needs to come in action. The projected incidence of Breast cancer in India with crude rate of 29.9 and a cumulative risk of developing breast cancer at age of 0-74 years is 1 in 29 [3]. An integrated approach like awareness and preventive programs, early screening programs for early detection and well facilities of treatment may help and sounds vital for reducing the rate of incidence and rate of mortality of breast cancer [4,5].

Breast cancer disease is complex and features specific clinical, histological and molecular types. Various factors affect in the development of breast cancer like the intrinsic factors (genetic mutations and family history) and extrinsic factors (living style, radiation, environment etc.) [6].Various subtypes of breast cancer are based on the molecular receptors and the expression of the genes. Molecular receptors include progesterone (PR), estrogen receptor (ER) and human epidermal growth factor receptor-2 (HER-2). The subtypes based on the receptors include luminal A, luminal B, basal–like, normal breast-like and HER-2 enriched [7]. Prevailing treatments like surgery, chemotherapy and radiation therapy have limitations as well as side effects due to lack of specificity which affects the healthy cells of the body [8-10]. Over time the therapies fail as the patient incurs resistance to radiation and chemotherapeutic drugs. Due to the failure of therapies the risk of metastasis and relapse increases [11-13] the considerable problem with present therapies is that they depend mainly on the phenotype of the tumor. Breast cancer has many molecular arrays which results for varieties of phenotype. So the genome editing which seems to target such genetic changes occurring in breast cancer can lead to a promising cancer therapy [14].

Targeted editing of genome has gained attention and serves as a powerful tool to study and modify genes. Modification of genes is carried out either by silencing of the gene or by adding of gene[15].There are various types of genome editing tools like zinc finger nucleases (ZFN), transcription activator-like effector nucleases (TALEN) and clustered regularly interspaced short palindromic repeat- Cas- associated nucleases (CRISPR-Cas9). Genome editing requires a delivery system to reach the targeted site. The categories of delivery ways include physical method (electroporation, microinjection or hydrodynamic injection) and viral vectors. The risk of cell damage increases in physical method [16]. Viral vectors are well known for their easy transfection but also have limitations due to possible immunogenicity, carcinogenicity and size limited packaging [17]. These limitations lead to the augmentation and application of different method like nanotechnological applications for new, effective and safeguarded cancer treatment or cancer nanomedicine. The developing importance of implementing nanotechnology for drug carrier is due to its particular features like drug deliver, pint-sized medical device and therapeutic nature of some of the nanomaterials [18-20].

Furthermore, this chapter describes molecular mechanism of breast cancer, current treatment modalities, prons and cons, programmed cells with genome editing, CRISPR-Cas 9 role in breast cancer, role of nanomedicine in breast cancer and mechanism of nanocarrier in delivery of CRISPR-Cas9 nucleases to the targeted site. Our aim focuses on the advanced nanotechnology based drug delivery system using genome editing for the treatment of breast cancer.

2. MOLECULAR MECHANISM OF BREAST CANCER

Mutation and gene expression plays a crucial role in the mechanism of breast cancer. The classification based on expression of genes is still developing and makes the characterization even more complex. The molecular receptors and gene expression based on subtypes are listed in the Table 1 [6, 21, 22]. Gene expression classification includes inherited mutation, pathogenic mutation, genes positively associated with breast cancer susceptibility and mutations that can be correlated with increasing breast cancer risk. Inherited mutation includes genes namely BRCA1 (Breast Cancer gene 1) and BRCA2 (Breast Cancer gene 2). BRCA1 and BRCA2 genes have lifetime risk of breast cancer development approximately 72% and 69% by the age of 80 respectively. Pathogenic mutations also confer the high risk of breast cancer development. Pathogenic mutation in genes include PTEN (Phosphatase and tensin homolog), TP53 (Tumor protein 53), CDH1 (Cadherin-1), PALB2 (Partner and Localizer of BRCA2) and STK11 (Serine/Threonine kinase 11). Positively associated genes with susceptibility of breast cancer are FGFR2 (Fibroblast growth factor receptor 2), TNRC9 (Trinucleotide-repeat-containing 9), MAP3K1 (Mitogen - activated protein kinase 1) and LSP1 (Lymhocyte specific protein 1). These positively associated genes are related to signaling and cell growth. Mutations that are correlated with breast cancer increasing risk include ATM (ATM Serine/Threonine kinase), BRIP1 (BRCA1 interacting protein C-terminal helicase 1), CHEK2 (Checkpoint Kinase2), MRE11A (Meiotic recombination 11 homolog A), MSH6 (mutS homolog 6), NBN (Nibrin gene), NF1 (Neurofibromatosis 1), PMS2 (PMS1 Homolog2, mismatch repair system component), RAD50 (DNA repair protein RAD50 gene), RAD51C (DNA repair protein RAD51 homolog C) and SEC23B. Large number of these genes plays a role in DNA damage repair mechanism pathway. The genome-wide studies of single nucleotide polymorphism have determined more than 125 loci association with breast cancer genetic perceptivity. [21, 23]

Table 1. Breast Cancer Molecular Subtypes Features (Modified Reference [6, 21, 22])

Molecular Subtype	**IHC Markers**	**Frequency**	**Gene Markers**	**Histological Grade**	**Prognosis**	**Distribution among BRCA1**	**Distribution among BRCA 2**
Luminal A	ER+, PR+, HER2-	50-60%	ESR1, CCND1, GATA3, TFF3, FOXA1,	Low (Grade I or II)	Excellent	3%	14%
Luminal B	ER+, PR+/-, HER2+	10-20%	ESR1, FOXA1, GATA3, LAPTM4B, TFF3	Intermediate – High (Grade II or III)	Intermediate - High	27%	73%
HER2-enriched	ER+/-, PR+/-, HER2+	10-15%	GRB7, ERBB2	High (Grade II or III)	Bad	6%	-
Basal-like	ER-, PR+/-, HER2+/-	10-20%	FABP7, LAMC2, KRT17, TRIM29	High (Grade III)	Bad	61%	9%
Normal Breast like	ER+/-, PR+/-, HER2-	5-10%	VIM, CD36, TIMP1, FABP4, MMP2/14, ITGA7	Low (Grade I or II)	Intermediate	3%	4%

3. Existing Treatment and Its Pros and Cons

There are several ways to treat breast cancer. This is mainly dependent on the type of breast cancer and how much far it has spread to other organs of the body. There can be use of more than one treatment to treat breast cancer patients. Some treatments are localized or local meaning that only the tumor will be treated without affecting the other part of the body [24-26].The flow chart of the treatment is mentioned in Figure 1.

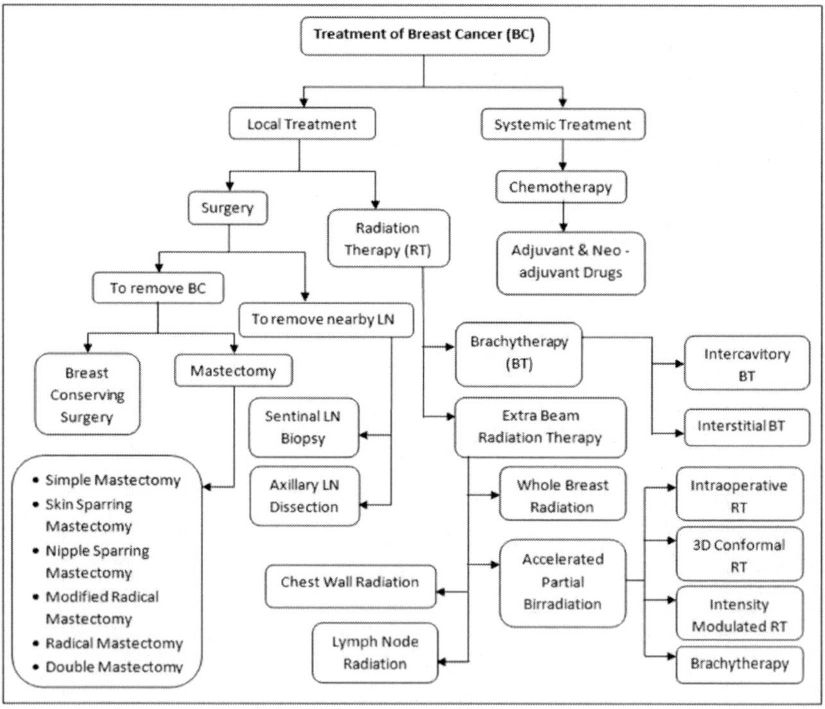

Figure 1. Various Treatment Strategies available for Breast Cancer.

3.1. Localized Treatment

3.1.1. Surgery

There are various types of surgery of patient suffering from breast cancer. The reason of surgery is also different and varies from patient to patient. Surgery can also be carried out just to relieve patient from the pain and discomfort. Surgery broadly divided into two namely; to remove breast cancer and to remove nearby lymph nodes. Surgery to remove breast cancer is furthered classified into two types that are Breast conserving surgery (BCS) and mastectomy. BCS is also termed as quadrantectomy, lumpectomy and partial or segmental mastectomy. In this type of surgery only the affected part of the breast is removed. Along with the tumor some normal tissues surrounding the cancer are removed to ensure the complete removal of the tumor. There are various side effects of breast conserving surgery which includes pain or tenderness, temporary swelling, hard scar or dimple formation, change of breast shape and nerve pain (neuropathic) in chest or armpit. Even lymphedema may occur if the axillary lymph node is not removed [24,26,27].

Mastectomy is a type of surgery in which the breast is removed entirely along with the nearby tissues. There are different types of mastectomy depending on the surgery namely; Simple mastectomy or total mastectomy (removes breast, nipples, areola and skin entirely), Skin sparring mastectomy (Skin is sparred and breast tissues, nipple and areola is removed), Nipple sparring mastectomy (Spares Nipple and skin and removes breast tissues and areola which helps to reconstruct the breast), Modified radical mastectomy (Simple mastectomy followed by removal of axillary lymph nodes from the armpit), Radical mastectomy (Same as modified radical mastectomy but also removes the pectoral tissues so not used widely) and Double mastectomy (Both the breast are removed to reduce the risk further). The side effects of mastectomy are pain or tenderness, swelling at the surgery area, hematoma, seroma, limited movement of shoulder and arm, numbness in upper arm area and chest and as with all surgery problems bleeding and infection at the site of operation are also the side effects [24, 29, 30].

To remove nearby lymph node surgery is performed to find out whether the breast cancer has spread to axillary lymph node under armpit. One or more lymph nodes are taken and tested in laboratory. Mainly there are two types Sentinel lymph node biopsy (One or more nodes are removed to reduce the risk of spreading of the breast cancer. Lymphedema can also avoided by it) and auxiliary lymph node dissection (Approximately less than 20 lymph nodes from the armpit are removed. This type of surgery was used earlier more frequently but is now only carried out in situation of metastasis of breast cancer). The common side effects of lymph node surgery are mainly lymphedema (swollen, tight and painful arm), limited movement of shoulder (a frozen shoulder) and upper arm and Numbness of the skin [24, 25].

3.2. Radiation Therapy

A treatment with high energy particle or rays used to kill the cancer cells is called as radiation therapy. Radiation is given to breast cancer along with other treatments like after BCS to reduce the risk of spreading to the lymph nodes and coming back of cancer in the same breast. It is also given along with mastectomy where the tumor was about 2 inches or the cancer has spread to nearby lymph nodes or other unreachable parts like bone and brain. The important types of radiation therapy for the treatment of cancer are external beam radiation (EBRT) and brachytherapy. In EBRT a beam of radiation is focused from outside the body. Depending on the type and schedules of treatment EBRT is further classified into whole breast radiation, accelerated partial breast radiation, chest wall radiation and lymph node radiation. EBRT side effects include damage to the arm nerves leading to numbness, weakens the shoulder and arm and lymphedemal. Sometimes ribs can also get damaged leading to permanent weakness of the ribs. In brachytherapy, which is also called as internal radiation, a radioactive pellets holding device is placed into breast tissues for a short period of time from where the tumor has been removed. There are two types of it intracavitary (Device is placed into the space, or small catheter is placed) and interstitial (Catheter is placed and device is placed in it for short period of time and is than removed). The side effects of brachytherapy include redness or bruising at the treatment site, pain and infection, fatty tissue damage and collection of fluid in the breast (Seroma) [24, 31-33].

3.3. Systematic Treatment

3.3.1. Chemotherapy

Chemotherapy also called as chemo is therapy which makes use of anticancer drugs which may be given orally or intravenously. The anticancer drug then reaches cancer cells by travelling through the blood stream or the lymphatic system. These drugs can be given before (neoadjuvant) or after (adjuvant) surgery. The drugs that are used for breast cancer are doxorubicin, epirubicin, paclitaxel, docetaxel, 5-fluorouracil, cyclophosphamide, carboplatin, vinorelbine, capecitabine, eribulin, ixabepilone and gemcitabine. These drugs are used in combination to treat breast cancer but in case of advanced stage single drug is used for treatment [35]. The common side effects of chemotherapy include diarrhea, mouth sores, hair fall, loss of weight, vomiting and nausea. Other major side effects of chemotherapy are menstrual changes, fertility issues, affects blood forming cells leading to infection due to low white blood cell count, Neuropathy (Nerve damage), foot-hand syndrome, chemo brain, increases risk of leukemia and fatigue [34].

3.4. Treatment of Breast Cancer by Stage

The localized treatment and the systematic treatments are used in various combinations to treat the various stages of breast cancer. Ductal carcinoma in site (DCIS) also called as stage 0 of breast cancer is mainly treated by carrying out breast conservation surgery (BCS) followed by radiation. If mastectomy is done then the nearby lymph nodes are checked to carry out sentinel lymph node biopsy (SLNB) or axillary lymph node dissection (ALND). Radiation is mainly not needed in case of mastectomy. For stage 1 breast cancer BCS and SNLB is followed by chemotherapeutic drugs depending on the size of the tumor. For stage 2 breast cancer mastectomy is done and nearby lymph nodes are checked for SLNB or ALND. The chemotherapy is given before or after the surgery depending on the size of the tumor. For stage 4 breast cancer the neoadjuvant chemo is given first to shrink the tumor enough to carry out BCS or mastectomy. Radiation is given after the surgery accompanied with adjuvant chemotherapy. Stage 4 breast cancers spreads beyond the breast and lymph nodes i.e., liver, lungs, bone and sometimes brain or the other organs. In certain cases surgery can be done or the radiation therapy is given to shrink the tumor followed by combinations of systematic therapy. It is mainly incurable. Recurrent breast cancer can be regional which is treated with localized treatment and can be distant which is treated like stage 4 depending on the health of the patient [24, 35].

Thus, the existing treatment of breast cancer has its own prons and cons. These treatments get costly as well as affect the health of the patient making them sometimes so weak to survive. Cancer itself is an unbearable pain to the patient and the family and the existing treatment is able to relive the pain only to some extent. So the new modalities are taken into account to lessen the pain even more and the cost. Various phytochemicals are looked into for the development of phytotherapy in breast cancer.

4. Programmed Cells with Genome Editing

DNA modifications involved in genome editing are mainly through replacement, deletion or insertion. The genome editing method involves endonucleases that makes double strand break in DNA. The breaks can be than repaired either by non-homologous end joining (NHEJ) or homology directed repair (HDR).

NHEJ is error prone as there the ligation takes place between the breaks by insertion or by deletion. HDR uses donor or template DNA for ligation which increases decisive changes that shall help gene to work properly (Figure 2)[14, 36, 37].

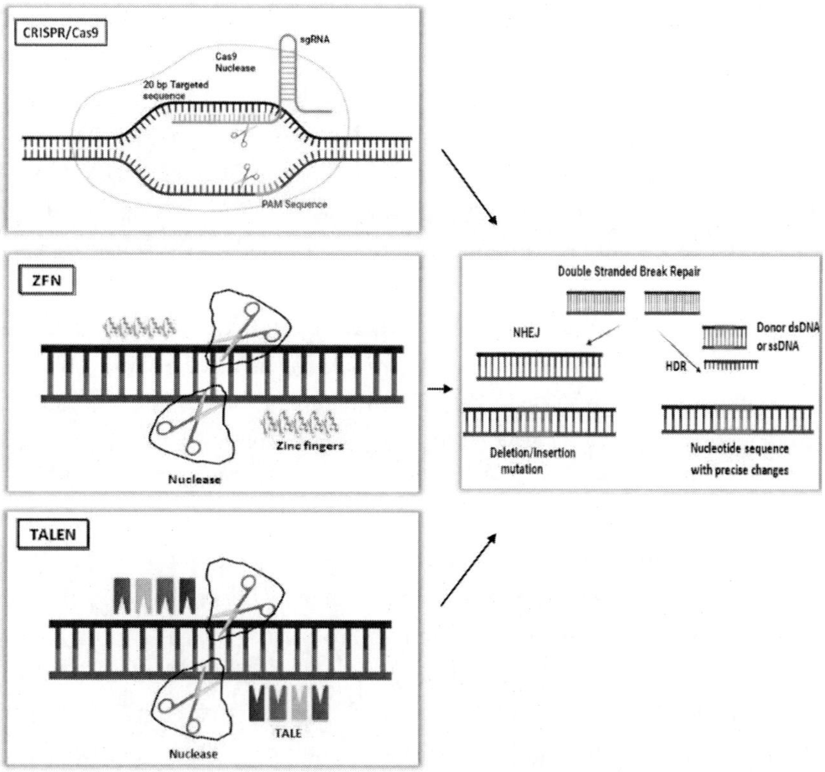

Figure 2. Genome editing by applying various nucleases and double strand break repair. NHEJ is prone to error and HDR is precised correction. (Modified from [14, 37]). Zinc finger nucleases (ZFN), transcription activator-like effector nucleases (TALEN), clustered regularly interspaced short palindromic repeat- Cas- associated nucleases (CRISPR-Cas9), homology-directed repair (HDR) and nonhomologous end joining (NHEJ).

Programmable nucleases are developed for genome editing systems like Zinc finger nucleases (ZFNs), transcription activator like effector nucleases (TALEN) and clustered regularly interspaced short palindromic repeats-cas9 (CRISPR-Cas9) system the possible treatment modalities. Genome editing comprises of DNA binding domain and DNA cleavage domain where DNA binding domain consists of specific sequence for targeting DNA site and cleavage domain consists of endonuclease to cleave double stranded DNA. ZFNs are prepared by combining the Zinc Finger (DNA binding domain) and Fok1 restriction endonuclease (DNA cleavage domain). TALENs are prepared by TALE (a DNA binding domain) and Fok1 restriction endonuclease (DNA cleavage domain). CRISPR-Cas9 has two components single stranded guide RNA (sgRNA) and Cas 9 endonuclease. sgRNA is designed to complement the targeted site of DNA. The design is such that it follows the sequence i.e., protospacer adjacent motif (PAM) which is necessary for cas9 compatibility to the targeted site of the DNA. PAM are short DNA sequences like NGG or NAG. CRISPR-cas9 helps in small insertion and deletion can with the help of highly integrated HDR. ZFN and TALEN require engineering of endonucleases every time when the target sequence changes. In CRISPR-cas9 the cas-0 remains the same always only the guide RNA sequence changes as per the target site. TALEN and ZFN are lot more expensive as well as laborious. CRISPR-cas 9 is easy as well as promising and multiple targets can be edited (Table2)[37-40].

CRISPR-Cas is mainly graded into two classes which is further classified into 6 types and several more subtypes. This classification is based on the recognition and cleavage mechanism. Class 1 proteins effects cleavage while class 2 protein focuses on genome editing. So class 1 is not used much extensively. Class 2 has 3 types namely type II, type V and type VI. Type II recognizes PAM site and utilizes Cas9 for cleavage. Type V recognizes PAM [41] and utilizes Cas12a and Cpf for cleavage. Class VI on contrarily recognizes protospacer flanking sequence (PFS) [42] and utilizes Cas13 for cleavage. Cas9 go through the conformational changes after binding to the sgRNA which helps to look for PAM sequence to cleave the targeted site three base pair upstream to

PAM [43]. The cleavage depends on the target site complimentary sequence and not just the PAM. Cas9 is found in *Streptococcus pyogenes* and is widely used for genome editing compared to other similar proteins derived from other species of bacteria. CRISPR-Cas9 is widely used for genome editing due to its simplicity, ease of design, cheaper and do not require protein engineering for targeting[14, 44].

Table 2. Comparison of genome editing platforms (Modified from [15, 37])

Platforms	CRISPR/Cas9	ZFN	TALEN
Source of Nuclease	Bacteria and archaea	Bacteria	Bacteria (*Xanthomonas* sp.)
Components	2 (sgRNA and Cas9)	2 (ZF and Fok1)	2 (TALE and Fok1)
Detection/ Recognition	sgRNA	Zinc finger protein	Repeat variable di residue region of TALE protein
Modifying component	Cas9	Fok1	Fok1
Sequence size of target (bp)	17-20 plus PAM sequence	9-18 (ZFN monomer) 18-36 (ZFN dimer)	14-20 (TALEN monomer) 28-40 (TALEN dimer)
Specificity	Low-Moderate (eSpCas9 shows reported improved specificity lately)	Low-moderate	Moderate
Patter of DSB	5' overhang bunt ends by SpCas9 and staggered cut by Cpf1	5' overhang staggered cut	Heterogeneous overhang staggered cut
Designing/ Engineering	Easy (requires standard cloning and oligosynthesis)	Difficult (Requires protein engineering)	Moderate (requires complex cloning)
Delivery modes	Viral transduction/ Electroporation[a]	Viral transduction/ Electroporation	Viral transduction/ Electroporation

[a]Nanoparticle carrier improves delivery. Zinc finger nucleases (ZFN), transcription activator-like effector nucleases (TALEN), clustered regularly interspaced short palindromic repeat- Cas-associated nucleases (CRISPR-Cas9).

5. Role and Therapeutic Ability of CRISPR-Cas9 in Breast Cancer

The development of cancer involves mainly two types of gene protooncogenes and tumor suppressor genes (TSGs). Protooncogenes when mutated helps tumor to grow and when TSGs are mutated there is uncontrolled growth leading to the genetic instability. CRISPR-Cas9 can target such mutations and hence shows better ability in treating breast cancer. The therapeutic ability of CRISPR-Cas9 includes targeting TSGs, oncogenes, DNA repair pathways, kinome and cellular adhesion. It also helps in altering the epigenetics effects and drug resistance [14].

5.1. Oncogenes and TSGs

Oncogenes like *PI3KCA, MYC* and HER2 [45] are involved in breast cancer. Knocking out of oncogene or the methylation of the binding sites of these genes slows down the process of proliferation and reduced growth of tumor. Numerous TSGs involved in breast cancer like *BRCA1*, *PTEN* and *BRCA2* [46, 47]. TSGs function is suppressed by defect in the regulation of transcriptional factors or the promoter region hypermethylation. To regain the function of TSGs is not as easy as knocking out the function of the gene. CRISPR-Cas9 helps to suppress the effects of hypermethylation or the improper functioning of the transcriptional genes. The approach of CRISPR in which dCas9 fuses with VP64, p65 and Rta (VPR) transcription activators which helps to activate TSGs that are deactivated due to improper functioning of transcriptional factors [48, 49]. The fusion of dCas9 with ten-eleven translocation (dCas9-TET) dioxygenases helps to demethylate the TSGs by converting 5'-methyl cytosine (5'-mC) to 5'- hydroxyl methyl cytosine (5'-hmc)[14].

5.2. DNA Repair Pathways

CRISPR when targets *PARP1* gene conjugated with platinum coated chemotherapies shows efficient cytotoxicity against breast cancer cells. PARP enzyme is involved in repair of single stranded breaks. So by targeting this DNA repair pathway and by upregulating the *PARP1* gene may work when BRCA fails [14].

5.3. Kinome Targeting

Another such target is a kinase protein which also refers to kinome. Cyclin dependent kinases (CDKs) helps to regulate the progression of cell cycle. CRSPR-Cas9 can knockout CDK2 resulting in cell cycle arrest and CDKs can serve as potential targets for the treatment [50]. *Cellular Adhesion.* Integrins are protein that helps cells to adhere to the extracellular matrix. When integrin is mutated it helps in metastasis resulting in tumor progression. CRISPR helps to knockout this protein and shows effective result in reducing tumor growth and metastasis ability of the breast cancer cells [14, 51].

5.4. Epigenetic Alteration

The epigenetics effects include i) DNA methylation, ii) non coding RNAs and iii) histone modifications. DNA methylation or hypermethylation is already discussed in this paragraph above. ii) non-coding RNAs: Non coding RNAs are transcribed from non-coding DNA. Mostly studied non coding RNAs are long non coding RNA and microRNA. Long non coding RNA effects mRNA by altering transcriptional factors and state of chromatin. Micro-RNA combines with its complementary mRNA and the mRNA is degraded or the translation is hindered. CRISPR-Cas9 knocks out long non coding RNA and microRNA at multiple sites making it dysfunctional. To avoid insertion or deletion CRISPR-cas9 targets multiple sites making it complete knockout. iii) Histone modifications: Histone helps in keeping DNA intake as the protein is positively charged and the negatively charged DNA can easily wound around the protein. Histone modifications may result in loose wounding resulting in more DNA transcription. The modification is mainly methylation or acetylation of the protein [52, 53]. Uncontrolled transcription may result in uncontrolled growth of the cell resulting into tumor formation or promotes tumor growth. So CRISPR-Cas9 can target overexpressed enzymes like acetyltransferases and demethylases by fusion of dCas9 to it. This binding helps to reverse the abnormal modifications and helps to activate TSGs [14].

5.5. Drug Resistance

Drug resistance can be due to single nucleotide polymorphism (SNP) or the efflux pumps which do not let the drug accumulate in the breast cancer cells in required amount [54] SNP and epigenetic changes can be edited using CRISPR-Cas9 as described earlier. Various transporters present on the cell like multidrug resistance associated protein 1 (MRP1), Breast cancer resistance protein (BCRP) and P-glycoprotein (P-gp) shows major role in drug resistance. CRISPR-cas9 mediated editing can help to knockout the function of these efflux pumps making it again sensitive towards the therapy [14].

6. Role of Nanomedicine in Breast Cancer Treatment

The rapid development of nanotechnology over the past years has invented many nanoparticles but only few are used in biomedical application and can meet the requirements of FDA for its clinical application. Nanoparticle to be used as biomaterial should be biocompatible and non toxic [55]. Almost 240 nanoparticles strategies were screened and it was concluded that to find a nanoparticle which suits all the desired characteristics is difficult to get. But there are these few therapeutic systems which are nanoparticle based and can be considered as promising treatment of cancer [56].

There has been a significant effort in the past to improvise the breast cancer treatment. Nanomedicines have properties like prolonged blood circulation, accumulation of drug at tumor site which helps to overcome the biological hurdles. The other properties like therapeutic drug stability, solubility is increased and there is control on release as the drug is either binding or is encapsulated in nanocarriers. The combination of drug can also be used to increase the anti-cancer effect [57, 58].The characteristics and role are summarized in Table 3.

Table 3. Summary of characteristics and role of nanocarriers [55-58]

Sr. No.	Drug Therapy challenges in Breast cancer	Characteristics of Nanocarriers	Role of Nanomedicine
1	Drug specificity insufficiency	Drug is released only at targeted sites helps to lower dose of drug	Passive or active targeting to increase drug level at tumor site and lessen the drug in nanocarrier
2	Short life and early clearance of drug	Considerable time of circulation	PEGylation extends the circulation time
3	Toxicity due to dose limiting	Control release of drug extrinsically	The release of drug is controlled by the nanocarrier
4	Resistance of drug at cellular and at tumor microenvironment level	Targets the required cells or tissues	Inhibits mechanism of drug efflux and targets the microenvironment of the tumor
5	Cancer stem cells are difficult to eradicate	Targeted carriers eradicate each cells	It targets the cancer stem cells
6	Sequence and dose schedule of multiple drugs is suboptimized	Carry more than one drug for combination therapy	Co-delivery of drugs is carefully optimized
7	Metastatic sites like the bones and brain are not accessible	Small size so can skip the immune system and reach the deeper areas	Nanocarreirs may improve the penetration inherently
8	Less solubility and stability in vivo of the drugs	Encapsulation makes the drug stable	Protects unstable drugs and help to solubilize.

7. Nanocarriers Entrapped CRISPR-Cas9 in Breast Cancer

Remedial potency of CRISPR-Cas9 depends on the competency of it to reach the targeted site with nominal biodegradation. The delivery of the components of the CRISPR-Cas9 is carried out in plasmid form, combinational (mRNA of Cas9 and sgRNA) form or as ribonucleoprotein form. Plasmid is easy to synthesis and is stable but the large size and the negative charge may create a problem in delivery [59]. Plasmid undergoes transcription and translation which elongates or delays the editing and off targeting also increases [60]. The combinational form involves RNA which complicates the delivery due to its instability [61]. The ribonucleoprotein is rapid and effective but the large size and net negative charge of the complex makes it difficult to enter the cell [62].

The delivery system for CRISPR-Cas9 various forms is broadly divided into three categories physical, viral and non-viral. Physical methods like microinjection, electroporation and hydrodynamic injection are applied for the delivery of the CRISPR-Cas9. In microinjection delivery system the microsized needle is used to insert the components into the cell. Electroporation is powerful method and macromolecules can enter any cell using this technique. These two physical techniques can cause a potential damage to the cells as well as there can be no specific delivery of the components. Hydrodynamic injections involve intravenous injection of the macromolecules. Target specificity also decreases in the technique. The second group i.e., delivery using viral vectors is specific; efficient as well as stability of the components can also be maintained. The main drawback of this technique is that use of viral vectors can induce immunogenicity; carcinogenicity; mutagenicity as well as loading macromolecules may not be possible [63-66].

Non-viral vector such as nanoparticles in particular, are gaining attention to deliver the components of gene editing. Nanoparticles range from 1 to 100 nanometer in size shows great ability to deliver the components as they show very low immunogenicity, large capacity to load the components due to high surface to area to volume ratio and is widely used as targeted therapy in treating cancer [14, 66]. Large loading capacity makes the components stable and loss of functionality or biodegradation of the components is decreased. Moreover, the circulation time increases due to low immunogenicity making its more target specific. Various nanoparticles are used for treating cancers. Hybrid nanocarriers are today's modern approach as it involves more than one nanoparticle and so to load components of CRISPR-cas9 gets easy and the therapy can get more effective. Various hybrid nanocarrier systems used in past recent years in various fields [67, 68]. Among these hybrid nanocarriers, hybrid nanoclay system is of importance due to its property like high surface area, cation exchange capacity, biodegradable, low cost, low toxicity and molecules intercalating capacity. Montmorillonite (MMT) is one of the constitutents of bentonite and is having wide application and can be used as delivery system for the treatment of cancer [69, 70]. Montmorillonite acts as a detoxifier and its non-toxicity has been proved hematologically, histopathologically and biochemically. An enormous deal of consideration is being given to MMT for its ability of controlled drug release, mucoadhesiveness and detoxification effect. MMT is been used in delivery of CRISPR-Cas9 for treating cervical cancer [71]. Zinc oxide nanoparticle have shown anticancer property and its application is been studied in many cancers including breast cancer [72]. MMT can also be used in combination with zinc oxide nanoparticle (FDA approved) as a hybrid nanocarrier to increase the target specificity for the delivery of components of CRISPR-Cas9 as well as enhancing the effect of the treatment.

FUTURE PROSPECTS AND CONCLUSION

Genome editing has been extensively studied in the past recent years for the treatment of various cancers. As discussed in this review there are many genome editing systems used for the treatment of breast cancer. After carrying out survey related to the ease of synthesis and target specific delivery of the components of the different genome editing system, CRISPR-Cas9 was scoring as best candidate to be used for treatment of cancer. Moreover, very preliminary work is done in area of treating breast cancer using CRISPR-Cas9 genome editing system. The work needs to be carried out extensively to overcome the problems and carry the CRISPR-Cas9 genome editing system to clinical trials. Later on the role and function of nanocarriers were discussed in breast cancer. Many nanocarriers are gaining importance for purpose of drug delivery in various cancers and hybrid nanocarriers shall suffice the needs of it. Zinc oxide has a FDA approval and Montmorillonite in gaining as nanoclay due to its outstanding properties of drug delivery. So far both the nanoparticles are not combined and used for breast cancer treatment. Thus, if a hybrid nanoparticle is prepared using Zinc oxide and Montmorillonite along with CRISPR-Cas9 than efforts can be made to increase the effect of the treatment. The concluding notes of this review states the use of hybrid nanocarriers along with CRISPR-Cas9 genome editing system for the proper and personalized treatment of the breast cancer for the well-being of the society.

REFERENCES

[1] Akram, Muhammad, Mehwish Iqbal, Muhammad Daniyal, and AsmatUllah Khan. "Awareness and current knowledge of breast cancer."*Biological research* 50, no. 1 (2017): 1-23.

[2] World Health Organization (2020). *Global Cancer Profile 2020.* https://gco.iarc.fr/today/data/factsheets/cancers/20-Breast-fact-sheet.pdf

[3] Mathur, Prashant, Krishnan Sathishkumar, Meesha Chaturvedi, Priyanka Das, Kondalli Lakshminarayana Sudarshan, Stephen Santhappan, Vinodh Nallasamy et al. "Cancer statistics, 2020: report from national cancer registry programme, India." *JCO Global Oncology* 6 (2020): 1063-1075.

[4] Chaturvedi, Meesha, K. Vaitheeswaran, K. Satishkumar, Priyanka Das, S. Stephen, and A. Nandakumar. "Time trends in breast cancer

among Indian women population: an analysis of population based cancer registry data." *Indian journal of surgical oncology* 6, no. 4 (2015): 427-434.

[5] Takiar, R. "Status of breast and cervix cancer in selected registries of India." *Ann Womens Health.* 2018; 2 (1) 1012 (2018).

[6] Xie, Mingsi, Jie Liu, RongTsao, Ziyuan Wang, Baoguo Sun, and Jing Wang. "Whole grain consumption for the prevention and treatment of breast cancer." *Nutrients* 11, no. 8 (2019): 1769.

[7] Dai, Xiaofeng, Hongye Cheng, Zhonghu Bai, and Jia Li. "Breast cancer cell line classification and its relevance with breast tumor subtyping." *Journal of Cancer* 8, no. 16 *(2017): 3131.*

[8] Diana Rodriguez (2010, February 9). *Cancer Treatment: Know Your options.* Retrieved from https://www.everydayhealth.com/cancer/understanding/treatment-methods.aspx

[9] American Society of Clinical Oncology (ASCO). *What is cancer surgery?* Accessed at https://www.cancer.net/navigating-cancer-care/how-cancer-treated/surgery/what-cancer-surgeryon October 2, 2019.

[10] National Cancer Institute (NCI). *Surgery to treat cancer.* Accessed at https://www.cancer.gov/about-cancer/treatment/types/surgeryon October 3, 2019.

[11] Brassil, Kelly J. Chemotherapy and immunotherapy guidelines and recommendations for practice. *Oncology Nursing Society*, 2019.

[12] Chu, Edward, and Vincent T. DeVita Jr. *Physicians' Cancer Chemotherapy Drug Manual 2020.* Jones & Bartlett Learning, 2019.

[13] Olsen MiKaela M, Naseman Ryan W. Chemotherapy. In Olsen MM, LeFebvre KB, Brassil KJ, eds. *Chemotherapy and Immunotherapy Guidelines and Recommendations for Practice.* Pittsburgh, PA: Oncology Nursing Society; 2019:61-90.

[14] Padayachee, Jananee, and Moganavelli Singh. "Therapeutic applications of CRISPR/Cas9 in breast cancer and delivery potential of gold nanomaterials." *Nanobiomedicine* 7 (2020): 1849543520983196.

[15] Guha, Tuhin Kumar, Alvan Wai, and Georg Hausner. "Programmable genome editing tools and their regulation for efficient genome engineering." *Computational and structural biotechnology journal* 15 (2017): 146-160.

[16] Glass, Zachary, Matthew Lee, Yamin Li, and Qiaobing Xu. "Engineering the delivery system for CRISPR-based genome editing." *Trends in biotechnology* 36, no. 2 (2018): 173-185.

[17] Shi, Jinjun, Philip W. Kantoff, Richard Wooster, and Omid C. Farokhzad. "Cancer nanomedicine: progress, challenges and opportunities." *Nature reviews cancer* 17, no. 1 (2017): 20.

[18] van der Meel, Roy, Einar Sulheim, Yang Shi, Fabian Kiessling, Willem JM Mulder, and Twan Lammers. "Smart cancer nanomedicine." *Nature nanotechnology* 14, no. 11 (2019): 1007-1017.

[19] Li, Ling, Shuo Hu, and Xiaoyuan Chen. "Non-viral delivery systems for CRISPR/Cas9-based genome editing: Challenges and opportunities." *Biomaterials* 171 (2018): 207-218.

[20] Mendes, Rita, Alexandra R. Fernandes, and Pedro V. Baptista. "Gold nanoparticle approach to the selective delivery of gene silencing in cancer—the case for combined delivery?." *Genes* 8, no. 3 (2017): 94.

[21] Tharmapalan, Pirashaanthy, Mathepan Mahendralingam, Hal K. Berman, and Rama Khokha. "Mammary stem cells and progenitors: targeting the roots of breast cancer for prevention." *The EMBO journal* 38, no. 14 (2019): e100852.

[22] Larsen, Martin J., Torben A. Kruse, Qihua Tan, Anne-VibekeLaenkholm, Martin Bak, Anne E. Lykkesfeldt, Kristina P. Sørensen et al. "Classifications within molecular subtypes enables identification of BRCA1/BRCA2 mutation carriers by RNA tumor profiling." *PloS one* 8, no. 5 (2013): e64268.

[23] Evans, D. Gareth, Sacha J. Howell, and Anthony Howell. "Personalized prevention in high risk individuals: Managing hormones and beyond." *The Breast* 39 (2018): 139-147.

[24] American Cancer Society (ACS) (2019, September 18). *Treating Breast Cancer*. Retrieved from: https://www.cancer.org/content/dam/CRC/PDF/Public/8581.00.pdf.

[25] Archana, Kumar, A., & Parashar, B. (2020). Breast Cancer: It's just a thing. *European Journal of Pharmaceutical and Medical Research*, 7(8), 241-250.

[26] Moo, Tracy-Ann, Rachel Sanford, Chau Dang, and Monica Morrow. "Overview of breast cancer therapy." *PET clinics* 13, no. 3 (2018): 339-354.

[27] Waks, Adrienne G., and Eric P. Winer. "Breast cancer treatment: a review." *JAMA* 321, no. 3 (2019): 288-300.

[28] Rossi, Giovanna, Zhaomei Mu, Alfred W. Rademaker, Laura K. Austin, Kimberly S. Strickland, Ricardo Lima Barros Costa, Rebecca J. Nagy et al. "Cell-free DNA and circulating tumor cells: comprehensive liquid biopsy analysis in advanced breast cancer." *Clinical Cancer Research* 24, no. 3 (2018): 560-568.

[29] Carbine, Nora E., Liz Lostumbo, Judi Wallace, and Henry Ko. "Risk-reducing mastectomy for the prevention of primary breast cancer." *Cochrane Database of Systematic Reviews* 4 (2018).

[30] Smith, Barbara L., Rong Tang, Upahvan Rai, Jennifer K. Plichta, Amy S. Colwell, Michele A. Gadd, Michelle C. Specht, William G. Austen Jr, and Suzanne B. Coopey. "Oncologic safety of nipple-sparing mastectomy in women with breast cancer." *Journal of theAmerican College of Surgeons* 225, no. 3 (2017): 361-365.

[31] Ajkay, Nicolas, Abigail E. Collett, Erica V. Bloomquist, Edward J. Gracely, Thomas G. Frazier, and Andrea V. Barrio. "A comparison of complication rates in early-stage breast cancer patients treated with brachytherapy versus whole-breast irradiation." *Annals of surgical oncology* 22, no. 4 (2015): 1140-1145.

[32] Castaneda, Serguei A., and Jon Strasser. "Updates in the treatment of breast cancer with radiotherapy." *Surgical Oncology Clinics* 26, no. 3 (2017): 371-382.

[33] Deng, Xinna, Haijiang Wu, Fei Gao, Ye Su, Qingxia Li, Shuzhen Liu, and Jianhui Cai. "Brachytherapy in the treatment of breast

cancer." *International journal of clinical oncology* 22, no. 4 (2017): 641-650.

[34] Furmaniak, Anna C., Matthias Menig, and Martina H. Markes. "Exercise for women receiving adjuvant therapy for breast cancer." *Cochrane Database of Systematic Reviews* 9 (2016).

[35] Henry, N. L., P. D. Shah, I. Haider, P. E. Freer, R. Jagsi, and M. S. Sabel. "Chapter 88: cancer of the breast." *Abeloff's clinical oncology,* 6th edn. Elsevier, Philadelphia, PA (2020).

[36] Zhan, Tianzuo, Niklas Rindtorff, Johannes Betge, Matthias P. Ebert, and Michael Boutros. "CRISPR/Cas9 for cancer research and therapy." In *Seminars in cancer biology*, vol. 55, pp. 106-119. Academic Press, 2019.

[37] Li, Hongyi, Yang Yang, Weiqi Hong, Mengyuan Huang, Min Wu, and Xia Zhao. "Applications of genome editing technology in the targeted therapy of human diseases: mechanisms, advances and prospects." *Signal transduction and targeted therapy* 5, no. 1 (2020): 1-23.

[38] Sternberg, Samuel H., Sy Redding, Martin Jinek, Eric C. Greene, and Jennifer A. Doudna. "DNA interrogation by the CRISPR RNA-guided endonuclease Cas9."*Nature* 507, no. 7490 (2014): 62-67.

[39] Cox, David Benjamin Turitz, Randall Jeffrey Platt, and Feng Zhang. "Therapeutic genome editing: prospects and challenges." *Nature medicine* 21, no. 2 (2015): 121-131.

[40] Paschon, David E., Stephanie Lussier, Tenzin Wangzor, Danny F. Xia, Patrick W. Li, Sarah J. Hinkley, Nicholas A. Scarlott et al. "Diversifying the structure of zinc finger nucleases for high-precision genome editing." *Nature communications* 10, no. 1 (2019): 1-12.

[41] Gleditzsch, Daniel, Patrick Pausch, Hanna Müller-Esparza, Ahsen Özcan, Xiaohan Guo, Gert Bange, and Lennart Randau. "PAM identification by CRISPR-Cas effector complexes: diversified mechanisms and structures." *RNA biology* 16, no. 4 (2019): 504-517.

[42] Burmistrz, Michal, Kamil Krakowski, and Agata Krawczyk-Balska. "RNA-targeting CRISPR–Cas systems and their applications." *International journal of molecular sciences* 21, no. 3 (2020): 1122.

[43] Le Rhun, Anaïs, Andrés Escalera-Maurer, Majda Bratovič, and Emmanuelle Charpentier. "CRISPR-Cas in Streptococcus pyogenes." *RNA biology* 16, no. 4 (2019): 380-389.

[44] Lone, Bilal Ahmad, Shibendra Kumar Lal Karna, Faiz Ahmad, NerinaShahi, and Yuba Raj Pokharel. "CRISPR/Cas9 system: a bacterial tailor for genomic engineering." *Genetics research international* 2018 (2018).

[45] Pranavathiyani, G., Raja Rajeswary Thanmalagan, Naorem Leimarembi Devi, and Amouda Venkatesan. "Integrated transcriptome interactome study of oncogenes and tumor suppressor genes in breast cancer." *Genes & diseases* 6, no. 1 (2019): 78-87.

[46] Hou, Sheng-Qi, Meng Ouyang, Andrew Brandmaier, HongboHao, and Wen H. Shen. "PTEN in the maintenance of genome integrity: From DNA replication to chromosome segregation." *Bioessays* 39, no. 10 (2017): 1700082.

[47] Gorodetska, Ielizaveta, IrynaKozeretska, and Anna Dubrovska. "BRCA genes: the role in genome stability, cancer stemness and therapy resistance." *Journal of Cancer* 10, no. 9 (2019): 2109.

[48] Moses, Colette, Fiona Nugent, Charlene Babra Waryah, Benjamin Garcia-Bloj, Alan R. Harvey, and Pilar Blancafort. "Activating PTEN tumor suppressor expression with the CRISPR/dCas9 system." *Molecular Therapy-Nucleic Acids* 14 (2019): 287-300.

[49] Li, Shuting, Yanwei Shen, Mengying Wang, Jiao Yang, MengLv, Pan Li, Zheling Chen, and Jin Yang. "Loss of PTEN expression in breast cancer: association with clinicopathological characteristics and prognosis." *Oncotarget* 8, no. 19 (2017): 32043.

[50] Liu, Houguang, Zheng Li, Shanshan Huo, Qiongling Wei, and Ling Ge. "Induction of G0/G1 phase arrest and apoptosis by CRISPR/Cas9-mediated knockout of CDK2 in A375 melanocytes." *Molecular and clinical oncology* 12, no. 1 (2020): 9-14.

[51] Xiao, Yajuan, Yunfei Li, Hua Tao, Brock Humphries, Aimin Li, Yiguo Jiang, Chengfeng Yang, Rongcheng Luo, and Zhishan Wang. "Integrin α5 down-regulation by miR-205 suppresses triple negative breast cancer stemness and metastasis by inhibiting the Src/Vav2/Rac1 pathway." *Cancer letters* 433 (2018): 199-209.

[52] Yen, Ching-Yu, Hurng-Wern Huang, Chih-Wen Shu, Ming-Feng Hou, Shyng-Shiou F. Yuan, Hui-Ru Wang, Yung-Ting Chang, Ammad Ahmad Farooqi, Jen-Yang Tang, and Hsueh-Wei Chang. "DNA methylation, histone acetylation and methylation of epigenetic modifications as a therapeutic approach for cancers." *Cancer letters* 373, no. 2 (2016): 185-192.

[53] Xiao, Bo, Shigang Yin, Yang Hu, Maoxin Sun, Jieqiong Wei, Zhenghui Huang, Yuhao Wen et al. "Epigenetic editing by CRISPR/dCas9 in *Plasmodium falciparum*." *Proceedings of the National Academy of Sciences* 116, no. 1 (2019): 255-260.

[54] Ha, Jong Seong, Juyoung Byun, and Dae-Ro Ahn. "Overcoming doxorubicin resistance of cancer cells by Cas9-mediated gene disruption." *Scientific reports* 6, no. 1 (2016): 1-7.

[55] Adair, James H., Mylisa P. Parette, Erhan I. Altınoglu, and Mark Kester. "Nanoparticulate alternatives for drug delivery."*ACS nano* 4, no. 9 (2010): 4967-4970.

[56] Tang, Xiaomeng, Welley S. Loc, Cheng Dong, Gail L. Matters, Peter J. Butler, Mark Kester, Craig Meyers, Yixing Jiang, and James H. Adair. "The use of nanoparticulates to treat breast cancer." *Nanomedicine* 12, no. 19 (2017): 2367-2388.

[57] Di Wu, Mengjie Si, Hui-Yi Xue, and Ho-Lun Wong. "Nanomedicine applications in the treatment of breast cancer: current state of the art." *International journal of nanomedicine* 12 (2017): 5879.

[58] Zhang, RuiXue, Ho Lun Wong, Hui Yi Xue, June Young Eoh, and Xiao Yu Wu. "Nanomedicine of synergistic drug combinations for cancer therapy–Strategies and perspectives." *Journal of Controlled Release* 240 (2016): 489-503.

[59] Ran, F. Ann, Le Cong, Winston X. Yan, David A. Scott, Jonathan S. Gootenberg, Andrea J. Kriz, Bernd Zetsche et al. "*In vivo* genome editing using Staphylococcus aureus Cas9." *Nature* 520, no. 7546 (2015): 186-191.

[60] Kim, Sojung, Daesik Kim, Seung Woo Cho, Jungeun Kim, and Jin-Soo Kim. "Highly efficient RNA-guided genome editing in human cells via delivery of purified Cas9 ribonucleoproteins." *Genome research* 24, no. 6 (2014): 1012-1019.

[61] Miller, Jason B., Shuyuan Zhang, Petra Kos, Hu Xiong, Kejin Zhou, Sofya S. Perelman, Hao Zhu, and Daniel J. Siegwart. "Non-viral CRISPR/Cas gene editing *in vitro* and *in vivo* enabled by synthetic nanoparticle co-delivery of Cas9 mRNA and sgRNA." *Angewandte Chemie International* Edition 56, no. 4 (2017): 1059-1063.

[62] Zuris, John A., David B. Thompson, Yilai Shu, John P. Guilinger, Jeffrey L. Bessen, Johnny H. Hu, Morgan L. Maeder, J. Keith Joung, Zheng-Yi Chen, and David R. Liu. "Cationic lipid-mediated delivery of proteins enables efficient protein-based genome editing *in vitro* and *in vivo*." *Nature biotechnology* 33, no. 1 (2015): 73-80.

[63] Chandrasekaran, Arun Pandian, Minjung Song, Kye-Seong Kim, and Suresh Ramakrishna. "Different methods of delivering CRISPR/Cas9 into cells." *Progress in molecular biology and translational science* 159 (2018): 157-176.

[64] Wang, Le, Wenfu Zheng, Shaoqin Liu, Bing Li, and Xingyu Jiang. "Delivery of CRISPR/Cas9 by novel strategies for gene therapy." *Chem Bio Chem* 20, no. 5 (2019): 634-643.

[65] Lino, Christopher A., Jason C. Harper, James P. Carney, and Jerilyn A. Timlin. "Delivering CRISPR: a review of the challenges and approaches." *Drug delivery* 25, no. 1 (2018): 1234-1257.

[66] Rahimi, Hossein, Marziyeh Salehiabar, Jalil Charmi, Murat Barsbay, Mohammadreza Ghaffarlou, Mahdi Roohi Razlighi, Soodabeh Davaran et al. "Harnessing nanoparticles for the efficient delivery of the CRISPR/Cas9 system." *Nano Today* 34 (2020): 100895.

[67] Senapati, Sudipta, Arun Kumar Mahanta, Sunil Kumar, and PralayMaiti. "Controlled drug delivery vehicles for cancer treatment and their performance." *Signal transduction and targeted therapy* 3, no. 1 (2018): 1-19.

[68] Mousa, Mohamed, Nicholas D. Evans, Richard OC Oreffo, and Jonathan I. Dawson. "Clay nanoparticles for regenerative medicine and biomaterial design: A review of clay bioactivity." *Biomaterials* 159 (2018): 204-214.

[69] Bee, Soo-Ling, M. A. A. Abdullah, Soo-Tueen Bee, Lee Tin Sin, and A. R. Rahmat. "Polymer nanocomposites based on silylated-montmorillonite: A review." *Progress in polymer scienc*e 85 (2018): 57-82.

[70] Tipa, Cezar, Maria T. Cidade, Tânia Vieira, Jorge Carvalho Silva, Paula IP Soares, and João Paulo Borges. "A New Long-Term Composite Drug Delivery System Based on Thermo-Responsive Hydrogel and Nanoclay." *Nanomaterials* 11, no. 1 (2021): 25.

[71] Niu, Gang, Zhuang Jin, Chong Zhang, Dan He, XueqinGao, Chenming Zou, Wei Zhang et al. "An effective vaginal gel to deliver CRISPR/Cas9 system encapsulated in poly (β-amino ester) nanoparticles for vaginal gene therapy." *E Bio Medicine* 58 (2020): 102897.

[72] Jin, Su-Eon, and Hyo-Eon Jin. "Synthesis, characterization, and three-dimensional structure generation of zinc oxide-based nanomedicine for biomedical applications." *Pharmaceutics* 11, no. 11 (2019): 575.

In: The Future of Metabolic Engineering ISBN: 978-1-68507-362-6
Editors: A. Sharma and D. Amin

Chapter 4

INVERTEBRATE AND VERTEBRATE CELL CULTURES IN BIO-THERAPEUTICS

Krutika Saurabh Abhyankar*, PhD
Division of Biomedical and Life Sciences, School of Science,
Navrachana University, Vadodara, India

ABSTRACT

Emerging and reemerging infectious diseases drive and result in the development of bio-therapeutics for human health care. Animal biotechnology-based innovative techniques and strategies for developing new animal cell lines from various organisms belonging to the kingdom Animalia are profoundly researched to produce safe and efficient bio-therapeutics to cure disease conditions. Demand for bio-therapeutics, including vaccines in pandemic conditions directs rigorous research and quality product development. To fulfill the demand, innovative ideas, unique approaches, and a vast array of substrates are utilized. However, the final product so obtained should be compatible with its bioactivity and should not lead to allergic reactions in the host. Therefore, in the present chapter, several invertebrate and vertebrate cell cultures developed from different tissues, and their application in bio-therapeutics, i.e., recombinant proteins and viral vaccines, is

* Corresponding Author's E-mail: bholekrutika@gmail.com.

emphasized. Some benefits of one cell line over another and the advantages of mammalian cell lines are also enclosed in the present chapter. At last, the chapter also reflects the different cell lines and their products, some approved by FDA and EMA and some in the clinical phase.

Keywords: recombinant proteins, vaccines, Sf-9, CHO, BHK, Vero

INTRODUCTION

Animal cell cultures are established for studying therapeutic proteins, testing anticancer agents, bio-pesticide activity, and testing the toxicity of chemical entities. For human and animal health care, many animal cell-derived products like virus-based vaccines, antibodies, interferons, growth factors, immunoregulators, etc. are already acknowledged for treating the disease's condition. In terms of diagnostic and therapeutic, animal cell lines possess an enormous potential as they can undergo post-translation modifications and express the protein with appropriate biological activity (Godia and Cairó 2002, 289). A variety of expression systems are manufactured to derive therapeutic proteins for the prevention or treatment of human and animal diseases (Wu et al., 2002, 63; Ruggiero et al., 2008, 75; Steinbach et al., 2002, 83). Using recombinant DNA technology in these varied expression systems based on mammalian, plant, insect, fungal, and bacterial cells and their transgenic organisms, helps to yield selective therapeutic proteins (Wu et al., 2002, 63; Chen et al., 2003, 785; Katagiri et al., 2003, 11897; Hammonds et al., 2007, 8036). On the other hand, looking at the global scenario increase in the cases of infectious diseases due to emerging viral infections, there is a huge demand for new vaccines. Thus to manufacture a bulk amount of these vaccines requires new strategies to improve the production of vaccines by utilizing varied cell substrates and should not compromise the potency and safety of the vaccine produced (AHN et al., 2008, S71). Therefore, the present chapter highlights few invertebrates and vertebrates cell lines and their role in producing bio-therapeutics. The cell lines in discussion includes insect-based cell lines, avian and mammalian-based cell lines.

INSECT-BASED CELL LINES

Insect-based cell lines are developed from mature and immature organism tissues. Tissues from embryos, midgets, fat bodies, hemocytes, ovaries, etc. are utilized to develop cell lines (Baines et al., 1994, 347; Goodman et al., 2001, 367; Belloncik et al., 2007, 245; Ahmed et al., 2014, 2). A vast array of cell lines is developed from order Lepidoptera, Diptera, Hymenoptera, Heteroptera, Dictyoptera, Orthoptera and Coleoptera (Lynn and Dwight 2001, 319). In the case of lepidopteran insects, varied metamorphic stages and distinct parts such as whole embryos, ovaries, fat body, testis, imaginal discs, dorsal vessel, etc. are considered for the development of cell lines. These cell lines are used as research tools in virology, in studies of signaling mechanisms to study insect immunity, hemocytes migration, test hypotheses about gene expression, and screening programs designed to discover new insecticide chemistries (Smagghe et al., 2009, 93). Currently, two animal vaccines are on the market, and several immuno-therapeutic and human vaccines are being developed and produced in SF21, SF9, express SF+, or High Five cells (Granados et al., 2007, 83). Using baculovirus expression vector system (i.e., adding any foreign gene) in insect cell line Porcilis PestiTM and BayovacTM veterinary vaccines against the classical swine fever virus is developed by Merck and Bayer. Further, veterinary vaccines protecting against porcine circovirus type 2 namely CircoFLEXTM (Boehringer Ingelheim) and Porcilis™ PCV are also developed using a baculovirus expression vector system (Cox and Manon 2012, 1759; Liu et al., 2013, 104). For human utilization, using baculovirus expression vector system in an insect cell line, Cervarix™ vaccine developed by GlaxoSmithKline is approved by FDA to protect against cervical cancer caused by a human papillomavirus (Cox and Manon 2012, 1759). Furthermore, the baculovirus expression vector system in insect cell lines has also helped develop therapeutic vaccines such as Provenge™ by Dendreon to treat advanced prostate cancer patients. Additionally, the FluBlok™ vaccine by Protein Sciences Corporation is reported possessing active immunization against disease caused by influenza virus subtypes A and type B (Liu et al., 2013, 104; Cox et al., 2011, S31). Furthermore, the therapeutic vaccine for treating diabetes type I Diamyd™ by Diamyd Medical is under Phase III clinical trial. Additionally, many proteins and virus-like particle based vaccines for treating Hepatitis E, Influenza, ParvovirusB-19, Ebola, HIV, Chikungunya virus, etc. are being screened at different phases of clinical trials (Agardh et al., 2009, 1363; Shrestha et al., 2007, 895; López-Macías et al., 2011, 7826; Bernstein et al., 2011, 7357; Sun et al., 2009, 12; Pillay et al., 2009, 1153; Metz et al., 2013, e2124).

On the other hand, mosquitoes belonging to the order Diptera are used for understanding arbovirus research. ATC-15 cell line is used to study around 50 viruses transmitted by mosquitoes and ticks (Singh 1972, 187). *C. bitaeniorhynchus* cell line for Japanese encephalitis virus and *Ae. albopictus* cell line or C6/36 cells for dengue viruses are used to isolate these viruses from mosquitoes collected during epidemic periods (Igarashi and Akira 1978, 531). Insect cell lines can also be used for the screening of insecticides or biocontrol compounds against pests, or to assess the potential risk to non-target organisms such as the honey bee (Walker et al., 2014, 1). Some of the general applications of Insect- based cell lines in bio-therapeutics are shown in Table 1.

The insect cell expression system may prove to be a valuable tool for producing therapeutic glycoproteins, including monoclonal antibodies, which are very valuable glycosylated therapeutic agents in oncology, inflammation and cardiovascular disease (HU and Yu-chen 2005, 405; Hervas-Stubbs et al., 2007, 2361). The utility of insect cell lines for protein production has grown from laboratory-scale experimental work to industrial applications (Elias et al., 2007, 225). The insect cell lines derived from order Lepidoptera and Diptera are mostly used for recombinant protein expression (Geisler et al., 2018, 25).

Table 1. Lepidopterian and dipertian derived Insect-based cell lines and their applications

Order	Name of the organism	Name of the cell line	Remarks
Lepidoptera	*Spodoptera frugiperda*	Sf-9	Used for studying the expression of foreign proteins additionally reported expressing human bone morphogenetic protein -2 using baculovirus/Sf-9 insect cell systems as therapeutic protein. Also, the Sf-9 cell line is highly utilized by pharmaceutical industries for drug discovery (Sriram et al., 1997, 181; Sehgal et al., 1998, 997; Palhan et al., 1995, 97; Maruoka et al., 1995, 957; Toya et al., 2002, 1087).
		Sf-21	Used for protein expression (Sriram et al., 1997, 181; Sehgal et al., 1998, 997; Palhan et al., 1995, 97)
	Trichoplusia ni	BTI TN5B1-4 (High Five cells)	Used for production of expressed protein at higher yield compared to Sf-9 and Sf-21 Cell lines (Granados et al., 2007, 83; Sriram et al., 1997, 181; Sehgal et al., 1998, 997; Palhan et al., 1995, 97; Ikonomou et al., 2003, 1)
		IAL-TND1	Reported to be Ecdysteroid-responsive cell lines and exposure of 20-hydroxyecdysone stimulates the synthesis of protein respectively (Lynn and Oberlander 1983, 591)
	Bombyx mori	BmN-4/ Bm5	Both the cell lines are used for foreign protein expression. Additionally, they are reported to be Ecdysteroid-responsive cell lines. (Sriram et al., 1997, 181; Sehgal et al., 1998, 997; Palhan et al., 1995, 97; Maeda and Susumu 2018, 167; Grace 1967, 613; Swevers et al., 2003, 134)
	Heliothis armigera	NIV-HA-1195/ NIV-HA-197	Used for testing the susceptibility to different baculoviruses and also reported to exhibit significant expression of Hepatitis C virus core protein (Sudeep et al., 2005, 725)
	Spodoptera. litura	Sl-1	Used to study the mechanism of cell cycle arrest and apoptosis by an herbal insecticide (Huang et al., 2011, 16)

Table 1. (Continued)

Order	Name of the organism	Name of the cell line	Remarks
Diptera	*Culex bitaenior-hynchus*	Unnamed	Used for understanding the Japanese encephalitis virus (Igarashi and Akira 1978, 531; Pant et al., 1982, 789)
	Aedes albopictus	C6/36	Used to understand the transfection, the mechanism of viral infection by chikungunya virus also used for elucidating information on the structure and characterization of the Denso virus (Singh 1967, 506; Chen et al., 2004, 123; Roberts et al., 2017, 1)
	Aedes aegypti	CCL-125	Used to study the permissive activity of Dengue virus infection and get information on insect cell and virus interaction (Wikan et al., 2009, 227)
		Aag-2	Used to understand mosquito immune system in response to infection with Sindbis virus (Barletta et al., 2012, 1)
		RML-12	Used to understand the interaction of varied viruses like dengue, West Nile virus, and Japanese encephalitis virus (Kuno and Goro 1983, 707; Sudeep et al., 2009, 491)

Many recombinant proteins of interest can be generated using the baculovirus expression vector system in Lepidopteran cell lines (Contreras-Gómez et al., 2014, 1; van Oers and Monique M 2011, S3). *Spodoptera frugiperda* derived cell line Sf2, Sf9 and Sf900+ (Vaughn et al., 1977, 213; Summers et al., 1987, 1; Smith et al., 2000, 1); *Trichoplusia ni*, derived cell lines such as TN-368, BTI-Tn5B1-4 and Tni PRO™ (Hink 1970, 466; Wickham et al., 1992, 391), and Bm-N (Volkman et al., 1982, 227) cell line derived from *Bombyx mori* are routinely used for expression of the protein of interest.

Few therapeutic proteins obtained using baculovirus expression vector system include tissue plasminogen activator, enzymes, viral and parasitic proteins (Wu et al., 1994, 55; Morais et al., 2001, 719; Metz et al., 2011, S16; Cruz et al., 1998, 149). The advantages of using the baculovirus-insect cell system are - is its wide utilization for the expression of various recombinant proteins. Its approach is simple and quick to manufacture milligram quantities of protein. But, the baculovirus-insect cell system has few drawbacks, such as the expressed protein can be driven by the delayed viral promoter, protein of interest peaks at a particular point, while cells are in the state of dying due to viral infection (Bernard et al., 1994, 139).

Dipteran insect cell lines are not susceptible to baculovirus infection and rather use plasmid-mediated recombinant protein expression method to express the protein of interest (Hink 1970, 466; McCarroll et al., 1997, 590; Lee et al., 2008, 9). *Drosophila melanogaster* derived cell lines S2 and S2R+ are mostly used for recombinant protein expression (Schneider and Imogene 1972, 353; Yanagawa et al., 1998, 32353). A high viability recombinant protein, namely human vascular cell adhesion protein (secreted protein) and dopamine D4 receptor (an integral membrane receptor) were expressed in the S2 Cell line (Bernard et al., 1994, 139).

FISH BASED CELL LINES

Fish cell lines serve as essential tools for pathological, toxicological and immunological studies. Fish cell lines have enormous applications in biomedical research, toxicology, gene regulation, gene expressions and gene transfer. The main advantage of cell culture is that cell lines allow higher control of conditions of experiments and, at the same time, reduces the variability of the in vivo responses that arise due to the responses of fish to stress and environmental influences. Fish cell lines have been developed from many tissues such as the ovary, fin, spleen, swim bladder, heart, liver, eye muscle, brain, and skin. Development of cell lines have been reported from important fish species like *Etroplus suratensis*, *Clarias batrachus* (Babu et al., 2011, 355), *Schizothorax richardsoni* (Goswami et al., 2013, 219), Catla catla (Chaudhary et al., 2014, 299), *Channa striatus*, *Puntius chelynoides* (Goswami et al., 2014, 206) and *Wallago attu* (Dubey et al., 2014, 915). The RTgill-W1 trout-derived gill cell line is helpful to understand fish health and understand mechanisms of toxicity, ranking the potencies of toxicants, and evaluating the toxicity of environmental samples. These cells are also valuable for high throughput toxicogenomic and toxic proteomic studies (Lee et al., 2009, 127).

AVIAN BASED CELL LINES

Avian embryos and avian cell lines are favorable for the development of safe and economic recombinant proteins and vaccines for human health care (Farzaneh et al., 2017). Vaccine development at the beginning of the century included avian embryonated eggs for virus inoculation (Goodpasture et al., 1931, 371). Avian egg development takes place externally in the calcareous shelled and this benefits researchers to explore them for developing transgenic models. These transgenic birds can further help produce recombinant proteins, create disease-resistant strains, and protect endangered species (Kagami and Hiroshi, 2016, 1065). In case of vaccine production, chick embryo fibroblast is efficient to start material. From 10-day old chick embryo head, wings and body cavity contents etc., are dissected and further mechanically/enzymatically homogenized and passaged (Hernandez et al., 2010, A-41). Chicken embryo-derived cells are reported to maintain genomic stability after repeated cell passage (Aubrit et al., 2015, 5905; Genzel and Yvonne 2015, 728). To treat diseases such as lymphomas and influenza, available cell lines CEF, DT95, DEF, DF-1, QT-35, QT-6, QM7 derived from chicken, quails, and ducks are commercially bestowed (Farzaneh et al., 2017, 1600598). For the development of measles and mumps vaccine, primary chicken embryo fibroblasts are considered (Lubiniecki and Anthony 1998, 139) also, looking at the current scenario, even avian eggs are employed for manufacturing seasonal flu vaccines (Gerdil and Catherine 2003, 1776). However, avian eggs derived cells possess certain drawbacks such as the risk of insufficient supply of vaccines; laborious process without consistency in the developed end-product, the manufacturing cost is high and the possibility to exhibit allergic responses due to egg components (Montomoli et al., 2012, 587; Audsley et al., 2008, 1483; Fedson and David 2008, 346). Therefore, to cope with the extensive demand for vaccines in human and animal health systems currently, avian-based cell lines are explored. Avian-derived cell lines are an alternative for virus production because they provide a simple, flexible system that can be controlled for an extended time with the capability for rapid vaccine production during a pandemic (Aubrit et al., 2015, 5905). Other options from the avian source include embryonic stem cells (ESCs) to develop continuous cell lines (Pain et al., 1996, 2339). EB66 avian ESC lines; are reported to efficiently propagate influenza, poxviruses, parainfluenza, and measles viruses. Additionally, EB66 is a good source for human and animal vaccines (Mehtali et al., 2015, 1; Olivier et al., 2010, 405). Other related cell lines can be derived from avian cultured PGCs (Primordial germ cells), and iPSCs (Induced pluripotent stem cells) are also known to be most efficient for the production of viral vaccines (Maruyama et al., 2012, 2754; Shittu et al., 2016, 24). Some of the avian-based cell lines and their application are depicted in Table 2.

Table 2. Avian-based cell lines and their applications

Order - Family	Name of the organism	Name of the cell line	Remarks
Galliformes-Phasianidae	Red junglefowl *Gallus gallus*	Primary cells Chicken embryo fibroblast (CEF)	Used for development of following vaccines Rabies (Rabipur®); TBE (FSME-Immun®, Encepur®); measles (Attenuvax®); mumps (Mumpsvax®) (Aubrit et al., 2015, 5905)
Anseriformes-Anatidae	Muscovy duck *Cairina moschata*	AGE1.CR®	Used for development of MVA-based vaccines and influenza virus (Lohr, et al., 2014, 1)
Anseriformes-Anatidae	Mallard *Anas platyrhynchos*	EB66®	Used for development of pandemic & seasonal flu; NDV; MVA recombinant vaccines (Endo et al., 2020, 551)

Transgenic avian animals can produce recombinant protein. The benefit of transgenic avian includes low maintenance steps. They are easy to rear and manage. Also, they have a short breeding time plus possess high productivity (Scott et al., 2010, 353; Nishijima et al., 2013, 207). Transgenic eggs with an ovalbumin OV promoter help produce a large quantity of recombinant protein (Ivarie and Robert 2006, 99). Few of the therapeutic proteins using OVA promoter are dansyl hapten mAb, prostate-specific membrane antigen mAb (Zhu et al., 2005, 1159), human interleukin 1 receptor antagonist (rhIL1RN) (Kwon et al., 2010, 1057), human extracellular superoxide dismutase (EC-SOD) (Byun et al., 2013, 404), human lysozyme, recombinant human lysozyme (rhYL) (Cao, et al., 2015, e0118626) and neutrophil defensin 4 (HNP4) (Liu et al., 2015, e0127922). Other promoters such as chicken β-actin are responsible for the production of antiprion single-chain Fv fused with the Fc region (scFv-Fc) (Kamihira et al., 2005, 10864); RSV promoter helps in human parathormone protein production (Lee et al., 2007, 2221). PGK Promoter in human erythropoietin protein production is useful (Kojima et al., 2014, 676; Koo et al., 2010, 437). Additionally, urokinase-type plasminogen activator (Lee et al., 2013, 2396), epidermal growth factors (Park et al., 2015, 2386), etc. are also obtained from egg white of transgenic avian animals.

Mammalian Based Cell Lines

To synthesize biopharmaceuticals products, one of the preferable systems includes mammalian cell line expression. It is adopted for the production of vaccines and bio-therapeutic proteins. Mammalian cell line helps in the production of large and complex proteins. Additionally, yielded proteins are glycosylated (after post-translation modification) and thus show similarity with proteins produced in humans (Durocher et al., 2009, 700; Ghaderi et al., 2012, 147; Swiech et al., 2012, 147). In vaccine production, utilization of mammalian cell lines exhibits benefits over traditional usage of embryonated eggs categorically in a pandemic situation. Pre-pandemic and pandemic time, the demand of vaccine is huge and to suffice the demand mammalian cell lines provides the substrate for virus growth rapidly helps in the production of vaccine and probably speeds up the production of desirable vaccine (Audsley et al., 2008, 1483).

Table 3. Mammalian-based cell lines and their applications

Order - Family	Name of the organism	Name of the cell line	Remarks
Primates - Hominidae	Human *Homo sapiens sapiens*	Diploid cell lines MRC5	Used for the manufacture of Varicella zoster virus-based vaccine (Varilrix®; Biopox®; ProQuad®); polioviruses (Poliovax®); rabies (Imovax®); hepatitis A (VAQTA®) (Aubrit et al., 2015, 5905)
Primates - Hominidae	Human *Homo sapiens sapiens*	Diploid cell line WI-38	Used for vaccine Rubella (Meruvax® II); adenoviruses; (Adenovirus Type 4 and Type 7 Vaccine, Live, Oral®) (Aubrit et al., 2015, 5905)
Carnivora-Canidae	Cocker spaniel *Canis lupus familiaris*	Continous cell line: MDCKa	Used for the manufacture of Seasonal flu (Optaflu®, Flucelvax®) vaccine (Aubrit et al., 2015, 5905)
Primates-Cercopithecidae	African green monkey *Chlorocebus sabaeus*	Continous cell line: Vero	Production of vaccines Pandemic (Celvapan®) & seasonal flu (Preflucel®); smallpox (ACAM2000®); JEV (Ixiaro®, IMOJEV®); polioviruses (OPV®, IMOVAX Polio®, PolioRIX®, Adacel®); rabies (VERORAB®, Abhayrab®), rotaviruses (RotaRIX®, RotaTeq®) (Aubrit et al., 2015, 5905)
Primates – Hominidae	Human *Homo sapiens sapiens*	Continous cell line: PER.C6®	Used for development of a vaccine for Influenza virus; West Nile virus (Aubrit et al., 2015, 5905)

In vaccine production, high-yield strains of the virus are allowed to infect a large number of selected mammalian cells and allowed to proliferate (Mabrouk et al., 2002, 125). Many researchers allow a virus to infect the cell growth medium along with the selected cells, although, another reported technique allows exchanging of cell culture medium before virus infection. However, to avoid any contamination, media should be devoid of animal-derived components i.e., especially serum (Genzel et al., 2009, 1681). Mammalian cell lines such as Madin-Darby canine kidney (MDCK), Vero (African green monkey kidney) cells, and human retina–derived cell line PER.C6 is wieldy used for the production of vaccines against influenza viral titers (Montomoli et al., 2012, 587). Some of the mammalian-based cell lines and their applications are shown in Table 3.

Mammalian cells are widely used for bio-therapeutic production since they possess heterologous expression systems due to the presence of enormous types of enzymes and a variety of cell organelles which allows the appropriate type of protein folding and post-translation modification (Barnes et al., 2003, 631). In this context, bio therapeutic proteins can aid in the treatment of varied disease conditions such as hematologic, autoimmune disorders, genetic disorders, hormonal imbalances, cancers etc. produced using non-human mammalian and human-based cell lines (Dumont et al., 2012, 3024). Majorly mammalian cell lines host for generation of recombinant protein includes Chinese hamster ovary (CHO) and nonsecreting (NS0) myeloma cells. Also, for the production of non-recombinant monoclonal antibodies, hybridoma cell lines are utilized commercially (Barnes et al., 2003, 631). Some of the attributes and usage of mammalian cell lines in vaccine and bio-therapeutic proteins production are discussed below:

i. Non-human mammalian cell lines for vaccine production:
   a) Madin-Darby canine kidney cells (MDCK): Original MDCK cell line is derived from the kidney of a normal adult cocker spaniel in 1958. Also, MDCK cells are adherently growing cells (Madin et al., 1958, 574). Globally, MDCK cells are frequently used in varied fields including public health programs, clinical virology, and research-based studies for cultivation, sensitivity and high productivity of different influenza virus-based vaccines (Fujii et al., 2002, 97; Oh et

al., 2008, 2189). MDCK cells are reported to be the preferred substrate for influenza viruses compared to embryonated eggs (Trusheim et al., 2007, 515). Also, Clinical studies on children, adults and the elderly upon injection by vaccine-derived from embryonated egg and MDCK cells showed no significant differences in immunogenicity and physiological systemic events (Reisinger et al., 2009, 849). Thus, MDCK cells can be considered to be doubtlessly safe (Gregersen and Jens-Peter 2008, 3297). Optaflu an intramuscular vaccine developed by Novartis vaccines possesses trivalent influenza active substances purified from influenza virus antigens of type A (H1N1 and H3N2) and type B (Doroshenko et al., 2009, 679). Provided, Novartis-based Celtura® vaccine against S-OIV H1N1 is also derived from MDCK 33016 cell line (Novartis 2009).

b) African green monkey kidney-derived Vero cells: Vero cells are derived from the kidney of an African green monkey (*Cercopithecus aethiops*). They were first successfully used for the production of inactivated poliovirus vaccine for human utilization; also Vero cells were subsequently used for the production of inactivated rabies vaccine (Montagnon et al., 1981, 55). Consequently, Vero cell lines are also utilized for the production of oral poliovirus vaccine (Sutter et al., 2018, 866). A comparative study on the immunogenicity of egg-derived and trivalent Vero cell-derived influenza virus vaccines showed a similar humoral immune response. Besides, the cellular response was higher in the case of vaccine-derived from the Vero cell line (Brühl et al., 2000, 1149). Many vaccines are reported to be derived using Vero cell line few among them are A/Vietnam/1203/2004 vaccine licensed as the first cell culture-derived from H5N1 (Barrett et al., 2009, 607). CELVAPAN developed against wild-type A/California/07/2009 H1N1 strain was licensed by the European regulatory authorities also developed using Vero cell line (Baxter International 2009). Using Vero cell lines, a replication-deficient vaccinia virus vector H5N1 live vaccine and dVV-HA5 expressing the hemagglutinin of a virulent Clade 1 H5N1 strain were also developed by Baxter International Inc. (Mayrhofer et al., 2009, 5192). Preflucel® a new seasonal influenza vaccine is also reported to be developed by Baxter's using Vero cell technology (Montomoli et al., 2012, 587).

c) Baby hamster kidney cells (BHK): BHK is an adherently growing and anchor-dependent mammalian cell line. It shows fibroblast growth pattern when provided standard cell culture conditions using fetal bovine serum (Cruz et al., 1998, 149; Hernandez et al., 2010, A-41). For virus antigen production BHK are susceptible to a range of viruses such as the virus of foot-and-mouth disease (FMD) (Guskey et al., 1975, 433).

ii. Human mammalian cell lines for vaccine production:

a) Human retina-derived PER.C6 cells: For the manufacture of viral vaccines, human diploid cells are constantly researched and studied (Petricciani et al., 2008, 359). The cell line is created from human embryonic retinal cells; they were further immortalized by transfection with the adenovirus E1 gene (Havenga et al., 2008, 273). PER.C6 cells were developed for specific applications i.e., for the development of human adenovirus vectors in vaccine development and gene therapy (Butler et al., 2014, 107). Consequently, PER.C6 cells are well characterized to fulfill regulatory and biosafety requirements (Genzel et al., 2009, 1681). At first, Pau et al. reported the potential of PER.C6 cells for influenza vaccine production (Pau et al., 2001, 2716). PER.C6 cells in both adherent and suspension states (using serum-free media) can generate virus seeds by reverse genetics (Koudstaal et al., 2009, 2558). The vaccine-derived from PER.C6 cells was Crucell NV the influenza vaccine (Crucell 2007). Cox and his co-workers conducted Phase I clinical trial of the influenza H7 virus vaccine grown on PER.C6 cells (Cox et al., 2009, 1889). The outcome of Phase I H7N1 vaccine study indicated that the vaccine was well tolerated, without any serious side effects. However, the vaccine was not licensed as it could not fulfill certain criteria of The Committee for Medicinal Products for Human Use as it was immunogenic poor and did not induce sufficient humoral immune (Olivier et al., 2010, 405).

b) Human embryonic kidney (HEK): HEK 293 cells can be easily grown in suspension and serum-free medium and extensively used to propagate vectored vaccines. Due to replication-defective adenovirus, the genome of these cells lacks E1 genes (E1A and E1B) (Hehir et al., 1996, 8459). It is now considered one of the best cell culture-based approaches for producing recombinant viral vectors and vaccines. Different viruses such as adenovirus or adeno-associated virus, retrovirus, lentivirus, reovirus, influenza etc. are studied using HEK293 cells. In addition, virus-like particles from rabies and influenza are also examined by HEK cells (Nadeau et al., 2003, 475; Fontana et al., 2014, 2799; Emma et al., 2013, 275; Pham et al., 2003, 332; Rodrigues et al., 2013, 131). HEK cells are quick and efficient in the replication of viral particles with little immunogenicity and substantial therapeutic activity. Besides, these cells can also actively help in viral vector amplification. Thus HEK 293 can be efficiently employed for adenovirus vaccine manufacturing for gene therapy, preventive inoculation, and cancer treatment (Nadeau et al., 2003, 475; Murakami et al., 2010, 196). HEK293-T cells are applied for the production of retroviral vectors and are also known to express simian virus 40 large T antigen. Thus, due to high titers of viral gene vectors, HEK293-T is used in gene therapy (Yamaguchi et al., 2003, 796).

iii. Non-human mammalian cell lines for Therapeutic proteins production:

a) Baby hamster kidney cells (BHK): Human coagulation factor VIII is a large plasma glycoprotein and crucial in the blood coagulation intrinsic pathway. The absence of this factor leads to Hemophilia A and thus, for treatment purposes either human plasma-derived VIII coagulation factor or recombinant coagulation factor is adopted. However, blood-borne pathogens contribute to issues of safety on the usage of human plasma-derived factors. Thus second option utilization of recombinant VIII coagulation factor is more preferable (Mei et al., 2006, 165). Furthermore, BHK-21 cells are useful in producing a factor VIII coagulation factor (Wurm and Florian 2004, 1393).

b) Murine myeloma cell lines (NS0 and Sp2/0): Recombinant antibodies manufactured using mammalian cell lines are currently in demand. To enhance the productivity in mammalian cell lines, studies on clonal differences to increase productivity (Smales et al., 2004, 474; Dinnis et al., 2006, 830) and strategies such as application of physiological conditions such as low temperature (Barnabé et al., 1994, 1235; Fogolín et al., 2004, 179), high osmolarity (Shen et al., 2006, 132; Sun et al., 2004, 576), and chemical treatment with sodium butyrate (Cherlet et al., 2000, 17; de Leon Gatti et al., 2002, 205) etc. are carried out. Arresting proliferating stage in the NS0 myeloma cell line increased the antibody production to 2-4 folds (Watanabe et al., 2002, 1; Ibarra et al., 2003, 224). Murine myeloma cells are chosen for the production of mAbs e.g., Palivizumab (humanized monoclonal antibody for the prevention of serious lower respiratory tract disease caused by a respiratory syncytial virus in infants at high risk) and Ofatumumab (anti-CD20 IgG1κ human monoclonal antibody as a treatment for chronic lymphocytic leukemia) (Ghaderi et al., 2012, 147;

Butler et al., 2014, 107; Barnes et al., 2000, 109; Fenton et al., 2004, 177; Zhang and Bodi 2009, 326).

c) Chinese Hamster Ovary Cell line (CHO): Currently, around 60- 70% of recombinant therapeutic proteins available in the market are derived from CHO cells (McAtee et al., 2014, 63). In 1986, the tissue plasminogen activator was approved to be the first recombinant bio-therapeutic protein produced in CHO cells (Kim et al., 2012, 917). The benefits of CHO cells for therapeutic protein production include high output, robustness and are safe (Meleady et al., 2011, 1). They are also known to exhibit post-translation modifications and glycosylation patterns identical to humans (Birch et al., 2006, 671). But some glycosylation patterns such as [2-6] sialyl transferase a [1-3/ 4] fucosyl transferases are seen in humans are not expressed by CHO cells. Additionally, CHO cells are reported to produce glycans (a-gal and NGNA) that are not expressed in humans. Expression of this glycan residue in the final product can be a source for elevated immune reaction and altered pharmacokinetics in humans (Ghaderi et al., 2012, 147; Bosques et al., 2010, 1153; Dietmair et al., 2012, e43394; Ghaderi et al., 2010, 863; Padler-Karavani et al., 2008, 818). Comparing CHO cell lines with other mammalian cells derived from mouse, i.e., NS0 cell line in terms of expressing recombinant antibodies, CHO possess dihydrofolate reductase and glutamine synthetase antibiotic-free vector systems. In contrast, NS0 cell lines lack antibiotic-free vector systems (Butler et al., 2005, 283).

iv. Human mammalian cell lines for therapeutic proteins production: Currently, human-based cell lines are extensively researched for the production of clinical products useful in biomedical research. An appropriate type of translation products and high productivity being achieved by using a human cell line for bio-therapeutic protein production. Several clonal isolates are studied and screened for their productivity and the one with specific clones with the highest yield of the specific therapeutic protein will be considered at industrial scale manufacturing (Berkner et al., 1993, 450).

a) Human retina-derived PER.C6®cells: the cells are derived from human embryonic retinal cells which are immortalized by transfection with the adenovirus E1 gene (Havenga et al., 2008, 273). PER.C6 is involved in the production of a high level of protein while associated with the production of human IgG (Jones et al., 2003, 163). Many products derived from the PER.C6 cell line are currently under Phase I and II clinical trials (Durocher et al., 2009, 700). A human IgG antibody against granulocyte-macrophage colony-stimulating factor (MOR103 mAb) for treating rheumatoid arthritis and multiple sclerosis condition is reported to show promising clinical trial outcomes (Behrens et al., 2015, 1058). Moreover, CL184 a combination of human IgG1 (l) and human IgG1(k) acts as an antibody against the rabies virus (Bakker et al., 2008, 5922; Nagarajan et al., 2014, 1).
b) Human embryonic kidney (HEK): Several factors contribute to the advantages of HEK293 cells, such as their ability to grow easily in suspension serum-free culture, they can reproduce rapidly, they are highly susceptible to varied transfection methods and most importantly they are tremendously efficient at bio therapeutic protein production (Swiech et al., 2012, 147; Dumont et al., 2016, 1110; Thomas et al., 2005, 187). Thus, due to the eminent adaptive nature of HEK cells to transient and stable transfection techniques and high transfection efficiency yields in the production of huge proteins (Büssow and Konrad 2015, 81; Hacker et al., 2016, 189). Another benefit of HEK293 cells is to have a greater capacity for g-carboxylation than CHO cells (Berkner et al., 1993, 450). Therefore, as a result, the FDA or the EMA approved five therapeutic proteins due to immense research and attributes of HEK293 cells (Dumont et al.,

2016, 1110). The first recombinant therapeutic protein based on human HEK293 cell lines was Drotrecogin alfa (Xigris ®; Eli Lilly Corporation, Indianapolis, IN) for the treatment of patients with sepsis and approved by FDA in 2001 and EMA in 2002 but later in 2011 Drotrecogin alfa was withdrawn by the company. The other four FDA or EMA approved biopharmaceutical products are efmoroctocog alfa (Eloctate ®), human- cl rhFVIII (Nuwiq ®; Octapharma, Lachen, Switzerland), eftrnonacog alfa (Alprolix ®), and dulaglutide (Trulicity TM) (Dumont et al., 2016, 1110; Swiech et al. 2017, 462). Dulaglutide is produced using HEK293- EBNA cells and is a Glucagon-1-like peptide (GLP-1) Fc fusion protein. It has been approved for the treatment of type 2 diabetes mellitus by FDA (Glaesner et al., 2010, 287; Sanford and Mark 2014, 2097). Eloctate ® and Alprolix ® helps in controlling and prevent bleeding episodes, perioperative management, and routine prophylaxis to prevent or reduce the frequency of bleeding episodes in people with hemophilia A and B, respectively (Ducore et al., 2014, 559; FDA 2019; Mahlangu et al., 2014, 317; Powell et al., 2013, 2313).

c) HT-1080: It is produced from fibrosarcoma and phenotypically appears as epithelial cells (Swiech et al., 2012, 147). Usages of HT-1080 cells lead to developing few therapeutic products such as Epoetin delta, Iduronate-2-sulfatase, Agalsidase alfa, and Velaglucerase alfa Shire (Swiech et al., 2012, 147). The first therapeutic protein, Epoetin delta (DYNEPO) was useful in the treatment of anemia secondary to chronic renal failure; however, due to commercial reasons, Epoetin delta was withdrawn by the company (Martin et al., 2007, 26). Iduronate-2-sulfatase (idursulfase; ELAPRASE) was approved by FDA in 2006 and EMA in 2007. It is licensed as enzyme replacement therapy and is essentially used for Hunter syndrome (mucopolysaccharidosis II), an X-linked lysosomal storage disorder treatment (Kim et al., 2013, 796). Therapeutic protein Agalsidase alfa (REPLAGAL; Shire Human Genetic Therapies, Danderyd, Sweden) is useful for treating Fabry disease (Shire Human Genetic Therapies 2010). At last, Velaglucerase alfa (VPRIV; Shire Human Genetic Therapies, Lexington, MA) was known to be helpful in the treatment of type 1 Gaucher disease (Swiech et al., 2012, 147; Shire Human Genetic Therapies, 2019).

d) Other reported cell lines: adopting amniocentesis procedure in humans a new cell line CAP was derived using amniocytes. The adenovirus types 5 E1 gene aids in the immortalization of amniocytes. Thus, the resultant cells can yield a good quantity of post-translated modified protein (Genzel et al., 2013, 111). Another HKB-11 cell line was created by HEK293-S and a human B-cell line (modified Burkitt lymphoma cells) fusion using polyethylene glycol (Durocher et al., 2009; Cho et al., 2003, 229; Picanco-Castro et al., 2013, 1373). HKB-11 cells are used for the production of a recombinant factor called VIII protein and tissue factor (Durocher et al., 2009). HKB-11 exhibited increased expression of human FVIII as compared to HEK293 and BHK21 cell lines (Mei et al., 2006, 165). Human hepatocellular carcinoma-derived HuH-7-CD4 clone is efficient to produce recombinant human factor IX with a correct human glycosylation profile. The post-translation modifications in terms of glycosylation, sialylation, phosphorylation, and sulfation profiles were reported to be similar to plasma-derived factor IX. Besides the resultant recombinant factor IX derived from the HuH-7-CD4 clone was much enhanced as compared to the CHO cells derived recombinant factor IX (Enjolras et al., 2013, 266). Moreover, HuH-7 cell line-derived mutant recombinant factor IX is known to have raised binding affinity for activated FVIII and enhanced clotting activity in an in-vivo experiment carried on mice (Perot et al., 2015, 1017).

"Designer" Cell Lines

The present approach has now shifted to develop a specific cell line with a designated cell substrate for one application and is well characterized. However, there is scope to explore the other potentials of the developed cell line in the future (Genzel and Yvonne 2015, 728). For the development of designer cell lines, certain factors such as the origin of cells source, tissue, or organ for the new cell line are considered. Besides, immortalization strategy and growth conditions are to be thoroughly planned for the establishment of the new cell line with a designated product of interest. For viral vaccines production some cell lines such as PER.C6, CAP, AGE1.CR, EB66 cells, PBS-1, QOR/2E11, SogE, MFF-8C1 cells are well established (Genzel and Yvonne 2015, 728). Thus, with a new zest of learning and research in the in-vitro forum, new characterized cell lines can surely help develop bio therapeutic proteins and vaccines.

Conclusion

The invertebrate and vertebrate cell lines from different developmental stages and organs are characterized and looked for their potential in obtaining new recombinant proteins of therapeutic use and viral vaccine for human health benefits. The cell lines are preferable due to their rapid growth, high output of required product, abundant protein yield and optimal investment. The final product of need is post-translationally similar to actual protein seen in humans and exhibits potent bioactivity. Thus transgenic insect cell lines, avian and mammalian based cell lines are accepted profoundly due to the safety of the protein product produced by cell lines in vitro systems. However, viral, bacterial and fungal contaminations during the manufacturing process need to be rigorously monitored. In terms of vaccine production, the primary method utilizes embryonated eggs, but the requirement for vaccine production increases to a huge fold during the pandemic situation. The supply of embryonated eggs could be limited, which can have a huge impact on vaccine production. In conclusion, large-scale cell culture-derived vaccine manufacturing can ease the substrate supply; one can continuously monitor the controlled product formed using closed system bioreactors and essentially reduce microbial contamination.

REFERENCES

[1] Agardh, C. D., K. F. Lynch, M. Palmer, Katarina Link, and Åke Lernmark. "GAD65 vaccination: 5 years of follow-up in a randomised dose-escalating study in adult-onset autoimmune diabetes." *Diabetologia* 52, no. 7 (2009): 1363-1368.

[2] Ahmed, Ibrahim, H. Huebner, and R. Buchholz. *Establishment and characterization of three new embryonic Spodoptera littoralis cell lines and testing their susceptibility to SpliMNPV*. PhD diss., Friedrich-Alexander-Universität Erlangen-Nürnberg (FAU), 2015.

[3] Aubrit, Françoise, Fabien Perugi, Arnaud Léon, Fabienne Guéhenneux, Patrick Champion-Arnaud, Mehdi Lahmar, and Klaus Schwamborn. "Cell substrates for the production of viral vaccines." *Vaccine* 33, no. 44 (2015): 5905-5912.

[4] Audsley, Jennifer M., and Gregory A. Tannock. "Cell-based influenza vaccines." *Drugs* 68, no. 11 (2008): 1483-1491.

[5] Babu, V. S., K. S. N. Nambi, V. Chandra, V. P. Ishaq Ahmed, R. Bhonde, and A. S. Sahul Hameed. "Establishment and characterization of a fin cell line from Indian walking catfish, Clarias batrachus (L.)." *Journal of fish diseases* 34, no. 5 (2011): 355-364.

[6] Baines, D., A. Brownwright, and J. L. Schwartz. "Establishment of primary and continuous cultures of epithelial cells from larval

lepidopteran midguts." *Journal of insect physiology* 40, no. 4 (1994): 347-357.

[7] Bakker, A. B. H., C. Python, C. J. Kissling, P. Pandya, W. E. Marissen, M. F. Brink, F. Lagerwerf et al. "First administration to humans of a monoclonal antibody cocktail against rabies virus: safety, tolerability, and neutralizing activity." *Vaccine* 26, no. 47 (2008): 5922-5927.

[8] Barletta, Ana Beatriz Ferreira, Maria Clara L. Nascimento Silva, and Marcos H. Ferreira Sorgine. "Validation of Aedes aegypti Aag-2 cells as a model for insect immune studies." *Parasites & vectors* 5, no. 1 (2012): 1-9.

[9] Barnabé, N., and M. Butler. "Effect of temperature on nucleotide pools and monoclonal antibody production in a mouse hybridoma." *Biotechnology and bioengineering* 44, no. 10 (1994): 1235-1245.

[10] Barnes, Louise M., Catherine M. Bentley, and Alan J. Dickson. "Advances in animal cell recombinant protein production: GS-NS0 expression system." *Cytotechnology* 32, no. 2 (2000): 109-123.

[11] Barnes, Louise M., Catherine M. Bentley, and Alan J. Dickson. "Stability of protein production from recombinant mammalian cells." *Biotechnology and bioengineering* 81, no. 6 (2003): 631-639.

[12] Barrett, P. Noel, Wolfgang Mundt, Otfried Kistner, and M. Keith Howard. "Vero cell platform in vaccine production: moving towards cell culture-based viral vaccines." *Expert review of vaccines* 8, no. 5 (2009): 607-618.

[13] Baxter International, Inc: *Baxter receives European Commission approval for CELVAPAN H1N1 pandemic influenza vaccine.* Press release 7 October 2009. www.evaluatepharma.com/Universal/View.aspx?type=Story&id=197126&sectionID=&isEPVantage=no.

[14] Behrens, Frank, Paul P. Tak, Mikkel Østergaard, Rumen Stoilov, Piotr Wiland, Thomas W. Huizinga, Vadym Y. Berenfus et al. "MOR103, a human monoclonal antibody to granulocyte–

macrophage colony-stimulating factor, in the treatment of patients with moderate rheumatoid arthritis: results of a phase Ib/IIa randomised, double-blind, placebo-controlled, dose-escalation trial." *Annals of the rheumatic diseases* 74, no. 6 (2015): 1058-1064.

[15] Belloncik, S., O. Petcharawan, M. Couillard, G. Charpentier, B. Larue, H. Guardado, S. Chareonsak, and S. Imanishi. "Development and characterization of a continuous cell line, AFKM-On-H, from hemocytes of the European corn borer *Ostrinia nubilalis* (Hübner) (Lepidoptera, Pyralidae)." *In Vitro Cellular & Developmental Biology-Animal* 43, no. 7 (2007): 245-254.

[16] Berkner, Kathleen L. "[26] Expression of recombinant vitamin K-dependent proteins in mammalian cells: Factors IX and VII." *Methods in enzymology* 222 (1993): 450-477.

[17] Bernard, A. R., T. A. Kost, L. Overton, C. Cavegn, J. Young, M. Bertrand, Z. Yahia-Cherif, C. Chabert, and A. Mills. "Recombinant protein expression in a *Drosophila* cell line: comparison with the baculovirus system." *Cytotechnology* 15, no. 1 (1994): 139-144.

[18] Bernstein, David I., Hana M. El Sahly, Wendy A. Keitel, Mark Wolff, Gina Simone, Claire Segawa, Susan Wong, Daniel Shelly, Neal S. Young, and Walla Dempsey. "Safety and immunogenicity of a candidate parvovirus B19 vaccine." *Vaccine* 29, no. 43 (2011): 7357-7363.

[19] Birch, John R., and Andrew J. Racher. "Antibody production." *Advanced drug delivery reviews* 58, no. 5-6 (2006): 671-685.

[20] Bosques, Carlos J., Brian E. Collins, James W. Meador, Hetal Sarvaiya, Jennifer L. Murphy, Guy DelloRusso, Dorota A. Bulik et al. "Chinese hamster ovary cells can produce galactose-α-1, 3-galactose antigens on proteins." *Nature biotechnology* 28, no. 11 (2010): 1153-1156.

[21] Brühl, Peter, Astrid Kerschbaum, Otfried Kistner, Noel Barrett, Friedrich Dorner, and Marijan Gerenčer. "Humoral and cell-

mediated immunity to Vero cell-derived influenza vaccine." *Vaccine* 19, no. 9-10 (2000): 1149-1158.

[22] Büssow, Konrad. "Stable mammalian producer cell lines for structural biology." *Current opinion in structural biology* 32 (2015): 81-90.

[23] Butler, Michael, and Maureen Spearman. "The choice of mammalian cell host and possibilities for glycosylation engineering." *Current opinion in biotechnology* 30 (2014): 107-112.

[24] Butler, Michael. "Animal cell cultures: recent achievements and perspectives in the production of biopharmaceuticals." *Applied microbiology and biotechnology* 68, no. 3 (2005): 283-291.

[25] Byun, Sung June, Mi-Ran Ji, Ye-Jin Jang, A-In Hwang, Hee Kyoung Chung, Jeom Sun Kim, Kyung-Woon Kim et al. "Human extracellular superoxide dismutase (EC-SOD) expression in transgenic chicken." *BMB reports* 46, no. 8 (2013): 404.

[26] Cao, Dainan, Hanyu Wu, Qingyuan Li, Yingmin Sun, Tongxin Liu, Jing Fei, Yaofeng Zhao, Sen Wu, Xiaoxiang Hu, and Ning Li. "Expression of recombinant human lysozyme in egg whites of transgenic hens." *PloS one* 10, no. 2 (2015): e0118626.

[27] Chaudhary, Dharmendra Kumar, Neeraj Sood, Gaurav Rathore, Pravata Kumar Pradhan, Peyush Punia, Naresh Kumar Agarwal, and Joy Krushna Jena. "Establishment and characterization of macrophage cell line from thymus of *Catla catla* (Hamilton, 1822)." *Aquaculture Research* 45, no. 2 (2014): 299-311.

[28] Chen, Jinghua, Michael B. Duncan, Kevin Carrick, R. Marshall Pope, and Jian Liu. "Biosynthesis of 3-O-sulfated heparan sulfate: unique substrate specificity of heparan sulfate 3-O-sulfotransferase isoform 5." *Glycobiology* 13, no. 11 (2003): 785-794.

[29] Chen, Senxiong, Lingpeng Cheng, Qinfen Zhang, Wei Lin, Xinying Lu, Jennifer Brannan, Z. Hong Zhou, and Jingqiang Zhang. "Genetic, biochemical, and structural characterization of a new densovirus isolated from a chronically infected Aedes albopictus C6/36 cell line." *Virology* 318, no. 1 (2004): 123-133.

[30] Cherlet, Marc, and Annie Marc. "Stimulation of monoclonal antibody production of hybridoma cells by butyrate: evaluation of a feeding strategy and characterization of cell behaviour." *Cytotechnology* 32, no. 1 (2000): 17-29.

[31] Cho, M. S., H. Yee, C. Brown, B. Mei, C. Mirenda, and S. Chan. "Versatile expression system for rapid and stable production of recombinant proteins." *Biotechnology progress* 19, no. 1 (2003): 229-232.

[32] Contreras-Gómez, A., A. Sánchez-Mirón, F. García-Camacho, E. Molina-Grima, and Y. Chisti. "Protein production using the baculovirus-insect cell expression system." *Biotechnology progress* 30, no. 1 (2014): 1-18.

[33] Cox, Manon MJ, and Yoshifumi Hashimoto. "A fast track influenza virus vaccine produced in insect cells." *Journal of invertebrate pathology* 107 (2011): S31-S41.

[34] Cox, Manon M. J. "Recombinant protein vaccines produced in insect cells." *Vaccine* 30, no. 10 (2012): 1759-1766.

[35] Cox, Rebecca J., Abdullah S. Madhun, Solveig Hauge, Haakon Sjursen, Diane Major, Mirjam Kuhne, Katja Höschler et al. "A phase I clinical trial of a PER. C6® cell grown influenza H7 virus vaccine." *Vaccine* 27, no. 13 (2009): 1889-1897.

[36] Crucell Holland BV: WO2007045674 (2007).

[37] Cruz, Pedro E., Cristina C. Peixoto, José L. Moreira, and Manuel JT Carrondo. "Production and quality analysis of Pr55gag particles produced in baculovirus-infected insect cells." *Journal of Chemical Technology & Biotechnology: International Research in Process, Environmental and Clean Technology* 72, no. 2 (1998): 149-158.

[38] de Leon Gatti, M., P. Hossler, and W. S. Hu. "Effect of sodium butyrate on gene expression of recombinant mammalian cells." In *abstracts of papers of the American chemical society,* vol. 224, pp. U205-u206. 1155 16th st, NW, Washington, DC 20036 USA: Amer chemical soc, 2002.

[39] Dietmair, Stefanie, Mark P. Hodson, Lake-Ee Quek, Nicholas E. Timmins, Peter Gray, and Lars K. Nielsen. *A multi-omics analysis of recombinant protein production in Hek293 cells.* (2012): e43394.

[40] Dinnis, Diane M., Scott H. Stansfield, Stefan Schlatter, C. Mark Smales, Daniel Alete, John R. Birch, Andrew J. Racher, Carol T. Marshall, Lars K. Nielsen, and David C. James. "Functional proteomic analysis of GS-NS0 murine myeloma cell lines with varying recombinant monoclonal antibody production rate." *Biotechnology and bioengineering* 94, no. 5 (2006): 830-841.

[41] Doroshenko, Alexander, and Scott A. Halperin. "Trivalent MDCK cell culture-derived influenza vaccine Optaflu® (Novartis Vaccines)." *Expert review of vaccines* 8, no. 6 (2009): 679-688.

[42] Dubey, Akhilesh, Mukunda Goswami, Kamalendra Yadav, and Bhagwati S. Sharma. "Development and characterization of a cell line WAF from freshwater shark *Wallago attu*." *Molecular biology reports* 41, no. 2 (2014): 915-924.

[43] Ducore, Jonathan M., Maricel G. Miguelino, and Jerry S. Powell. "Alprolix (recombinant Factor IX Fc fusion protein): extended half-life product for the prophylaxis and treatment of hemophilia B." *Expert review of hematology* 7, no. 5 (2014): 559-571.

[44] Dumont, Jennifer A., Tongyao Liu, Susan C. Low, Xin Zhang, George Kamphaus, Paul Sakorafas, Cara Fraley et al. "Prolonged activity of a recombinant factor VIII-Fc fusion protein in hemophilia A mice and dogs." *Blood, the Journal of the American Society of Hematology* 119, no. 13 (2012): 3024-3030.

[45] Dumont, Jennifer, Don Euwart, Baisong Mei, Scott Estes, and Rashmi Kshirsagar. "Human cell lines for biopharmaceutical manufacturing: history, status, and future perspectives." *Critical reviews in biotechnology* 36, no. 6 (2016): 1110-1122.

[46] Durocher, Yves, and Michael Butler. "Expression systems for therapeutic glycoprotein production." *Current opinion in biotechnology* 20, no. 6 (2009): 700-707.

[47] Durocher, Yves, and Michael Butler. "Expression systems for therapeutic glycoprotein production." *Current opinion in biotechnology* 20, no. 6 (2009): 700-707.

[48] Elias, Cynthia B., Barbara Jardin, and Amine Kamen. "Recombinant protein production in large-scale agitated bioreactors using the baculovirus expression vector system." In *Baculovirus and Insect Cell Expression Protocols,* pp. 225-245. Humana Press, 2007.

[49] Emma, Petiot, and Amine Kamen. "Real-time monitoring of influenza virus production kinetics in HEK293 cell cultures." *Biotechnology progress* 29, no. 1 (2013): 275-284.

[50] Endo, Masafumi, Mitsuyoshi Tanishima, Kayo Ibaragi, Kenshi Hayashida, Tadashi Fukuda, Tetsuro Tanabe, Takeshi Naruse, Yoichiro Kino, and Kohji Ueda. "Clinical phase II and III studies of an AS03-adjuvanted H5N1 influenza vaccine produced in an EB66® cell culture platform." *Influenza and Other Respiratory Viruses* 14, no. 5 (2020): 551-563.

[51] Enjolras, Nathalie, Yesim Dargaud, Eloïse Pérot, Florine Guillaume, Michel Becchi, and Claude Négrier. "Human hepatoma cell line HuH-7 is an effective cellular system to produce recombinant factor IX with improved post-translational modifications." *Thrombosis research* 130, no. 5 (2012): e266-e273.

[52] *Factor, Antihemophilic, and Antihemorrhagic Blood Coagulation Factor VIII.* "ELOCTATE®." https://www.fda.gov/vaccines-blood-biologics/approved-blood-products/eloctate last modified September 2019.

[53] Farzaneh, Maryam, Seyedeh-Nafiseh Hassani, Paul Mozdziak, and Hossein Baharvand. "Avian embryos and related cell lines: A convenient platform for recombinant proteins and vaccine production." *Biotechnology journal* 12, no. 5 (2017): 1600598.

[54] Fedson, David S. "New technologies for meeting the global demand for pandemic influenza vaccines." *Biologicals* 36, no. 6 (2008): 346-349.

[55] Fenton, Caroline, Lesley J. Scott, and Greg L. Plosker. "Palivizumab." *Pediatric Drugs* 6, no. 3 (2004): 177-197.

[56] Fogolín, Mariela Bollati, Roland Wagner, Marina Etcheverrigaray, and Ricardo Kratje. "Impact of temperature reduction and expression of yeast pyruvate carboxylase on hGM-CSF-producing CHO cells." *Journal of biotechnology* 109, no. 1-2 (2004): 179-191.

[57] Fontana, Diego, Ricardo Kratje, Marina Etcheverrigaray, and Claudio Prieto. "Rabies virus-like particles expressed in HEK293 cells." *Vaccine* 32, no. 24 (2014): 2799-2804.

[58] Fujii, Hayato, Hiroshi Takahashi, Takaaki Ohyama, Kinuyo Hattori, and Shigeto Suzuki. "Evaluation of the school health surveillance system for influenza, Tokyo, 1999-2000." *Japanese journal of infectious diseases* 55, no. 3 (2002): 97-99.

[59] Geisler, Christoph, and Donald L. Jarvis. "Adventitious viruses in insect cell lines used for recombinant protein expression." *Protein expression and purification* 144 (2018): 25-32.

[60] Genzel, Yvonne, and Udo Reichl. "Continuous cell lines as a production system for influenza vaccines." *Expert review of vaccines* 8, no. 12 (2009): 1681-1692.

[61] Genzel, Yvonne, Ilona Behrendt, Jana Rödig, Erdmann Rapp, Claudia Kueppers, Stefan Kochanek, Gudrun Schiedner, and Udo Reichl. "CAP, a new human suspension cell line for influenza virus production." *Applied microbiology and biotechnology* 97, no. 1 (2013): 111-122.

[62] Genzel, Yvonne. "Designing cell lines for viral vaccine production: where do we stand?." *Biotechnology journal* 10, no. 5 (2015): 728-740.

[63] Gerdil, Catherine. "The annual production cycle for influenza vaccine." *Vaccine* 21, no. 16 (2003): 1776-1779.

[64] Ghaderi, Darius, Mai Zhang, Nancy Hurtado-Ziola, and Ajit Varki. "Production platforms for biotherapeutic glycoproteins. Occurrence, impact, and challenges of non-human sialylation."

Biotechnology and Genetic Engineering Reviews 28, no. 1 (2012): 147-176.

[65] Ghaderi, Darius, Rachel E. Taylor, Vered Padler-Karavani, Sandra Diaz, and Ajit Varki. "Implications of the presence of N-glycolylneuraminic acid in recombinant therapeutic glycoproteins." *Nature biotechnology* 28, no. 8 (2010): 863-867.

[66] Glaesner, Wolfgang, Andrew Mark Vick, Rohn Millican, Bernice Ellis, Sheng-Hung Tschang, Yu Tian, Krister Bokvist et al. "Engineering and characterization of the long-acting glucagon-like peptide-1 analogue LY2189265, an Fc fusion protein." *Diabetes/metabolism research and reviews* 26, no. 4 (2010): 287-296.

[67] Godia, F., and J. J. Cairó. "Metabolic engineering of animal cells." *Bioprocess and Biosystems Engineering* 24, no. 5 (2002): 289-298.

[68] Goodman, Cynthia L., Galal N. El Sayed, Arthur H. McIntosh, James J. Grasela, and Brad Stiles. "Establishment and characterization of insect cell lines from 10 lepidopteran species." *In Vitro Cellular & Developmental Biology-Animal* 37, no. 6 (2001): 367-373.

[69] Goodpasture, E. W., Alice M. Woodruff, and G. J. Buddingh. "The cultivation of vaccine and other viruses in the chorioallantoic membrane of chick embryos." *Science* 74, no. 1919 (1931): 371-372.

[70] Goswami, M., B. S. Sharma, Kamalendra Yadav, S. N. Bahuguna, and W. S. Lakra. "Establishment and characterization of a piscean PCF cell line for toxicity and gene expression studies as in vitro model." *Tissue and Cell* 46, no. 3 (2014): 206-212.

[71] Goswami, M., B. S. Sharma, S. N. Bahuguna, N. S. Nagpure, and W. S. Lakra. "A SRCF cell line from snowtrout, Schizothorax richardsonii: development and characterization." *Tissue and Cell* 45, no. 3 (2013): 219-226.

[72] Grace, T. D. C. "Establishment of a line of cells from the silkworm Bombyx mori." *Nature* 216, no. 5115 (1967): 613-613.

[73] Granados, Robert R., Guoxun Li, and G. W. Blissard. "Insect cell culture and biotechnology." *Virologica Sinica* 22, no. 2 (2007): 83-93.

[74] Gregersen, Jens-Peter. "A risk-assessment model to rate the occurrence and relevance of adventitious agents in the production of influenza vaccines." *Vaccine* 26, no. 26 (2008): 3297-3304.

[75] Guskey, Louis E., and Howard M. Jenkin. "Adaptation of BHK-21 cells to growth in shaker culture and subsequent challenge by Japanese encephalitis virus." *Applied microbiology* 30, no. 3 (1975): 433-438.

[76] Hacker, David L., and Sowmya Balasubramanian. "Recombinant protein production from stable mammalian cell lines and pools." *Current opinion in structural biology* 38 (2016): 129-136.

[77] Hammonds, Jason, Xuemin Chen, Xiugen Zhang, Francis Lee, and Paul Spearman. "Advances in methods for the production, purification, and characterization of HIV-1 Gag–Env pseudovirion vaccines." *Vaccine* 25, no. 47 (2007): 8036-8048.

[78] Havenga, M. J. E., L. Holterman, I. Melis, S. Smits, J. Kaspers, E. Heemskerk, R. van der Vlugt et al. "Serum-free transient protein production system based on adenoviral vector and PER. C6 technology: High yield and preserved bioactivity." *Biotechnology and bioengineering* 100, no. 2 (2008): 273-283.

[79] Hehir, Kathleen M., Donna Armentano, Lisa M. Cardoza, Terry L. Choquette, Patricia B. Berthelette, Gary A. White, Larry A. Couture et al. "Molecular characterization of replication-competent variants of adenovirus vectors and genome modifications to prevent their occurrence." *Journal of Virology* 70, no. 12 (1996): 8459-8467.

[80] Hernandez, Raquel, and Dennis T. Brown. "Growth and maintenance of chick embryo fibroblasts (CEF)." *Current protocols in microbiology* 17, no. 1 (2010): A-4I.

[81] Hervas-Stubbs, Sandra, Paloma Rueda, Lissette Lopez, and Claude Leclerc. "Insect baculoviruses strongly potentiate adaptive immune

responses by inducing type I IFN." *The Journal of Immunology* 178, no. 4 (2007): 2361-2369.

[82] Hink, W. F. "Established insect cell line from the cabbage looper, Trichoplusia ni." *Nature* 226, no. 5244 (1970): 466-467.

[83] HU, Yu-chen. "Baculovirus as a highly efficient expression vector in insect and mammalian cells." *Acta Pharmacologica Sinica* 26, no. 4 (2005): 405-416.

[84] Huang, Jing-Fei, Ke-Juan Shui, Hai-Yi Li, Mei-Ying Hu, and Guo-Hua Zhong. "Antiproliferative effect of azadirachtin A on Spodoptera litura Sl-1 cell line through cell cycle arrest and apoptosis induced by up-regulation of p53." *Pesticide biochemistry and physiology* 99, no. 1 (2011): 16-24.

[85] Ibarra, Neysi, Shikiko Watanabe, Jin-Xiu Bi, John Shuttleworth, and Mohamed Al-Rubeai. "Modulation of cell cycle for enhancement of antibody productivity in perfusion culture of NS0 cells." *Biotechnology progress* 19, no. 1 (2003): 224-228.

[86] Igarashi, Akira. "Isolation of a Singh's Aedes albopictus cell clone sensitive to Dengue and Chikungunya viruses." *Journal of General Virology* 40, no. 3 (1978): 531-544.

[87] Ikonomou, L., Y. J. Schneider, and S. N. Agathos. "Insect cell culture for industrial production of recombinant proteins." *Applied microbiology and biotechnology* 62, no. 1 (2003): 1-20.

[88] Ivarie, Robert. "Competitive bioreactor hens on the horizon." *TRENDS in Biotechnology* 24, no. 3 (2006): 99-101.

[89] Jones, David, Nathalie Kroos, Regina Anema, Bart Van Montfort, Andre Vooys, Sven van der Kraats, Esmeralda van der Helm et al. "High-level expression of recombinant IgG in the human cell line PER. C6." *Biotechnology progress* 19, no. 1 (2003): 163-168.

[90] Kagami, Hiroshi. "Perspectives on avian stem cells for poultry breeding." *Animal Science Journal* 87, no. 9 (2016): 1065-1075.

[91] Kamihira, Masamichi, Ken-ichiro Ono, Kazuhisa Esaka, Ken-ichi Nishijima, Ryoko Kigaku, Hiroyuki Komatsu, Takashi Yamashita, Kenji Kyogoku, and Shinji Iijima. "High-level expression of single-chain Fv-Fc fusion protein in serum and egg white of

genetically manipulated chickens by using a retroviral vector." *Journal of virology* 79, no. 17 (2005): 10864-10874.

[92] Katagiri, Yasuhiro, Shelesa A. Brew, and Kenneth C. Ingham. "All six modules of the gelatin-binding domain of fibronectin are required for full affinity." *Journal of Biological Chemistry* 278, no. 14 (2003): 11897-11902.

[93] Kim, J., M. R. Park, D. S. Kim, J. O. Lee, S. H. Maeng, S. Y. Cho, Y. Han, K. Ahn, and D. K. Jin. "IgE-mediated anaphylaxis and allergic reactions to idursulfase in patients with H unter syndrome." *Allergy* 68, no. 6 (2013): 796-802.

[94] Kim, Jee Yon, Yeon-Gu Kim, and Gyun Min Lee. "CHO cells in biotechnology for production of recombinant proteins: current state and further potential." *Applied microbiology and biotechnology* 93, no. 3 (2012): 917-930.

[95] Kojima, Yusuke, Jun Wakita, Yujin Inayoshi, Ryosuke Suzuki, Yuki Yamada, Hidenori Kaneoka, Ken-ichi Nishijima, and Shinji Iijima. "Galactosylation of human erythropoietin produced by chimeric chickens expressing galactosyltransferase." *Journal of bioscience and bioengineering* 117, no. 6 (2014): 676-679.

[96] Koo, Bon Chul, Mo Sun Kwon, Hyuna Lee, Minjee Kim, Dohyang Kim, Ji Yeol Roh, Yoon-Yub Park et al. "Tetracycline-dependent expression of the human erythropoietin gene in transgenic chickens." *Transgenic research* 19, no. 3 (2010): 437-447.

[97] Koudstaal, W., L. Hartgroves, M. Havenga, I. Legastelois, C. Ophorst, M. Sieuwerts, D. Zuijdgeest et al. "Suitability of PER. C6® cells to generate epidemic and pandemic influenza vaccine strains by reverse genetics." *Vaccine* 27, no. 19 (2009): 2588-2593.

[98] Kuno, Goro. "Cultivation of mosquito cell lines in serum-free media and their effects on dengue virus replication." *In vitro* 19, no. 9 (1983): 707-713.

[99] Kwon, Se Chang, Jin Won Choi, Hyun-Jun Jang, Sang Su Shin, Seul Ki Lee, Tae Sub Park, In Young Choi, Gwan Sun Lee, Gwonhwa Song, and Jae Yong Han. "Production of biofunctional recombinant human interleukin 1 receptor antagonist (rhIL1RN)

from transgenic quail egg white." *Biology of reproduction* 82, no. 6 (2010): 1057-1064.

[100] Lee, Jae Man, Masateru Takahashi, Hiroaki Mon, Hitoshi Mitsunobu, Katsumi Koga, Yutaka Kawaguchi, Yumiko Nakajima, and Takahiro Kusakabe. "Construction of gene expression systems in insect cell lines using promoters from the silkworm, *Bombyx mori*." *Journal of biotechnology* 133, no. 1 (2008): 9-17.

[101] Lee, L. E. J., V. R. Dayeh, Kirstin Schirmer, and N. C. Bols. "Applications and potential uses of fish gill cell lines: examples with RTgill-W1." *In Vitro Cellular & Developmental Biology-Animal* 45, no. 3-4 (2009): 127-134.

[102] Lee, S. H., M. K. Gupta, D. W. Han, S. Y. Han, S. J. Uhm, T. Kim, and H. T. Lee. "Development of transgenic chickens expressing human parathormone under the control of a ubiquitous promoter by using a retrovirus vector system." *Poultry science* 86, no. 10 (2007): 2221-2227.

[103] Lee, Sung Ho, Mukesh Kumar Gupta, Young Tae Ho, Teoan Kim, and Hoon Taek Lee. "Transgenic chickens expressing human urokinase-type plasminogen activator." *Poultry science* 92, no. 9 (2013): 2396-2403.

[104] Liu, Fuxiao, Xiaodong Wu, Lin Li, Zengshan Liu, and Zhiliang Wang. "Use of baculovirus expression system for generation of virus-like particles: successes and challenges." *Protein expression and purification* 90, no. 2 (2013): 104-116.

[105] Liu, Tongxin, Hanyu Wu, Dainan Cao, Qingyuan Li, Yaqiong Zhang, Ning Li, and Xiaoxiang Hu. "Oviduct-specific expression of human neutrophil defensin 4 in lentivirally generated transgenic chickens." *PloS one* 10, no. 5 (2015): e0127922.

[106] Lohr, Verena, Oliver Hädicke, Yvonne Genzel, Ingo Jordan, Heino Büntemeyer, Steffen Klamt, and Udo Reichl. "The avian cell line AGE1. CR. pIX characterized by metabolic flux analysis." *BMC biotechnology* 14, no. 1 (2014): 1-14.

[107] López-Macías, Constantino, Eduardo Ferat-Osorio, Alejandra Tenorio-Calvo, Armando Isibasi, Juan Talavera, Oscar Arteaga-Ruiz, Lourdes Arriaga-Pizano et al. "Safety and immunogenicity of a virus-like particle pandemic influenza A (H1N1) 2009 vaccine in a blinded, randomized, placebo-controlled trial of adults in Mexico." *Vaccine* 29, no. 44 (2011): 7826-7834.

[108] Lubiniecki, Anthony S. "Historical reflections on cell culture engineering." In *Cell Culture Engineering* VI, pp. 139-145. Springer, Dordrecht, 1998.

[109] Lynn, D. E., and H. Oberlander. "The establishment of cell lines from imaginal wing discs of *Spodoptera frugiperda* and *Plodia interpunctella*." *Journal of Insect Physiology* 29, no. 7 (1983): 591-596.

[110] Lynn, Dwight E. "Novel techniques to establish new insect cell lines." In *Vitro Cellular & Developmental Biology-Animal* 37, no. 6 (2001): 319-321.

[111] Mabrouk, T., and R. W. Ellis. "Influenza vaccine technologies and the use of the cell-culture process (cell-culture influenza vaccine)." *Developments in biologicals* 110 (2002): 125-134.

[112] Madin, S. H., and N. B. Darby Jr. "Established kidney cell lines of normal adult bovine and ovine origin." *Proceedings of the Society for Experimental Biology and Medicine* 98, no. 3 (1958): 574-576.

[113] Maeda, Susumu. "Gene transfer vectors of a baculovirus, Bombyx mori nuclear polyhedrosis virus, and their use for expression of foreign genes in insect cells." In *Invertebrate cell system applications,* pp. 167-181. CRC Press, 2018.

[114] Mahlangu, Johnny, Jerry S. Powell, Margaret V. Ragni, Pratima Chowdary, Neil C. Josephson, Ingrid Pabinger, Hideji Hanabusa et al. "Phase 3 study of recombinant factor VIII Fc fusion protein in severe hemophilia A." *Blood, The Journal of the American Society of Hematology* 123, no. 3 (2014): 317-325.

[115] Martin, K. J. "The first human cell line-derived erythropoietin, epoetin-delta (Dynepo), in the management of anemia in patients

with chronic kidney disease." *Clinical nephrology* 68, no. 1 (2007): 26-31.

[116] Maruoka, Yutaka, Shinichiro Oida, Tadahiro Iimura, Kohsuke Takeda, Izumi Asahina, Shoji Enomoto, and Satoshi Sasaki. "Production of functional human bone morphogenetic protein-2 using a baculovirus/Sf-9 insect cell system." *Biochemistry and molecular biology international* 35, no. 5 (1995): 957-963.

[117] Maruyama, Tetsuo, and Yasunori Yoshimura. "Stem cell theory for the pathogenesis of endometriosis." *Front Biosci* 4, no. 8 (2012): 2754-2763.

[118] Mayrhofer, Josef, Sogue Coulibaly, Annett Hessel, Georg W. Holzer, Michael Schwendinger, Peter Brühl, Marijan Gerencer et al. "Nonreplicating vaccinia virus vectors expressing the H5 influenza virus hemagglutinin produced in modified Vero cells induce robust protection." *Journal of virology* 83, no. 10 (2009): 5192-5203.

[119] McAtee, Allison G., Neil Templeton, and Jamey D. Young. "Role of Chinese hamster ovary central carbon metabolism in controlling the quality of secreted biotherapeutic proteins." *Pharmaceutical Bioprocessing* 2, no. 1 (2014): 63-74.

[120] McCarroll, Lynn, and Linda A. King. "Stable insect cell cultures for recombinant protein production." *Current opinion in biotechnology* 8, no. 5 (1997): 590-594.

[121] Mehtali, Majid, Patrick Champion-Arnaud, and Arnaud Leon. *Production of viral vaccines in suspension on avian embryonic derived stem cell lines*. U.S. Patent 9,040,296, issued May 26, 2015.

[122] Mei, Baisong, Yaoqi Chen, Jianmin Chen, Clark Q. Pan, and John E. Murphy. "Expression of human coagulation factor VIII in a human hybrid cell line, HKB11." *Molecular Biotechnology* 34, no. 2 (2006): 165-178.

[123] Meleady, Paula, Padraig Doolan, Michael Henry, Niall Barron, Joanne Keenan, Finbar O'Sullivan, Colin Clarke et al. "Sustained productivity in recombinant Chinese hamster ovary (CHO) cell

lines: proteome analysis of the molecular basis for a process-related phenotype." *BMC biotechnology* 11, no. 1 (2011): 1-11.

[124] Metz, Stefan W., and Gorben P. Pijlman. "Arbovirus vaccines; opportunities for the baculovirus-insect cell expression system." *Journal of Invertebrate Pathology* 107 (2011): S16-S30.

[125] Metz, Stefan W., Joy Gardner, Corinne Geertsema, Thuy T. Le, Lucas Goh, Just M. Vlak, Andreas Suhrbier, and Gorben P. Pijlman. "Effective chikungunya virus-like particle vaccine produced in insect cells." *PLoS Negl Trop Dis 7,* no. 3 (2013): e2124.

[126] Montagnon, B. J., B. Fanget, and A. J. Nicolas. "The large-scale cultivation of VERO cells in micro-carrier culture for virus vaccine production. Preliminary results for killed poliovirus vaccine." *Developments in biological standardization* 47 (1981): 55-64.

[127] Montomoli, Emanuele, Baharak Khadang, Simona Piccirella, Claudia Trombetta, Elisa Mennitto, Ilaria Manini, Valerio Stanzani, and Giulia Lapini. "Cell culture-derived influenza vaccines from Vero cells: a new horizon for vaccine production." *Expert review of vaccines* 11, no. 5 (2012): 587-594.

[128] MORAIS, Vanessa A., Jacinta SERPA, Angelina S. PALMA, Teresa Costa, Luís Maranga, and Júlia Costa. "Expression and characterization of recombinant human α-3/4-fucosyltransferase III from *Spodoptera frugiperda* (Sf9) and *Trichoplusia ni* (Tn) cells using the baculovirus expression system." *Biochemical Journal* 353, no. 3 (2001): 719-725.

[129] Murakami, Miho, Hideyo Ugai, Minghui Wang, Natalya Belousova, Paul Dent, Paul B. Fisher, Joel N. Glasgow, Maaike Everts, and David T. Curiel. "An adenoviral vector expressing human adenovirus 5 and 3 fiber proteins for targeting heterogeneous cell populations." *Virology* 407, no. 2 (2010): 196-205.

[130] Nadeau, Isabelle, and Amine Kamen. "Production of adenovirus vector for gene therapy." *Biotechnology advances* 20, no. 7-8 (2003): 475-489.

[131] Nagarajan, Thirumeni, Wilfred E. Marissen, and Charles E. Rupprecht. "Monoclonal antibodies for the prevention of rabies: theory and clinical practice." *Antibody Technology Journal* 4 (2014): 1-12.

[132] Nishijima, Ken-ichi, and Shinji Iijima. *"Transgenic chickens." Development, growth & differentiation* 55, no. 1 (2013): 207-216.

[133] Novartis AG. *Novartis receives regulatory approval in Germany for Celtura, a cell culture-based influenza A (H1N1) pandemic vaccine* November 05 (2009). www.novartis.com/newsroom/ media releases/en/2009/1352736.shtml.

[134] Oh, Ding Yuan, Ian G. Barr, Jenny A. Mosse, and Karen L. Laurie. "MDCK-SIAT1 cells show improved isolation rates for recent human influenza viruses compared to conventional MDCK cells." *Journal of clinical microbiology* 46, no. 7 (2008): 2189-2194.

[135] Olivier, Stéphane, Marine Jacoby, Cédric Brillon, Sylvana Bouletreau, Thomas Mollet, Olivier Nerriere, Audrey Angel et al. "EB66 cell line, a duck embryonic stem cell-derived substrate for the industrial production of therapeutic monoclonal antibodies with enhanced ADCC activity." In *MAbs,* vol. 2, no. 4, pp. 405-415. Taylor & Francis, 2010.

[136] Padler-Karavani, Vered, Hai Yu, Hongzhi Cao, Harshal Chokhawala, Felix Karp, Nissi Varki, Xi Chen, and Ajit Varki. "Diversity in specificity, abundance, and composition of anti-Neu5Gc antibodies in normal humans: potential implications for disease." *Glycobiology* 18, no. 10 (2008): 818-830.

[137] Pain, Bertrand, M. E. Clark, Mi Shen, H. Nakazawa, M. Sakurai, Jacques Samarut, and R. J. Etches. "Long-term in vitro culture and characterisation of avian embryonic stem cells with multiple morphogenetic potentialities." *Development* 122, no. 8 (1996): 2339-2348.

[138] Palhan, V. B., S. Sumathy, and K. P. Gopinathan. "Baculovirus mediated high-level expression of luciferase in silkworm cells and larvae." *Biotechniques* 19, no. 1 (1995): 97-98.

[139] Pant, U., K. Banerjee, S. A. Athawale, and V. Dhanda. "Susceptibility of Culex bitaeniorhynchus cell line to some arboviruses." *Indian journal of medical research* 76 (1982): 789-794.

[140] Park, Tae Sub, Hyo Gun Lee, Jong Kook Moon, Hong Jo Lee, Jong Won Yoon, Bit Na Rae Yun, Sang-Chul Kang et al. "Deposition of bioactive human epidermal growth factor in the egg white of transgenic hens using an oviduct-specific minisynthetic promoter." *The FASEB Journal* 29, no. 6 (2015): 2386-2396.

[141] Pau, M. G., C. Ophorst, Martin H. Koldijk, G. Schouten, M. Mehtali, and Fons Uytdehaag. "The human cell line PER. C6 provides a new manufacturing system for the production of influenza vaccines." *Vaccine* 19, no. 17-19 (2001): 2716-2721.

[142] Perot, Eloïse, Nathalie Enjolras, Sandra Le Quellec, Alice Indalecio, Jonathan Girard, Claude Negrier, and Yesim Dargaud. "Expression and characterization of a novel human recombinant factor IX molecule with enhanced in vitro and in vivo clotting activity." *Thrombosis research* 135, no. 5 (2015): 1017-1024.

[143] Petricciani, John, and Rebecca Sheets. "An overview of animal cell substrates for biological products." *Biologicals* 36, no. 6 (2008): 359-362.

[144] Pham, Phuong Lan, Sylvie Perret, Huyen Chau Doan, Brian Cass, Gilles St-Laurent, Amine Kamen, and Yves Durocher. "Large-scale transient transfection of serum-free suspension-growing HEK293 EBNA1 cells: Peptone additives improve cell growth and transfection efficiency." *Biotechnology and bioengineering* 84, no. 3 (2003): 332-342.

[145] Picanco-Castro, Virginia, Rafael Tage Biaggio, Dimas Tadeu Cova, and Kamilla Swiech. "Production of recombinant therapeutic proteins in human cells: current achievements and future perspectives." *Protein and peptide letters* 20, no. 12 (2013): 1373-1381.

[146] Pillay, Sirika, Ann Meyers, Anna-Lise Williamson, and Edward P. Rybicki. "Optimization of chimeric HIV-1 virus-like particle

production in a baculovirus-insect cell expression system." *Biotechnology progress* 25, no. 4 (2009): 1153-1160.

[147] Powell, Jerry S., K. John Pasi, Margaret V. Ragni, Margareth C. Ozelo, Leonard A. Valentino, Johnny N. Mahlangu, Neil C. Josephson et al. "Phase 3 study of recombinant factor IX Fc fusion protein in hemophilia B." *New England Journal of Medicine* 369, no. 24 (2013): 2313-2323.

[148] Reisinger, Keith S., Stanley L. Block, Allen Izu, Nicola Groth, and Sandra J. Holmes. "Subunit influenza vaccines produced from cell culture or in embryonated chicken eggs: comparison of safety, reactogenicity, and immunogenicity." *The Journal of infectious diseases* 200, no. 6 (2009): 849-857.

[149] Roberts, Grace C., Carsten Zothner, Roland Remenyi, Andres Merits, Nicola J. Stonehouse, and Mark Harris. "Evaluation of a range of mammalian and mosquito cell lines for use in Chikungunya virus research." *Scientific reports* 7, no. 1 (2017): 1-13.

[150] Rodrigues, A. F., A. S. Formas-Oliveira, V. S. Bandeira, P. M. Alves, W. S. Hu, and A. S. Coroadinha. "Metabolic pathways recruited in the production of a recombinant enveloped virus: Mining targets for process and cell engineering." *Metabolic engineering* 20 (2013): 131-145.

[151] Ruggiero, Florence, and Manuel Koch. "Making recombinant extracellular matrix proteins." *Methods* 45, no. 1 (2008): 75-85.

[152] Sanford, Mark. "Dulaglutide: first global approval." *Drugs* 74, no. 17 (2014): 2097-2103.

[153] Schneider, Imogene. "Cell lines derived from late embryonic stages of *Drosophila melanogaster*." *Development* 27, no. 2 (1972): 353-365.

[154] Scott, Benjamin B., Tarciso A. Velho, Shuyin Sim, and Carlos Lois. "Applications of avian transgenesis." *ILAR journal* 51, no. 4 (2010): 353-361.

[155] Sehgal, Deepak, and Karumathil P. Gopinathan. "Recombinant *Bombyx mori* nucleopolyhedrovirus harboring green fluorescent protein." *Biotechniques* 25, no. 6 (1998): 997-1006.

[156] Shen, Duan, and Susan T. Sharfstein. "Genome-wide analysis of the transcriptional response of murine hybridomas to osmotic shock." *Biotechnology and bioengineering* 93, no. 1 (2006): 132-145.

[157] Shire Human Genetic Therapies."VPRIV (velaglucerase)." (2019).

[158] Shittu, Ismaila, Ziying Zhu, Yangqing Lu, Jessica M. Hutcheson, Steven L. Stice, Franklin D. West, Meritxell Donadeu et al. "Development, characterization and optimization of a new suspension chicken-induced pluripotent cell line for the production of Newcastle disease vaccine." *Biologicals* 44, no. 1 (2016): 24-32.

[159] Shrestha, Mrigendra Prasad, Robert McNair Scott, Durga Man Joshi, Mammen P. Mammen Jr, Gyan Bahadur Thapa, Narbada Thapa, Khin Saw Aye Myint et al. "Safety and efficacy of a recombinant hepatitis E vaccine." *New England Journal of Medicine* 356, no. 9 (2007): 895-903.

[160] Singh, K. R. P. "Cell cultures derived from larvae of *Aedes albopictus* (Skuse) and *Aedes aegypti* (L.)." *Current Science* 36, no. 19 (1967): 506-508.

[161] Singh, K. R. P. "Growth of arboviruses in arthropod tissue culture." *Advances in virus research* 17 (1972): 187-206.

[162] Smagghe, Guy, Cynthia L. Goodman, and David Stanley. "Insect cell culture and applications to research and pest management." *In Vitro Cellular & Developmental Biology-Animal* 45, no. 3-4 (2009): 93-105.

[163] Smales, Christopher Mark, Diane M. Dinnis, Scott H. Stansfield, D. Alete, E. A. Sage, John R. Birch, Andrew J. Racher, Carol T. Marshall, and David C. James. "Comparative proteomic analysis of GS-NS0 murine myeloma cell lines with varying recombinant monoclonal antibody production rate." *Biotechnology and bioengineering* 88, no. 4 (2004): 474-488.

[164] Smith, Gale E., Harald G. Foellmer, John Knell, James Debartolomeis, and Andrei I. Voznesensky. *Spodoptera frugiperda single cell suspension cell line in serum-free media, methods of producing and using.* U.S. Patent 6,103,526, issued August 15, 2000.

[165] Sriram, Satyanarayana, Vikas B. Palhan, and Karumathil P. Gopinathan. "Heterologous promoter recognition leading to high-level expression of cloned foreign genes in *Bombyx mori* cell lines and larvae." *Gen*e 190, no. 1 (1997): 181-189.

[166] Steinbach, Falko, Susanne Mauel, and Ilka Beier. "Recombinant equine interferons: expression cloning and biological activity." *Veterinary immunology and immunopathology* 84, no. 1-2 (2002): 83-95.

[167] Sudeep, A. B., D. T. Mourya, and A. C. Mishra. "Insect cell culture in research: Indian scenario." *Indian Journal of Medical Research* 121, no. 6 (2005): 725- 738.

[168] Sudeep, A. B., Deepti Parashar, Ramesh S. Jadi, Atanu Basu, Chetan Mokashi, Vidya A. Arankalle, and Akhilesh C. Mishra. "Establishment and characterization of a new *Aedes aegypti* (L.) (Diptera: Culicidae) cell line with special emphasis on virus susceptibility." *In Vitro Cellular & Developmental Biology-Animal* 45, no. 9 (2009): 491-495.

[169] Summers, Max D., and Gale E. Smith. *"A manual of methods for baculovirus vectors and insect cell culture procedures."* (1987):1-56.

[170] Sun, Yuliang, Ricardo Carrion Jr, Ling Ye, Zhiyuan Wen, Young-Tae Ro, Kathleen Brasky, Anysha E. Ticer et al. "Protection against lethal challenge by Ebola virus-like particles produced in insect cells." *Virology* 383, no. 1 (2009): 12-21.

[171] Sun, Zhe, Rui Zhou, Shuyan Liang, Kathleen M. McNeeley, and Susan T. Sharfstein. "Hyperosmotic stress in murine hybridoma cells: effects on antibody transcription, translation, posttranslational processing, and the cell cycle." *Biotechnology progress* 20, no. 2 (2004): 576-589.

[172] Sutter, Roland W., Olen M. Kew, Stephen L. Cochi, and R. Bruce Aylward. "Poliovirus vaccine–live." In *Plotkin's vaccines,* pp. 866-917. Elsevier, 2018.

[173] Swevers, L., L. Kravariti, S. Ciolfi, M. Xenou-Kokoletsi, G. Wong, N. Ragousis, G. Smagghe, Y. Nakagawa, V. Mazomenos, and K. Iatrou. "A high-throughput screening system for fast detection of ecdysteroid mimetic and antagonistic substances using transformed *Bombyx mori*-derived cell lines." *FASEB J* 17, no. 14 (2003): 134-136.

[174] Swiech, Kamilla, Virgínia Picanço-Castro, and Dimas Tadeu Covas. "Human cells: new platform for recombinant therapeutic protein production." *Protein expression and purification* 84, no. 1 (2012): 147-153.

[175] Swiech, Kamilla, Virgínia Picanço-Castro, and Dimas Tadeu Covas. "Production of recombinant coagulation factors: Are humans the best host cells?." *Bioengineered* 8, no. 5 (2017): 462-470.

[176] Therapies, Shire Human Genetic. *Replagal: Summary of product characteristics.* Danderyd, Sweden: Shire Human Genetic Therapies (2010).

[177] Thomas, Philip, and Trevor G. Smart. "HEK293 cell line: a vehicle for the expression of recombinant proteins." *Journal of pharmacological and toxicological methods* 51, no. 3 (2005): 187-200.

[178] Toya, Tetsuya, Hiroshi Fukasawa, Akio Masui, and Yasuyuki Endo. "Potent and selective partial ecdysone agonist activity of chromafenozide in Sf9 cells." *Biochemical and biophysical research communications* 292, no. 4 (2002): 1087-1091.

[179] Trusheim, H., B. Roth, R. Wilms, S. Jung, R. Muth, M. Eggers, M. Enders et al. "The MDCK 33016-PF cell line is not only suitable for the production of cell-based influenza vaccine but is also an ideal substrate for influenza virus isolation." *Abstr. Options Control of Influenza VI*, Toronto, Canada, abstr (2007): P515.

[180] van Oers, Monique M. "Opportunities and challenges for the baculovirus expression system." *Journal of invertebrate pathology* 107 (2011): S3-S15.

[181] Vaughn, J. L., R. H. Goodwin, G. J. Tompkins, and P. J. I. V. McCawley. "The establishment of two cell lines from the *insectspodoptera frugiperda* (lepidoptera; noctuidae)." *In vitro* 13, no. 4 (1977): 213-217.

[182] Volkman, Loy E., and Phyllis A. Goldsmith. "Generalized immunoassay for Autographa californica nuclear polyhedrosis virus infectivity in vitro." *Applied and Environmental Microbiology* 44, no. 1 (1982): 227-233.

[183] Walker, Thomas, Claire L. Jeffries, Karen L. Mansfield, and Nicholas Johnson. "Mosquito cell lines: history, isolation, availability and application to assess the threat of arboviral transmission in the United Kingdom." *Parasites & vectors* 7, no. 1 (2014): 1-9.

[184] Watanabe, Shikiko, John Shuttleworth, and Mohamed Al-Rubeai. "Regulation of cell cycle and productivity in NS0 cells by the over-expression of p21CIP1." *Biotechnology and bioengineering* 77, no. 1 (2002): 1-7.

[185] Wickham, T. J., T. Davis, R. R. Granados, M. L. Shuler, and H. A. Wood. "Screening of insect cell lines for the production of recombinant proteins and infectious virus in the baculovirus expression system." *Biotechnology progress* 8, no. 5 (1992): 391-396.

[186] Wikan, Nitwara, Atichat Kuadkitkan, and Duncan R. Smith. "The Aedes aegypti cell line CCL-125 is dengue virus permissive." *Journal of virological methods* 157, no. 2 (2009): 227-230.

[187] Wu, Donglai, Kenji Murakami, Nihong Liu, Yasuo Inoshima, Takashi Yokoyama, Takehiro Kokuho, Shigeki Inumaru et al. "Expression of biologically active recombinant equine interferon-γ by two different baculovirus gene expression systems using insect cells and silkworm larvae." *Cytokine* 20, no. 2 (2002): 63-69.

[188] Wu, Suh-Chin, Donald L. Jarvis, Bruce E. Dale, and James C. Liao. "Heterologous protein expression affects the death kinetics of baculovirus-infected insect cell cultures: A quantitative study by use of n-target theory." *Biotechnology progress* 10, no. 1 (1994): 55-59.

[189] Wurm, Florian M. "Production of recombinant protein therapeutics in cultivated mammalian cells." *Nature biotechnology* 22, no. 11 (2004): 1393-1398.

[190] Yamaguchi, Kanji, Katsuhiko Itoh, Naoki Ohnishi, Yoshito Itoh, Christopher Baum, Takashi Tsuji, Toshikazu Nagao, Hiroaki Higashitsuji, Takeshi Okanoue, and Jun Fujita. "Engineered long terminal repeats of retroviral vectors enhance transgene expression in hepatocytes in vitro and in vivo." *Molecular Therapy* 8, no. 5 (2003): 796-803.

[191] Yanagawa, Shin-ichi, Jong-Seo Lee, and Akinori Ishimoto. "Identification and characterization of a novel line of Drosophila Schneider S2 cells that respond to wingless signaling." *Journal of Biological Chemistry* 273, no. 48 (1998): 32353-32359.

[192] Zhang, Bodi. "Ofatumumab." In *MAbs,* vol. 1, no. 4, pp. 326-331. Taylor & Francis, 2009.

[193] Zhu, Lei, Marie-Cecile Van de Lavoir, Jenny Albanese, David O. Beenhouwer, Pina M. Cardarelli, Severino Cuison, David F. Deng et al. "Production of human monoclonal antibody in eggs of chimeric chickens." *Nature biotechnology* 23, no. 9 (2005): 1159-1169.

In: The Future of Metabolic Engineering ISBN: 978-1-68507-362-6
Editors: A. Sharma and D. Amin

Chapter 5

An Insight in the Trends Timeline of Metabolic Engineering in Microbes and Its Reflection on Future Scope

***Ketankumar J. Panchal**, PhD**
Department of Animal Biotechnology,
College of Veterinary Science and Animal Husbandry,
Kamdhenu University, Anand, Gujarat, India

Keywords: metabolic engineering, flux, pathways, synthetic biology, gene, protein, co-factor

* Corresponding Author's E-mail: ketan9589@gmail.com.

1. Introduction

The food and beverage industries are the primary focus for the exploitation of microbes. Through methods such as microbial fermentation, various chemicals and fuels can be produced. The use of microbial fermentation is advantageous as it is scalable and has a lower carbon footprint. In the year 1973, Cohen, Boyer and co-workers (Cohen et al. 1973) were the first to report the successful genetic engineering of *Escherichia coli* (*E. coli*). After that, it became apparent that the microorganisms can be engineered into chemical factories. The most notable example of such is the production of human insulin by recombinant *E. coli*. In the same decade, researchers started investigations to understand the fundamental question of the functioning of metabolic networks like in the case of human insulin. This led to different processes such as thermodynamic analysis, kinetic control and design of genetic circuits, etc. This further development led to investigations in classical fermentation to apply metabolic engineering to optimize different processes. The use of classical approaches such as mutagenesis and screening have since been replaced by more advanced rational approaches such as introduction of directed genetic changes through recombinant DNA technology for strain improvement. In 1991 Bailey (Bailey 1991) discussed this newly emerging science and coined the term 'metabolic engineering' for the first time. During the early stages of method development, it harnessed applications from applied molecular biology and with time, new analytical and cloning techniques have further advanced methods and collectively it is possible to introduce the genetic changes at a cellular level. However, for smooth functioning of a cell after the introduction of genetic modification, integrating analysis of cellular function and genetic engineering is also involved. Initially, genetic engineering and metabolic engineering were considered to be the same but later metabolic engineering was differentiated by its distinct focus to study how the metabolic network works, apparently with the help of genetic engineering and metabolic engineering.

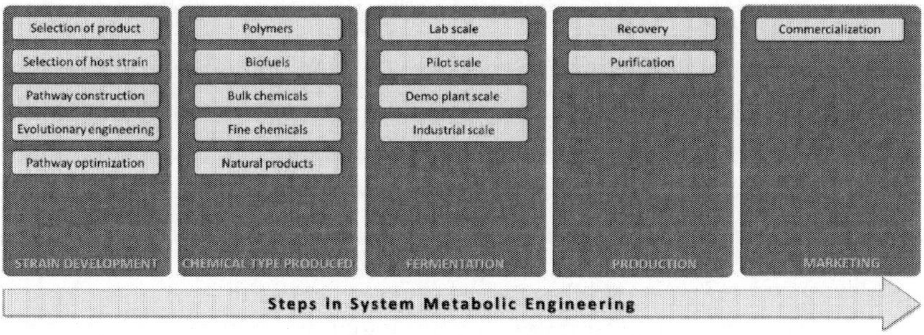

Figure 1. Flowchart of steps in the metabolic engineering.

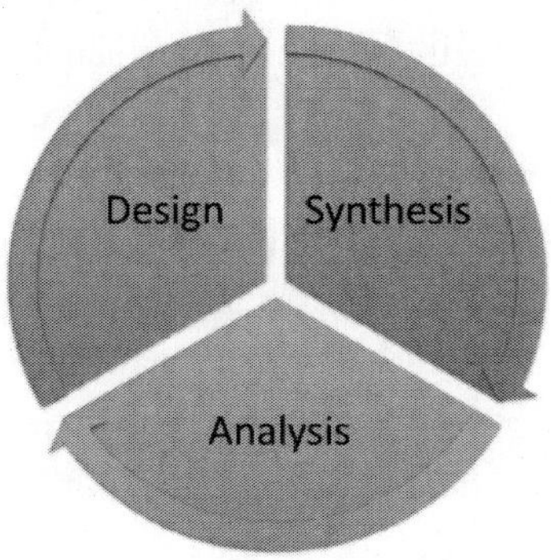

Figure 2. Cycle of strain improvement for production of target molecule.

As a key addition to the industry, metabolic engineering aims to construct microbes for cost effective production of pharmaceuticals, fuels and chemicals (Figure 1). It is not as simple as simply stitching a few genes together to create a small production factory. Instead, it encompasses overlapping fields in the scope of synthetic biology. Once successful, production could be possible by introducing the desired changes into the genome, but the process will not be cost effective until titre, productivity and yield factors are optimized. The timeline for creating a pathway that produces the desired product could be a month but to develop on an industrial scale production could take much longer. This underlines the importance in metabolic engineering to understand the baseline factors like construction, pathway design and optimization. These factors encompass not only metabolic engineering and molecular biology, but also chemical reaction engineering, biochemistry, graph theory and optimization. Different methodologies have been strategized over a period time for prokaryotes and eukaryotes that include metabolic flux rewiring (Qin et al. 2015), adaptive evolution (Reyes, Gomez, and Kao 2014), genome editing (Si, Xiao, and Zhao 2015, Cobb, Wang, and Zhao 2015), engineering the cell tolerance (Anfelt et al. 2013) and compartmentalization (Avalos, Fink, and Stephanopoulos 2013). This significant progress that has been accomplished is for production of high-value and high-volume chemicals. The *in-silico* technology has provided momentum to development of metabolic engineering cycle (design -> build -> test -> learn) by introducing the virtual system modelling and prediction of pathways. The process for effective strain selection during production of a molecule of interest, there is a cycle (Figure 2) which involves the synthesis of recombinant strain, followed by its performance analysis compared to original strain and then the design of next target. These processes run several rounds in a closed loop for continuous improvement of cellular properties until the target production is reached. Depending on the aim, the starting point could be any location in this cycle. This book chapter describes the trend in metabolic engineering from past to present and looking forward to the future.

2. Current Trends

2.1. Pathway Assessment and Enumeration for a Desired Product from a Specific Substrate

The fundamental question to be addressed while targeting the overproduction of a product of interest is 'which pathway to target?.' The chosen pathway must be thermodynamically balanced in order to produce the desired substrate in the host. If the pathway has a multistep network, it is essential to check all of them and to divert the intermediate in a single direction. The analysis of a pathway gives crucial information on how to achieve the maximum expected yield, as well as suggesting which bottleneck to bypass and assist with identification of intermediates.

2.2. Comprehension and Conception of Metabolic Control Supported by Metabolic Flux

Upon establishment of a possible pathway, the next step is to analyse the pathway operation and its control rate to propose the rational modification for a production phenotype (Nielsen 2001). Metabolic flux analysis (MFA) is the primary method for analysing identified metabolic networks. As an input for the MFA calculation, substrate uptake rate and product secretion rate values are provided, while as a result the biochemical reaction diagram is shown in the form of a diagram along with calculations and estimate of the steady state rate for each reaction in the diagram. Such map can help to pinpoint the critical pathway (Vallino and Stephanopoulos 1994), identify unusual pathways (Nissen et al. 1997) and gives theoretical information for the production of target molecules with the help of complex integrated pathway, including the intermediates and cofactors in it (Vallino and Stephanopoulos 2000). These form the bases to select the target to manipulate at the genetic level considering the given flux flexibility from the base state. Enzymes and branch points in the networks are either rigid or flexible for easy modulation. The experimental data on fluxes and their response to system perturbation is used to determine the flux control coefficient (FCC), concentration control coefficient (CCC) and elastic coefficient (EC). The FCC is the degree of control of an enzyme exerted on metabolite flux, CCC is an enzyme control over the metabolite concentration and EC defines the capability of an enzyme to change its rate in response to system perturbation. Collectively, FCC, CCC and EC can direct the genetic engineering by providing rational application of how to divert the flux to a specified product. Suppose in a simple pathway if one enzyme has higher FCC value, then it can limit the rate of reaction. Overproduction of this particular enzyme will lead to increase in the flux through the pathway; if the coefficient does not vary significantly between enzymes, then the pathway control is interpreted as disturbed.

2.3. *In Silico* Methods for Metabolic Engineering

For a selection of targets in metabolic engineering, metabolic control analysis (MCA) depends on accurate flux determination. However, this cannot help to achieve higher resolution in the case of simple pathways, but internal, bidirectional reaction can (Wiechert 2001). Techniques of isotopes like 13C – a stable isotope are used to get more insight; cells are grown on isotope substrate and when equilibrium condition is achieved, GC-MS/NMR (gas chromatography-mass spectroscopy/ nuclear magnetic resonance) is used to measure the isotopes in the metabolites. With the help of an algorithm based on isotopomer distribution, intracellular fluxes are determined more accurately; the progressive development of such algorithm started back in late 90s and further refined later (Schmidt, Nielsen, and Villadsen 1999, Antoniewicz, Kelleher, and Stephanopoulos 2007). Cost and the expertise required for these techniques are inherent problems. Alternately, flux balance analysis (FBA) balances as used in MFA with the help of genome-scale metabolic stoichiometric model (GSM) data. In this, rather than utilizing the isotopes, equations are formulated with an objective function to be either minimized or maximized (Raman and Chandra 2009). This method is advantageous as less date is required, *in silico* perturbations are possible and refinement of the available data with new omics data which continually accumulate in the database (Lee, Park, and Kim 2011). However, one cannot deny that all formulated fluxes are not real and there may chance of missing the important distinct. Despite that, for Bacillus subtilis (Oh et al. 2007), Arabidopsis thaliana (de Oliveira Dal'Molin et al. 2010) and others, GSM have been developed.

2.4. Metabolic Engineering from the Perspective of Rational Approach

Metabolic engineering relies on many factors such as pruning of genes to eliminate side reactions, measurement of intermediates to assess dysfunction, kinetic bottleneck identification, metabolic pathway regulation/kinetics and lastly the understanding of network topography. Almost two decades ago, initial efforts were made to manipulate metabolism rationally to direct the flux towards a desired product. Many reports (Park, Sinskey, and Stephanopoulos 1997, Nakamura and Whited 2003, Bro et al. 2006, Kötter and Ciriacy 1993) can advocate the success story of this approach. Most of them have constructed pathways through joint expression of selected and constituent genes followed by its inhibition, overproduction, modification, or knockout aiming the best pathway performance.

2.5. Connective Methods for Metabolic Engineering

The rational approach and modulation of pre-selected pathway genes are sufficient for target molecule production. All methods used in metabolic engineering select for the pathways that convert a substrate into a product as well as flux associated and regulatory mechanism. However, for best case scenario the individual method is insufficient to optimize the targeted production. For every step up in the production other factors like a metabolic burden on cells and toxicity of a product requires attention. These factors could interconnect by a complex network of genes which again needs to be manipulated. Poor understanding of these genes calls out the use of a rational approach. The cycling of randomized mutation and screening of improved phenotypes could help. Penicillin production and ethanol production (Jin et al. 2005) is the evidence of use of combinatorial methods.

3. PARAMETERS

3.1. Product

The alternate of fossil fuels has always been a topic of top priority for a more sustainable society. In order to reduce greenhouse gases, biofuels can be used as an alternative to petroleum-based fuel in mechanical instruments for which electrification is not a viable option. The best strategy for waste management is utilizing the agricultural waste as feedstock to produce fuel chemicals from the microbial cells. Microbial cell factories have already produced the diols, organic acids, fatty acids, higher alcohols, terpenoids, phenolics and alkaloids (Liu and Nielsen 2019a) from a variety of hosts such as *Saccharomyces cerevisiae, Escherichia coli, Pichia pastoris, Aspergillus niger* and *Bacillus subtilis*. For production of these biofuels alternates, five metabolic pathways namely, polyketide synthetic pathway, CoA-dependent reverse β-oxidation pathway, isoprenoid pathway, 2-keto acid pathway and fatty acid synthetic pathway have been advanced (Liu and Nielsen 2019a). *S. cerevisiae* platform was reported to produce cytosolic fatty acid (Liu and Nielsen 2019a) and high-level extracellular accumulation of free fatty acids (Zhou et al. 2016); deletion of FAA4 and FAA1 genes has made secretion of free fatty acid possible, parallel overexpression of ACC1 gene has resulted in high volume production of free fatty acids in the fed-batch fermentation. Of course, this is the eukaryotic biofuel production, which needs further optimization in terms of production titre and yield. The application of fatty acid is dependent on its carbon chain length. C7-C13 - medium chain fatty acid can be used in jet and gasoline; more than C22 for fatty waxes (Liu and Nielsen 2019b). As per the use and demand, now the attention has been given to the production of fatty acids of desired carbon length like C14 from *E. coli* (Fernandez-Moya et al. 2015), C22-C26 from *S. cerevisiae* (Yu et al. 2017). Other strategies like compartmentalization, novel pathway, transport engineering and protein engineering play crucial role in increasing the production of different fatty acids with varied carbon chain length.

Alkanes, fatty alcohol and fatty acid ethyl ester are the oleochemicals and the biosynthesis of which have captured great attention recently. To produce the branch chain alcohol, an intermediate in the amino acid biosynthesis i.e., 2-keto acid and its pathway can be recruited. Use of *E. coli* for the production of 1-octanol and 1-heptanol (Marcheschi et al. 2012) via protein engineering, protein-substrate modelling apart from the metabolic engineering can be used. Attempts were made by Mak et al. (Mak et al. 2015) to modify the substrate specificity to produce the long chain alcohol via a 2-keto acid pathway. Other enzymes like aldehyde-deformylating oxygenase and acetyl-transferase were also incorporated for alkanes and ester production (Yuan, Mishra, and Ching 2016, Zhang et al. 2016). Factors like redox imbalance, intolerance and inefficiency of carbon limit the performance of this pathway, but its accurate control could help in production of demanding new products.

Isoprenoid pathway is also recognized as a promising pathway for the production of biofuels due to its diverse alcohols and hydrocarbon end products. This pathway adds dimethylallyl pyrophosphate or isopentenyl pyrophosphate to elongate the carbon chain. One of such end products is farnesene that has wide application area such as cosmetics, food additives and pharmaceutics (Kang and Lee 2016). The introduction of heterologous enzyme named farnesene synthase and a few non-native enzymes to the *S. cerevisiae* system can enhance the production of farnesene by reduction in oxygen consumption. Farnesene is an example of viable strategy in utilization of rewired central carbon metabolism for maximum production of chemicals in demand.

For a long time herbal medicine has relied on secondary plant metabolites for therapeutic purposes. However, these metabolites have also been found in flavourings, colourings and texturings (Suástegui and Shao 2016). These occur as alkaloids, phenols and terpenoids in nature; extraction of these highly valued products from the plant source often yields low production and the extraction process is often very complicated (Milke et al. 2018). Plant tissue culture has been evaluated as an alternative (O'Neill and Kelly 2017). However, both of these are time consuming process, with low success ratio for genetic manipulation and possibility of yield variability. Replication of the production pathway with the help from a synthetic pathway circuit in microbial host is argued as a viable option.

One of the phenolic compounds, cinnamic acid, has wide variety application that includes anti-inflammatory agent, antimicrobial, food and flavour and is produced using S. cerevisiae (Gottardi et al. 2017). Stilbene family phenolic compound, resveratrol has anti-cancer property. This compound is produced via de novo synthetic pathway from the yeast (Xiong et al. 2017). *S. cerevisiae* is becoming the cell factory for the production of flavonoids and anthocyanins (Levisson et al. 2018). Alkaloids have been used for mostly therapeutic use. They are high value-added compounds found to be produced using *S. cerevisiae* and that has lowered the production cost. BIA is benzylisoquinoline alkaloids is produced by yeast fed on the intermediates (Trenchard and Smolke 2015) while *S. cerevisiae* used for its production with the help of *de novo* synthetic pathway and increase production was achieved by the supply of tyrosine. Similarly, alternative of production of this compound was also reported by researchers (Trenchard et al. 2015, DeLoache et al. 2015).

3.2. Methodology (Tools)

As described in the introduction, strain improvement requires the availability of proper tools for introduction of genetic modification and its analysis using metabolic engineering. The rapid expansion of such tools is due to recent developments in genomics and analytics. Genetic engineering, sequencing and bioinformatics have facilitated the genomics part while nuclear magnetic resonance (MNR), two-dimensional gel electrophoresis, matrix-assisted laser desorption/ionisation–time of flight (MALDI-TOF), gas chromatography - mass spectrometry (GC-MS) and DNA chips have facilitated the analytics part.

Starting material for the introduction of strain improvement could be the strain and the vector with reasonable transformation efficiency. It can be exemplified by *E. coli* and *S. cerevisiae*, for which high copy number vectors are available that enable the rapid integration of DNA efficiently. Furthermore, the autotropic type of strain availability for these microorganisms also facilitates the genetic engineering significantly. High frequency of homologous recombination in *S. cerevisiae* enables directed DNA integration in chromosome with high frequency. Similarly, Jones and Keasling (Jones and Keasling 1998) have developed a low copy number stable plasmids for rapid cloning of different genes. The genetic engineering could be cumbersome in case of low transformation efficiency. The second tool of molecular biology is to access of promoters. The name itself is suggestive of its function in gene expression regulation. Sometimes it is required to have overexpression of a gene of interest. The strong promoter could be constitutive like glycolytic promoter of glyceraldehyde-3-phosphate dehydrogenase gene or regulated promoter GAL7 promoter of *S. cerevisiae*. Third in raw molecular biology tool is the use of molecular cassette for gene disruption. Most of the time the strain improvement in the lab of a known strain with introduced genetic changes does enable its comparison of strains. For that strain specific disruption cassettes have been developed. loxP-kanMX-loxP cassette is known for its gene disruption function in *S. cerevisiae* (Güldener et al. 1996). Dominant genes can be looped out with the help of this cassette and thus can be used for other gene disruption also. The added advantage of this cassette is that it can disrupt the gene of interest on polyploid strain with repetitive use. Alternately, sequential gene disruption from the haploid strain can be possible, but the introduction of desired genetic changes will be impossible (Klein et al. 1999, Rønnow et al. 1999). Forth is use of RNA-antisense techniques. Researches (Christensen 1994, Zheng, Kobayashi, and Takeuchi 1998) have successfully silenced gene expression using this technique in filamentous fungi. A fifth technique is the use of gene shuffling through gene cloning in an organism of interest from source organism to get a protein with altered properties (Patten, Howard, and Stemmer 1997, Stemmer 1994a, Stemmer 1994b). A sixth idea is the use of error-prone PCR to apply directed evolution (Melnikov and Youngman 1999, Zhao et al. 1998). This technique has possibility to tailor enzyme of desired properties.

The functioning of a cell is determined by the active metabolic network, which is a makeup of the proteome. The proteome is controlled at transcription and translation level and so do the fluxes indirectly (Klein, Olsson, and Nielsen 1998). A key to decode the metabolic engineering mystery is to understand the cellular function and there are several different techniques available to do that. All of them are flagged under one title – "metabolic pathway analysis." This analysis comprised of three steps; (1) structure identification of metabolic network, (2) flux quantification through the metabolic network branch and (3) control structure identification. The metabolic pathways involve enzymes and co-factors. To evaluate or identify any metabolic network structure, already available literature could help to check the presence and absence of specific enzyme using an enzyme activity assay in different species. The majority of industrial important metabolic routes have been already identified. However, in many cases where the complete metabolic network is not identified, enzyme activity and co-factor identification could help to confirm the right pathway (Pedersen, Carlsen, and Nielsen 1999). Recently, the applicability of enzyme assay is limited to identification of pathway to study the microorganism. Alternately, the radiolabelled substrates are used especially ^{13}C-labelled and subsequently other intracellular metabolites (Christensen and Nielsen 1999b) duly analysed by NMR (Christensen and Nielsen 1999a) and GC-MS (Wendisch, de Graaf, and Sahm 1997). After identification of metabolic network structure, it is followed by the quantification of fluxome which is quantification of fluxes through the different branches in the network. Fluxome approach works on the concept of metabolites balancing. Assumption of steady states of metabolite concentration can be set up to obtain the set of algebraic equation relating to fluxes. The repetition of such exercise or use of liner programming, it is possible to calculate the fluxes through other branches. Many researchers have used this approach in different microbial systems (Jørgensen et al. 1995, Nissen et al. 1997, Pramanik and Keasling 1997, Pedersen, Carlsen, and Nielsen 1999,

Varma, Boesch, and Palsson 1993). Despite its simplicity, estimation of fluxes depends on NADH and NADPH co-factors balance is its limitation. It is important that the reaction must involve these co-factors within the cell since it is unlikely that all reactions are involving these co-factors. In such case the estimation of metabolic flux could be poor. However, it can be balanced by use of radio-labelled metabolite and analysed by NMR and GC-MS (Zupke and Stephanopoulos 1994). The last step is to know how the distribution of flux is controlled and by doing so, one can modulate flux and thus changes in cellular performance. It requires information of kinetics and thermodynamic constraints. The information on how enzyme as well as level of metabolites around the network is regulated could provide the sketch of *in vivo* regulation. Rapid quenching and efficient analytical procedure can be a handful. Like the use of cold methanol for rapid inactivation of cellular metabolism and at the same time measure different metabolites using sensitive, accurate analytical methods. The concept of metabolic analysis also implies to protein secretion pathways and signal transduction pathways.

4. Future Trends

The application of metabolic engineering will be expected to increase dramatically in the future. Synthetic DNA technology (synthetic biology approach) has made it possible to create the heterologous expression of a gene of any kind in the host cell. In addition to that, the high throughput sequencing and allied -omics technology backed by bioinformatic analysis has made it possible for the development of new and novel metabolic engineering tools for both standard as well as canonical organisms. The reduction cost of sequencing has added an advantage of using diverse capabilities of microorganisms from all branches and it is now not limited to a select few model lab organisms. The combination of these two developments has created infinite possibilities in term of pathway construction and production of any product. Use of synthetic

DNA could be more useful to those microbes who lack the component and artificial introduction of required synthetic component has enabled the capabilities of synthesizing and expressing the cellular components. The days are not far away when science will be practicing the synthetic cell exercise.

To date the basic principle and technologies that defined metabolic engineering are already developed and many of them are at their intellectual maturity. If that is the case, then what could be the future. The answer lies in the integration of developed with developing methodological advancement. The future prospects based on current methodology conceptualized and administered overlapped with other allied fields such as molecular biology, synthetic biology and systems biology are deciphered as follows.

4.1. Alternate Production Machinery for Metabolic Engineering

Table 1. Non-traditional hosts tried so far by different researchers

	Organism	**Product(s)**	**Reference**
1	*Bacillus subtilis*	Carotenoids	(Yoshida, Ueda, and Maeda 2009)
2	*Bacillus licheniformis*	2,3-Butanediol	(Wang et al. 2012)
3	*Synechococcus elongatus*	1-Butanol	(Lan and Liao 2011)
6	*Geobacillus thermoglucosidasius*	Ethanol	(Taylor et al. 2009)
9	*Thermophiles* (general)	Various	(Frock and Kelly 2012)
10	*Methylomonas* sp. strain 16a	Carotenoids	(Sharpe et al. 2007)
11	*Bacillus methanolicus*	L-Lysine	(Jakobsen et al. 2009)
12	Algae (general)	Various	(Lü, Sheahan, and Fu 2011)
13	Plants (general)	Various	(Wu and Chappell 2008)

To date *E. coli* and *S. cerevisiae* are the model organisms belonging to prokaryotes and eukaryotes, respectively. This is because they are resourced to experiment and fetch the fundamental biological phenomena or to sequence as well as to introduce the genetic modification. As a result of that, a plethora of information available for them pertaining to their genome sequence, physiology, metabolism and genetics. However, other niche of organisms are also explored, although on a small scale for metabolic engineering (Table 1). There could be niches explored such as optimum growth in a specific environment like thermophiles, use of alternate feedstock, naturally produce and tolerate desired chemical like solventogenic microorganisms. Thermophiles could be of industrial interest because of their nature of growing in higher temperature, allowing to take advantage of higher productivity of thermostable enzymes. In addition to that, it inhibits the growth of mesophilic microorganism as contaminants and also facilitate the faster recovery in case of volatile compounds. *Geobacillus thermoglucosidasius* and *Thermoanaerobacter saccharolyticum* have been engineered for ethanol production using derived sugar at elevated temperature (Shaw et al. 2008, Cripps et al. 2009). Adoption of *Clostridium* for production of iso-butanol is a working example, because it produces such solvents through acetone-butanol-ethanol pathway and such solvent are toxic to *E. coli* which makes *Clostridium* perfect host for production of such solvent (Higashide et al. 2011, Tracy et al. 2012). The production of 1-butanol from carbon dioxide is the first of its kind gas to liquid production by *Synechococcus*, a cyanobacterium [100]. Similarly, *B. subtilits* for carotinoids [159], *B. licheniformis* for 2,3-butanediol [162] and *B. methanolicus* for L-lysine [168]. Alternately, attempts of copying the pathway characteristics to model organisms *E. coli* has been met with limited success [99]. Access of genomics, development of new bioinformatics tools and progress of molecular biology is a less daunting barrier for these organisms as an alternate host. In this way, metabolic engineering can be applied efficiently to a non-restricted group of microorganisms allowing to capitalize the huge biological diversity. This can raise the question of choice of host to produce the molecule of interest. The decision in the choice of the host organism is mainly affected by the complexity of the desired trait. Because for the production of desired compound the complexity of pathway makes its host more efficient. Also, if the desired route can be distilled by a single gene or small set of gene, then more facile approach can port the capability into easy manipulable system. Such condition could be more advantageous if expression is done in the heterologous host in absence of native regulatory system. All in all, this information serves the motive and highlight of use of metabolic engineering in non-native microorganisms.

4.2. Advancement in Methods for Correcting Production Titre

The traditional method deals with overexpression or knocking out of desired gene/enzyme from the pathway to modulate its activity to optimize the strain. With the advancement of molecular biology, improved and finely tuned methodologies are being developed to regulate the pathway flux. One of such approach is to regulate enzyme via spatial organization which reduce the disturbance of inhibitory intermediates, its siphoning to alternate routes (Martin et al. 2003). Three methods, namely fusion proteins, scaffold binding and subcellular localization are described.

Fusion proteins: In this approach, genes of two or more proteins are fused by a linker sequence to generate single polypeptide with multiple functionalities. The linker region provides enough space for protein folding and to retain the same functionality. This can be beneficial to the proteins those are involved in a sequential process that helps to increase their concentration locally. Use of the fusion protein is very useful in a process where the functional pathways involve protein-protein interaction. A study of the use of such approach has yielded nearly 15-fold expression of desired product as compared to individual one (Zhang et al. 2006). In depth analysis revealed that it was because of the close proximity of active site rather than the catalytic efficiency of enzymes (Wang et al. 2011). Such study has exemplified the potential of fusion protein to improve concentration and yield of the desired product.

Scaffold binding: This is a sort of extended version of fusion protein where the enzyme is attached to a scaffold and it has numerous advantages over the use of fusion protein. One of such is the accommodation of a number of enzymes on a flexible scaffold. Scaffold structure has DNA or RNA molecules to which the enzymes can attach via zinc-finger domain (Collins et al. 2003). To obtain the high-throughput results, it is essential to maintain the ratio and spatial orientation of scaffold architecture at individual step. In vivo, up to 77-fold improvement has been achieved with optimum scaffold architecture (Dueber et al. 2009). Instead of facilitating the protein-protein interaction in case of use of fusion protein, the scaffold helps to increase the protein-protein interaction rate.

Subcellular localization: This can be visualized as a more complex version of scaffold protein binding approach which involve the localization of the enzyme to a cellular organelle. The protein holding capacity is much more than that of the scaffold and also mass transfer across the barrier. Prevention is helpful to cease the toxic effect of a pathway intermediates. Prokaryotes lack the organelles, but they do have microcompartments which are proteinaceous in nature and known as BMCs (bacterial microcompartments). Use of BMCs for production of target molecule is in its primitive stage. While in the case of eukaryotes, it is favourable to apply this approach for improved pathway productivity. Organelles such as peroxisome, vacuoles, mitochondria and chloroplast possess various friendly environments, those are thermodynamically favourable for the enzyme allowing them to perform an enzymatic reaction in the cell (Feldman and Sigman 1983). Use of this approach may push to attempt for generating new organelle with all desired functionalities in the near future.

4.3. Concept of *De Novo*

De novo pathway construction: Styrene production is native example of the *de novo* pathway usage; the ability to produce non-renewable fossils derived product. No natural production pathway known for styrene which was accomplished with the introduction of phenylalanine ammonia lyase encoding genes (McKenna and Nielsen 2011). Prediction of pathways using the computational power gives opportunity to facilitate the production of novel compound. Production of 1,4-butanediol in *E. coli* can be another example concept of using *de novo* pathways. However, to be a fruitful pathway; it requires optimal thermodynamic feasibility, number of non-native steps and pathway length. No matter how non-natural the chemical is, it can be producing even if it requires a large number of enzymatic steps from the toolbox of metabolic engineering.

De novo generation of enzyme: Conventionally, naturally occurring enzyme that catalyse reaction is identified and the process of its improvement is conceptualized when it has potential industrial application. In the improvement of the enzyme, mainly site directed mutagenesis or directed evolution methods are used. However, there is a possibility that no enzyme is available naturally or not known so far to catalyse the reaction of interest. As the field of metabolic engineering has continued to grow and has reached into a new level where hypothesised molecule can be synthesised. This can hopefully apply to make an entirely new enzyme of our interest that can catalyse the any virtual reaction, although this era is in its infancy, indeed the process is multistep to produce the *de novo* enzyme. They are idealizing the active site, mapping transition state energy of intermediates, screening for compatible scaffolds and many more. There could be the possibility that after generation of *de novo* enzyme, it may give low catalytic activity or no activity. In the case of low activity there could be the possibility for improvement (Woolston, Edgar, and Stephanopoulos 2013).

4.4. Concept of Systemic Control Genetic Circuit

The modulation of heterologous pathway uses static promoters for continuous expression despite of state of cellular environment. Such forced activity may hinder the overproduction of a compound for a longer period. So, it is required to generate flexible regulation in a cellular environment in order to overproduce a compound optimally.

Construction of a DSRS (dynamic sensor regulator system) has recently made significant advancement in this field. This approach is used in production of fatty acid ethyl ester (Zhang, Carothers, and Keasling 2012). Enabling of system level control can be expected from this approach. Most of the time, its applicability is based on replicating the functional circuits in the biological systems.

4.5. Attention to Synthesis of Unnatural Products

At the onset, overproduction of natural compound from one organism in order to overproduce the desired final product from another organism; this approach was an important step in early time of metabolic engineering. The next major paradigm can be achieved if production of non-natural compound involved, which is not known to be produced naturally in any living organism. Gateway to this can hopefully produce any imaginable compound through metabolic engineering. To date, the production of many non-natural compounds have been attempted by various researchers such as styrene (McKenna and Nielsen 2011), phenol (Wierckx et al. 2005) and long chain alcohols (Zhang et al. 2008).

5. Going Forward

In a decade, the field of metabolic engineering has come far since the first success story of genetic manipulation in a single gene. In recent times, it has been demonstrated how multiple gene manipulation can be exploited to achieve the targeted goal. Integration of various approaches to refine the research and validate the findings have been described. The current and future rapid development in genomics and functional genomics will guide to the cataloguing new pathways and co-factors. Furthermore, mining of public repository for the reported annotated genes may provide a base to formulate new biosynthetic pathways.

Conclusion

Metabolic engineering field is aged almost two decades and it is still in its progressive mode. As shown in this chapter, microbial metabolic engineering has direct contribution in improvement of industrial fermentation processes. Concept of manipulation of a single target at the onset has reached to a state where effective manipulation of any imaginable multiple targets is possible at present. Usage of combination of many fields has facilitated such transformation. Ever increasing the interest towards the production of target products can be considered as a driving force. Still, more efforts to refine the advancement of technology and a better of understating of fundamental biology which still limits us. Robust frameworks of analytic developments relies on growing our understanding in fundamental as well as applied research. Bioproduction of high value and high-volume of the product of interest with the help of metabolic engineering using well characterized host is vital. The developmental pace of biology will form the basis for additional applications of metabolic engineering for the improvement of microorganisms which in turn will be exploited for production, therefore having a significant contribution in society in the future.

References

Anfelt, Josefine, Björn Hallström, Jens Nielsen, Mathias Uhlén, and Elton P Hudson. 2013. "Using transcriptomics to improve butanol tolerance of Synechocystis sp. strain PCC 6803." *Applied and environmental microbiology* 79 (23):7419-7427.

Antoniewicz, Maciek R, Joanne K Kelleher, and Gregory Stephanopoulos. 2007. "Elementary metabolite units (EMU): a novel framework for modeling isotopic distributions." *Metabolic engineering* 9 (1):68-86.

Avalos, José L, Gerald R Fink, and Gregory Stephanopoulos. 2013. "Compartmentalization of metabolic pathways in yeast mitochondria improves the production of branched-chain alcohols." *Nature biotechnology* 31 (4):335-341.

Bailey, James E. 1991. "Toward a science of metabolic engineering." *Science* 252 (5013):1668-1675.

Bro, Christoffer, Birgitte Regenberg, Jochen Förster, and Jens Nielsen. 2006. "In silico aided metabolic engineering of *Saccharomyces cerevisiae* for improved bioethanol production." *Metabolic engineering* 8 (2):102-111.

Christensen, Bjarke, and Jens Nielsen. 1999a. "Isotopomer analysis using GC-MS." *Metabolic engineering* 1 (4):282-290.

Christensen, Bjarke, and Jens Nielsen. 1999b. "Metabolic network analysis." *Bioanalysis and Biosensors for Bioprocess Monitoring*: 209-231.

Christensen, Tove. 1994. "Application: *Aspergillus oryzae* as a host for production of industrial enzymes." In *The Genus Aspergillus*, 251-259. Springer.

Cobb, Ryan E, Yajie Wang, and Huimin Zhao. 2015. "High-efficiency multiplex genome editing of Streptomyces species using an engineered CRISPR/Cas system." *ACS synthetic biology* 4 (6):723-728.

Cohen, Stanley N, Annie CY Chang, Herbert W Boyer, and Robert B Helling. 1973. "Construction of biologically functional bacterial plasmids *in vitro*." *Proceedings of the National Academy of Sciences* 70 (11):3240-3244.

Collins, Cynthia H, Yohei Yokobayashi, Daisuke Umeno, and Frances H Arnold. 2003. "Engineering proteins that bind, move, make and break DNA." *Current opinion in biotechnology* 14 (4):371-378.

Cripps, RE, K Eley, David J Leak, B Rudd, M Taylor, M Todd, S Boakes, S Martin, and T Atkinson. 2009. "Metabolic engineering of *Geobacillus thermoglucosidasius* for high yield ethanol production." *Metabolic engineering* 11 (6):398-408.

de Oliveira Dal'Molin, Cristiana Gomes, Lake-Ee Quek, Robin William Palfreyman, Stevens Michael Brumbley, and Lars Keld Nielsen. 2010. "AraGEM, a genome-scale reconstruction of the primary metabolic network in Arabidopsis." *Plant physiology* 152 (2):579-589.

DeLoache, William C, Zachary N Russ, Lauren Narcross, Andrew M Gonzales, Vincent JJ Martin, and John E Dueber. 2015. "An enzyme-coupled biosensor enables (S)-reticuline production in yeast from glucose." *Nature chemical biology* 11 (7):465-471.

Dueber, John E, Gabriel C Wu, G Reza Malmirchegini, Tae Seok Moon, Christopher J Petzold, Adeeti V Ullal, Kristala LJ Prather, and Jay D Keasling. 2009. "Synthetic protein scaffolds provide modular control over metabolic flux." *Nature biotechnology* 27 (8):753-759.

Feldman, RI, and David S Sigman. 1983. "The synthesis of ATP by the membrane-bound ATP synthase complex from medium 32Pi under completely uncoupled conditions." *Journal of Biological Chemistry* 258 (20):12178-12183.

Fernandez-Moya, Ruben, Christopher Leber, Javier Cardenas, and Nancy A Da Silva. 2015. "Functional replacement of the *Saccharomyces cerevisiae* fatty acid synthase with a bacterial type II system allows flexible product profiles." *Biotechnology and bioengineering* 112 (12):2618-2623.

Frock, Andrew D, and Robert M Kelly. 2012. "Extreme thermophiles: moving beyond single-enzyme biocatalysis." *Current opinion in chemical engineering* 1 (4):363-372.

Gottardi, Manuela, Jan Dines Knudsen, Lydie Prado, Mislav Oreb, Paola Branduardi, and Eckhard Boles. 2017. "De novo biosynthesis of trans-cinnamic acid derivatives in Saccharomyces cerevisiae." *Applied microbiology and biotechnology* 101 (12):4883-4893.

Güldener, Ulrich, Susanne Heck, Thomas Fiedler, Jens Beinhauer, and Johannes H Hegemann. 1996. "A new efficient gene disruption cassette for repeated use in budding yeast." *Nucleic acids research* 24 (13):2519-2524.

Higashide, Wendy, Yongchao Li, Yunfeng Yang, and James C Liao. 2011. "Metabolic engineering of *Clostridium cellulolyticum* for production of isobutanol from cellulose." *Applied and environmental microbiology* 77 (8):2727-2733.

Jakobsen, Øyvind M, Trygve Brautaset, Kristin F Degnes, Tonje MB Heggeset, Simone Balzer, Michael C Flickinger, Svein Valla, and Trond E Ellingsen. 2009. "Overexpression of wild-type aspartokinase increases L-lysine production in the thermotolerant methylotrophic bacterium *Bacillus methanolicus*." *Applied and environmental microbiology* 75 (3):652-661.

Jin, Yong-Su, Hal Alper, Yea-Tyng Yang, and Gregory Stephanopoulos. 2005. "Improvement of xylose uptake and ethanol production in recombinant *Saccharomyces cerevisiae* through an inverse metabolic engineering approach." *Applied and environmental microbiology* 71 (12):8249-8256.

Jones, Kristala L, and Jay D Keasling. 1998. "Construction and characterization of F plasmid-based expression vectors." *Biotechnology and bioengineering* 59 (6):659-665.

Jørgensen, Henrik, Jens Nielsen, John Villadsen, and Henrik Møllgaard. 1995. "Metabolic flux distributions in *Penicillium chrysogenum* during fed-batch cultivations." *Biotechnology and Bioengineering* 46 (2):117-131.

Kang, Aram, and Taek Soon Lee. 2016. "Secondary Metabolism for Isoprenoid-based Biofuels." In *Biotechnology for Biofuel Production and Optimization*, 35-71. Elsevier.

Klein, Christopher JL, Lisbeth Olsson, and Jens Nielsen. 1998. "Glucose control in *Saccharomyces cerevisiae*: the role of Mig1 in metabolic functions." *Microbiology* 144 (1):13-24.

Klein, ChristopherJ L, JakobJ Rasmussen, Birgitte Rønnow, Lisbeth Olsson, and Jens Nielsen. 1999. "Investigation of the impact of MIG1 and MIG2 on the physiology of *Saccharomyces cerevisiae*." *Journal of biotechnology* 68 (2-3):197-212.

Kötter, Peter, and Michael Ciriacy. 1993. "Xylose fermentation by *Saccharomyces cerevisiae*." *Applied microbiology and biotechnology* 38 (6):776-783.

Lan, Ethan I, and James C Liao. 2011. "Metabolic engineering of cyanobacteria for 1-butanol production from carbon dioxide." *Metabolic engineering* 13 (4):353-363.

Lee, Sang Yup, Jong Myoung Park, and Tae Yong Kim. 2011. "Application of metabolic flux analysis in metabolic engineering." *Methods in enzymology* 498:67-93.

Levisson, Mark, Constantinos Patinios, Sascha Hein, Philip A de Groot, Jean-Marc Daran, Robert D Hall, Stefan Martens, and Jules Beekwilder. 2018. "Engineering de novo anthocyanin production in *Saccharomyces cerevisiae*." *Microbial cell factories* 17 (1):1-16.

Liu, Yi, and Jens Nielsen. 2019a. "Recent trends in metabolic engineering of microbial chemical factories." *Current opinion in biotechnology* 60:188-197.

Liu, Yi, and Jens %J Current opinion in biotechnology Nielsen. 2019b. "*Recent trends in metabolic engineering of microbial chemical factories*." 60:188-197.

Lü, Jing, Con Sheahan, and Pengcheng Fu. 2011. "Metabolic engineering of algae for fourth generation biofuels production." *Energy & Environmental Science* 4 (7):2451-2466.

Mak, Wai Shun, Stephen Tran, Ryan Marcheschi, Steve Bertolani, James Thompson, David Baker, James C Liao, and Justin B Siegel. 2015. "Integrative genomic mining for enzyme function to enable engineering of a non-natural biosynthetic pathway." *Nature communications* 6 (1):1-10.

Marcheschi, Ryan J, Han Li, Kechun Zhang, Elizabeth L Noey, Seonah Kim, Asha Chaubey, KN Houk, and James C Liao. 2012. "A synthetic recursive "+ 1" pathway for carbon chain elongation." *ACS chemical biology* 7 (4):689-697.

Martin, Vincent JJ, Douglas J Pitera, Sydnor T Withers, Jack D Newman, and Jay D Keasling. 2003. "Engineering a mevalonate pathway in Escherichia coli for production of terpenoids." *Nature biotechnology* 21 (7):796-802.

McKenna, Rebekah, and David R Nielsen. 2011. "Styrene biosynthesis from glucose by engineered *E. coli.*" *Metabolic engineering* 13 (5):544-554.

Melnikov, Alexandre, and Philip J Youngman. 1999. "Random mutagenesis by recombinational capture of PCR products in *Bacillus subtilis* and *Acinetobacter calcoaceticus*." *Nucleic acids research* 27 (4):1056-1062.

Milke, Lars, Jennifer Aschenbrenner, Jan Marienhagen, and Nicolai Kallscheuer. 2018. "Production of plant-derived polyphenols in microorganisms: current state and perspectives." *Applied microbiology and biotechnology* 102 (4):1575-1585.

Nakamura, Charles E, and Gregory M Whited. 2003. "Metabolic engineering for the microbial production of 1, 3-propanediol." *Current opinion in biotechnology* 14 (5):454-459.

Nielsen, Jens. 2001. "Metabolic engineering." *Applied microbiology and biotechnology* 55 (3):263-283.

Nissen, Torben L, Ulrik Schulze, Jens Nielsen, and John Villadsen. 1997. "Flux distributions in anaerobic, glucose-limited continuous cultures of *Saccharomyces cerevisiae*." *Microbiology* 143 (1):203-218.

O'Neill, Ellis C, and Steven Kelly. 2017. "Engineering biosynthesis of high-value compounds in photosynthetic organisms." *Critical reviews in biotechnology* 37 (6):779-802.

Oh, You-Kwan, Bernhard O Palsson, Sung M Park, Christophe H Schilling, and Radhakrishnan Mahadevan. 2007. "Genome-scale reconstruction of metabolic network in *Bacillus subtilis* based on high-throughput phenotyping and gene essentiality data." *Journal of Biological Chemistry* 282 (39):28791-28799.

Park, SM, AJ Sinskey, and Gregory Stephanopoulos. 1997. "Metabolic and physiological studies of *Corynebacterium glutamicum* mutants." *Biotechnology and bioengineering* 55 (6):864-879.

Patten, Phillip A, Russell J Howard, and Willem PC Stemmer. 1997. "Applications of DNA shuffling to pharmaceuticals and vaccines." *Current Opinion in Biotechnology* 8 (6):724-733.

Pedersen, Henrik, Morten Carlsen, and Jens Nielsen. 1999. "Identification of enzymes and quantification of metabolic fluxes in the wild type and in a recombinant *Aspergillus oryzae* strain." *Applied and environmental microbiology* 65 (1):11-19.

Pramanik, J, and JD Keasling. 1997. "Stoichiometric model of *Escherichia coli* metabolism: Incorporation of growth-rate dependent biomass composition and mechanistic energy requirements." *Biotechnology and bioengineering* 56 (4):398-421.

Qin, Jiufu, Yongjin J Zhou, Anastasia Krivoruchko, Mingtao Huang, Lifang Liu, Sakda Khoomrung, Verena Siewers, Bo Jiang, and Jens Nielsen. 2015. "Modular pathway rewiring of *Saccharomyces cerevisiae* enables high-level production of L-ornithine." *Nature communications* 6 (1):1-11.

Raman, Karthik, and Nagasuma Chandra. 2009. "Flux balance analysis of biological systems: applications and challenges." *Briefings in bioinformatics* 10 (4):435-449.

Reyes, Luis H, Jose M Gomez, and Katy C Kao. 2014. "Improving carotenoids production in yeast via adaptive laboratory evolution." *Metabolic engineering* 21:26-33.

Rønnow, Birgitte, Lisbeth Olsson, Jens Nielsen, and Jørn Dalgaard Mikkelsen. 1999. "Derepression of galactose metabolism in melibiase producing bakers' and distillers' yeast." *Journal of biotechnology* 72 (3):213-228.

Schmidt, Karsten, Jens Nielsen, and John Villadsen. 1999. "Quantitative analysis of metabolic fluxes in *Escherichia coli*, using two-dimensional NMR spectroscopy and complete isotopomer models." *Journal of Biotechnology* 71 (1-3):175-189.

Sharpe, Pamela L, Deana DiCosimo, Melissa D Bosak, Kyle Knoke, Luan Tao, Qiong Cheng, and Rick W Ye. 2007. "Use of transposon promoter-probe vectors in the metabolic engineering of the obligate methanotroph *Methylomonas* sp. strain 16a for enhanced C40 carotenoid synthesis." *Applied and environmental microbiology* 73 (6):1721-1728.

Shaw, A Joe, Kara K Podkaminer, Sunil G Desai, John S Bardsley, Stephen R Rogers, Philip G Thorne, David A Hogsett, and Lee R Lynd. 2008. "Metabolic engineering of a thermophilic bacterium to produce ethanol at high yield." *Proceedings of the National Academy of Sciences* 105 (37):13769-13774.

Si, Tong, Han Xiao, and Huimin Zhao. 2015. "Rapid prototyping of microbial cell factories via genome-scale engineering." *Biotechnology advances* 33 (7):1420-1432.

Stemmer, Willem P. 1994a. "DNA shuffling by random fragmentation and reassembly: in vitro recombination for molecular evolution." *Proceedings of the National Academy of Sciences* 91 (22):10747-10751.

Stemmer, Willem PC. 1994b. "Rapid evolution of a protein in vitro by DNA shuffling." *Nature* 370 (6488):389-391.

Suástegui, Miguel, and Zengyi Shao. 2016. "Yeast factories for the production of aromatic compounds: from building blocks to plant secondary metabolites." *Journal of Industrial Microbiology and Biotechnology* 43 (11):1611-1624.

Taylor, Mark P, Kirsten L Eley, Steve Martin, Marla I Tuffin, Stephanie G Burton, and Donald A Cowan. 2009. "Thermophilic ethanologenesis: future prospects for second-generation bioethanol production." *Trends in biotechnology* 27 (7):398-405.

Tracy, Bryan P, Shawn W Jones, Alan G Fast, Dinesh C Indurthi, and Eleftherios T Papoutsakis. 2012. "Clostridia: the importance of their exceptional substrate and metabolite diversity for biofuel and biorefinery applications." *Current opinion in biotechnology* 23 (3):364-381.

Trenchard, Isis J, Michael S Siddiqui, Kate Thodey, and Christina D Smolke. 2015. "De novo production of the key branch point benzylisoquinoline alkaloid reticuline in yeast." *Metabolic engineering* 31:74-83.

Trenchard, Isis J, and Christina D Smolke. 2015. "Engineering strategies for the fermentative production of plant alkaloids in yeast." *Metabolic engineering* 30:96-104.

Vallino, Joseph J, and Gregory Stephanopoulos. 1994. "Carbon flux distributions at the pyruvate branch point in *Corynebacterium glutamicum* during lysine overproduction." *Biotechnology progress* 10 (3):320-326.

Vallino, Joseph J, and Gregory Stephanopoulos. 2000. "Metabolic flux distributions in *Corynebacterium glutamicum* during growth and lysine overproduction." *Biotechnology and bioengineering* 67 (6):872-885.

Varma, Amit, Brian W Boesch, and Bernhard O Palsson. 1993. "Stoichiometric interpretation of *Escherichia coli* glucose catabolism under various oxygenation rates." *Applied and environmental microbiology* 59 (8):2465-2473.

Wang, Qingzhao, Tao Chen, Xueming Zhao, and Jauhleene Chamu. 2012. "Metabolic engineering of thermophilic *Bacillus licheniformis* for chiral pure D-2, 3-butanediol production." *Biotechnology and bioengineering* 109 (7):1610-1621.

Wang, Yechun, Hankuil Yi, Melissa Wang, Oliver Yu, and Joseph M Jez. 2011. "Structural and kinetic analysis of the unnatural fusion protein 4-coumaroyl-CoA ligase:: stilbene synthase." *Journal of the American Chemical Society* 133 (51):20684-20687.

Wendisch, Volker F, Albert A de Graaf, and Hermann Sahm. 1997. "Accurate Determination of13C Enrichments in Nonprotonated Carbon Atoms of Isotopically Enriched Amino Acids by1H Nuclear Magnetic Resonance." *Analytical biochemistry* 245 (2):196-202.

Wiechert, Wolfgang. 2001. "13C metabolic flux analysis." *Metabolic engineering* 3 (3):195-206.

Wierckx, Nick JP, Hendrik Ballerstedt, Jan AM de Bont, and Jan Wery. 2005. "Engineering of solvent-tolerant *Pseudomonas putida* S12 for bioproduction of phenol from glucose." *Applied and environmental microbiology* 71 (12):8221-8227.

Woolston, Benjamin M, Steven Edgar, and Gregory Stephanopoulos. 2013. "Metabolic engineering: past and future." *Annual review of chemical and biomolecular engineering* 4:259-288.

Wu, Shuiqin, and Joe Chappell. 2008. "Metabolic engineering of natural products in plants; tools of the trade and challenges for the future." *Current Opinion in Biotechnology* 19 (2):145-152.

Xiong, Dandan, Shikun Lu, Jieyuan Wu, Chaoning Liang, Wei Wang, Wenzhao Wang, Jian-Ming Jin, and Shuang-Yan Tang. 2017.

"Improving key enzyme activity in phenylpropanoid pathway with a designed biosensor." *Metabolic engineering* 40:115-123.

Yoshida, Kazuyuki, Shunsaku Ueda, and Isamu Maeda. 2009. "Carotenoid production in *Bacillus subtilis* achieved by metabolic engineering." *Biotechnology letters* 31 (11):1789-1793.

Yu, Tao, Yongjin J Zhou, Leonie Wenning, Quanli Liu, Anastasia Krivoruchko, Verena Siewers, Jens Nielsen, and Florian David. 2017. "Metabolic engineering of *Saccharomyces cerevisiae* for production of very long chain fatty acid-derived chemicals." *Nature communications* 8 (1):1-11.

Yuan, Jifeng, Pranjul Mishra, and Chi Bun Ching. 2016. "Metabolically engineered *Saccharomyces cerevisiae* for branched-chain ester productions." *Journal of biotechnology* 239:90-97.

Zhang, Fuzhong, James M Carothers, and Jay D Keasling. 2012. "Design of a dynamic sensor-regulator system for production of chemicals and fuels derived from fatty acids." *Nature biotechnology* 30 (4):354-359.

Zhang, Kechun, Michael R Sawaya, David S Eisenberg, and James C Liao. 2008. "Expanding metabolism for biosynthesis of non-natural alcohols." *Proceedings of the National Academy of Sciences* 105 (52):20653-20658.

Zhang, Lei, Yajing Liang, Wei Wu, Xiaoming Tan, and Xuefeng Lu. 2016. "Microbial synthesis of propane by engineering valine pathway and aldehyde-deformylating oxygenase." *Biotechnology for biofuels* 9 (1):1-9.

Zhang, Yansheng, Song-Zhe Li, Jia Li, Xiangqing Pan, Rebecca E Cahoon, Jan G Jaworski, Xuemin Wang, Joseph M Jez, Feng Chen, and Oliver Yu. 2006. "Using unnatural protein fusions to engineer resveratrol biosynthesis in yeast and mammalian cells." *Journal of the American Chemical Society* 128 (40):13030-13031.

Zhao, Huimin, Lori Giver, Zhixin Shao, Joseph A Affholter, and Frances H Arnold. 1998. "Molecular evolution by staggered extension process (StEP) in vitro recombination." *Nature biotechnology* 16 (3):258-261.

Zheng, XF, Y Kobayashi, and M Takeuchi. 1998. "Construction of a low-serine-type-carboxypeptidase-producing mutant of *Aspergillus oryzae* by the expression of antisense RNA and its use as a host for heterologous protein secretion." *Applied microbiology and biotechnology* 49 (1):39-44.

Zhou, Yongjin J, Nicolaas A Buijs, Zhiwei Zhu, Jiufu Qin, Verena Siewers, and Jens Nielsen. 2016. "Production of fatty acid-derived oleochemicals and biofuels by synthetic yeast cell factories." *Nature communications* 7 (1):1-9.

Zupke, C, and G Stephanopoulos. 1994. "Modeling of isotope distributions and intracellular fluxes in metabolic networks using atom mapping matrixes." *Biotechnology Progress* 10 (5):489-498.

In: The Future of Metabolic Engineering ISBN: 978-1-68507-362-6
Editors: A. Sharma and D. Amin

Chapter 6

METABOLIC ENGINEERING APPROACH FOR OPTIMIZATION OF MICROALGAE BIOREFINERY USING *PHAEODACTYLUM TRICORNUTUM* AS MODEL ORGANISM

Caroline Schmitz, Claudia M. Bauer and Marcelo Maraschin*

Plant Morphogenesis and Biochemistry Laboratory,
Federal University of Santa Catarina
Florianópolis, Santa Catarina State, Brazil

ABSTRACT

In the search for raw materials from renewable sources, microalgae cultivation systems have characteristics suitable for the bioeconomy area, such as having a biochemical matrix with multiproducts of high economic value and a high rate of biomass growth. However, despite the potential of microalgae, their production currently is not economically viable unless a set of compounds is obtained simultaneously following a biorefinery concept. The advantages of using microalgae biorefinery and its integration with the economy make

* Corresponding Author's E-mail: m.maraschin@ufsc.br.

it an important target in the development of processes optimized for industrial purposes, since it uses a profitable and sustainable approach that makes the most of the potential of biomass as source of bioactives. Despite the numerous advantages of using the biorefinery process, research on metabolic engineering applied to model species is still scarce. In this chapter, the main approaches in systems biology and genetic engineering applied to the microalgae *P. tricornutum* are reviewed, aiming at to clarify the state of the art of searching for strains that produce multiple compounds of economic interest. Finally, to find targets for metabolic modulation, computational tools such as *in silico* metabolic models and metabolic analysis with machine learning will be presented, comparing the resulting solutions, and discussing the methodological advantages and disadvantages of each experimental approach.

INTRODUCTION

Modern society faces the problem of an increasing demand for consumption and production that affects the environment and is one of the main intensifiers of climate change and the scarcity of natural resources. The development of processes aligned with the circular bioeconomy, attentive to social, economic and environmental issues, has sought to meet this current need (Ubando, Felix and Chen, 2020). Circular economy seeks to provide sustainable alternatives to the current economic model where biomass has its production optimized to reduce losses and its impact on the environment. In this context, microalga species stands out as alternative sources of biomasses with superior benefits when compared to terrestrial organisms. For example, the possibility of increasing their metabolic rate and biomass production, the existence of several biochemical profiles with high concentrations of commercial biomolecules, the prospect of bio-sequestration of carbon dioxide and photosynthetic efficiency, besides not creating competition for food crops lands that can be grown in degraded environments (Chew et al., 2017).

Microalgae are photosynthetic organisms widely distributed in aquatic ecosystems developing a complex metabolic diversity among their species, standing out for having a biochemical matrix rich in bioactive compounds (Milicevic et al., 2014, Desbois et al., 2008), such as antimicrobial, antioxidant and stimulant of plant growth and development. The increase interest in works related to microalgae as sources of biochemical and pharmaceutical products relies on the diversity of their bioactive compounds such as carbohydrates, proteins, polyphenols, phlorotannins, pigments, unsaturated fatty acids, and sterols (Ventura et al., 2017). However, despite the high economic value of some compounds accumulated by microalgae, their commercialization faces technical problems. Although the cultivation of microalgae requires simple conditions (i.e., temperature control, nutrients, and a light source), the process is still not economically sustainable (Tredici et al., 2015), unless multiple compounds of high economic value present in the cell's microalgae are explored simultaneously in a biorefinery process (Carrijo et al., 2015).

The concept of biorefinery has been closely linked to the bioeconomy and has priorities also related to it. Although the concept of biorefinery is not new, research and development in microalgae biorefinery are still in the early stages (De Jong, Langeveld and Van Ree, 2013; Peck et al., 2009). This might be noted in the search for an appropriate definition. The simultaneous acquisition of multiple alga's biomass compounds is related to the concept of biorefinery and is intended as a system considered financially and environmentally sustainable. Although the term biorefinery is associated with the crude oil refineries which rely on obtaining energy through the oxidation of carbon chains, microalga biorefineries involve differentiated technologies aiming to explore complex organic compounds derived from the metabolic functions of these microorganisms, e.g., proteins, carbohydrates, lipids, and pigments (Schiano di Visconte et al., 2019). In fact, using the concept of biorefinery, a relatively larger number of classes of compounds has been extracted from microalgae, making their exploitation on large-scale cultivation systems more profitable. Thus, for the development of biotechnological processes that align with the concept of biorefinery, it is necessary to have knowledge about the biochemical matrix of microalga species, i.e., their metabolomes, and how their metabolites oscillate under certain environmental conditions and over the cell cultivation.

On the other hand, the knowledge on microalga metabolism is the basis for the development and application of metabolic engineering tools that could help improving biotechnological processes for obtaining high commercial value bioproducts thereof. Nowadays, a complete and accurate mapping of microalgae' metabolic pathways is far from being elucidated considering its complexity due to the dependence on external conditions and variations in the genotypes that occur even among species. In this way, metabolic engineering not only makes use of the already established routes of metabolism, but also contributes to fill in gaps through predictions based on computational mathematics and experimental data, which may be validated in later experiments. Bioinformatics tools are fundamental for the optimization of biotechnological systems to produce metabolites with high market value since they can save time, costs and *in vitro* experiments through predictions returned by *in silico* models. The results afforded can be doubly advantageous since the *in silico* model itself is biochemically validated in the laboratory and can be improved by the results obtained and used in future works involving the species and related problems.

Metabolic engineering techniques have become increasingly useful to improve the yield and productivity of microalga biorefineries, through the integration of knowledge and applications in the areas of systems, evolutionary and synthetic biology, rationally optimizing cellular performance. The biosynthesis of target compounds is highly influenced by the culture conditions to which cells are exposed and by inducing the synthesis of such compounds one can generate significant gains in yield of microalga cell cultures. Indeed, the use of compounds that stimulate certain desired physiological responses in the biorefinery concept can also mitigate expenses, making it possible to grow microalgae on a commercial scale.

In view of the metabolic engineering applied to microalgae biorefinery, the diatom *Phaeodactylum tricornutum* is widely used as a biological model. This marine species has advantages for biotechnological uses such as its sequenced genome (Bowler et al., 2008), potential for commercial-scale cultivation (Benavides et al., 2013), and biomass with a high content of compounds with high economic value (Bellou et al., 2014). These characteristics are highly appreciated when looking for a model organism to optimize production processes. The biorefinery concept seeks at maximum the use of the biomass produced by a given species, and *P. tricornutum* has a great potential to synthesize various compounds of high value for the society such as fatty acids, fucoxanthin, and chrysolaminarin (Gao et al., 2017). The publication of the complete *P. tricornutum*'s genome sequence in 2008 made possible to expand information to explore bioinformatics tools that combine the information already available, enabling a greater understanding of the metabolome response. *P. tricornutum*'s genome is approximately 27.4 megabases (Mb) in size, serving as a model system for genetic and biochemical investigations. Additionally, the methodologies used and validated for the species can later be applied to other diatoms or even adapted to organisms of different taxonomic classifications.

Metabolic engineering has played an important role in the development of strategies to increase the biosynthesis of multiple compounds by a single strain, modulating the metabolism to drive metabolic flows to target compounds. To create and apply these strategies it is essential to know how the target compound oscillate, and this understanding is currently facilitated using metabolomics and computational tools, which help to elucidate metabolic profiles on a systems biology base. The development of machine learning tools and the increase of software dedicated to open access databases have driven the ways of finding targets for metabolic engineering applied to the optimization of biological production processes. Within this scope, this review brings a synthesis of the metabolic engineering strategies applied to the production of interest molecules by cells that can be applied to microalgae biorefinery systems. At first, a brief introduction of the metabolic engineering strategies targeted to the biosynthesis and accumulation of interest molecules by microalgae biorefinery process will be given, using *Phaeodactylum tricornutum* microalga as a study model. Subsequently, two ways to find potential pathways to modulate the metabolic flow to increase the biosynthesis of the economically valuable eicosapentaenoic acid (EPA) and fucoxanthin will be explored. Finally, to find targets for metabolic modulation, computational tools such as *in silico* metabolic models and metabolomic analysis with machine learning will be presented, following the comparison of the resulting solutions, and discussing the methodological advantages and disadvantages of each experimental approach.

METABOLIC ENGINEERING STRATEGIES APPLIED TO *PHAEODACTYLUM TRICORNUTUM*

Despite the potential for producing compounds by the species *P. tricornutum*, the area of metabolic engineering in microalgae is still underdeveloped when compared to other groups of organisms. However, it has been growing with advances in the molecular area such as genome sequencing and editing, understanding of the cell's biochemical and the selection of mutant strains (Fu et al., 2019). To develop strategies for optimizing the biorefinery process it is useful to develop research in two stages. At first, seeking to understand the cellular system as a whole and thus finding bottlenecks that will be targeted for optimization, and, in a second approach, to optimize strategies to guide the metabolic flow of the system to increase biosynthesis of the target compounds (Huang and Dauboussi, 2017).

The application of systems biology using approaches known as omics (e.g., metabolomics, transcriptomics, genomics, and fluxomics) has been growing since the publication of microalgae species genomes, allowing a more comprehensive description of the metabolism of microalgae and its regulation. Omics sciences are suitable tools for elucidating potential targets for metabolic engineering when seeking to optimize the production of compounds.

Experiments in systems biology use holistic approaches where no hypothesis is previously known or prescribed, but all data is acquired usually through high-throughput experiments coupled with quantitative and modeling analysis, to provide an insight into the processes that contribute to the organization and cell dynamics (Westerhoff and Palsson, 2004, Huang and Dauboussi, 2017). Genetic engineering techniques opened possibilities to redesign the system to improve the metabolic potential of cells (Fu et al., 2019).

P. tricornutum is one of the most suitable models in the context of industrial production of commercial interest compounds when aimed at biorefineries, as it has a considerable number of studies related to the metabolism of xanthophylls, lipids, and carbohydrates (Table 1). Engineering strategies to improve the production process are proving effective for this species. There is a tendency to apply metabolic engineering in the central carbon flow metabolism and further directing it to the routes of interest using nuclease addition methodologies such as TALE and CRISPR/Cas9 or carrying out gene knock-out techniques (Fu et al., 2019).

The simplest way to produce a compound is to dedicate the cultivation conditions to an increase in biomass production, i.e., knowing that a given organism naturally produces a compound of interest, cultivation conditions directed to increased biomass production lead to obtain greater amounts of such a compound. For this purpose, the selection of high producing strains can also be applied. Bioprospecting to identify strains with desirable characteristics can be done in oceans, ponds, rivers and other aquatic reservoirs, laboratories. Additionally, random chemical mutagenesis experiments can also be performed before a screening process to select the modified strains with the desirable phenotype, i.e., improved biosynthesis and accumulation rates of a biotechnologically relevant compound (Schiano di Visconte et al., 2019).

Table 1. System biology and metabolic engineering strategies applied to *Phaeodactylum tricornutum* for production of compounds with high commercial value

Strategy	Methods	Target	Author
	genome-scale metabolic network	carbohydrates	Kroth et al., 2(
	genome-scale metabolic network	genes and metabolites	Kim et al., 2(
	genome-scale metabolic network	lipids and GPR associations	Levering et al., 2(
	proteomics and genomics	lipid in dark stress	Bai et al., 2(
	proteomics and genomics	gene annotation	Yang et al., 2(
	genomics	gene annotation	Rastogi et al., 2(
	genomics, fluxomics and metabolomics	protein and lipid	Guerra et al., 2(
systems biology	fluxomics (FBA)	fucoxanthin	Bauer et al., 2(
	fluxomics (FBA)	TAGs and nitrogen starvation	Levitan et al., 2(
	proteomics	TAGs and nitrogen starvation	Bai et al., 2(
	proteomics	TAGs and nitrogen starvation	Longworth et al., 2(
	proteomics	TAGs and nitrogen starvation	Ge et al., 2(
	transcriptomics, proteomics and metabolomics	TAGs and nitrogen starvation	Remmers et al., 2(
	transcriptomic and metabolomic	cell response to nitrogen deprivation	Alipanah et al., 2(
mutagenesis	UVC radiation	carotenoids	Yi et al. 2(
	chemical (EMS and NTG)	carotenoids	Yi et al. 2(
	FCP A gene promoter	chlorophyll a/c and fucoxanthin	Watanabe et al., 2(
	Vde, Vdr and Zep3 overexpressed	carotenoids	Manfellotto et al., 2(
	CrGES expressed	geraniol	George et al., 2(
	AtLUS and LjLUS expressed	lupeol, betulin and betulinic acid	D'adamo et al., 2(
	EF2 and luciferase gene	luciferase protein	Seo et al., 2(

Strategy	Methods	Target	Author
genetic engineering	G6PDH overexpressed and antisense knockdown	lipids and growth rate	Wu et al., 2019
	PtPEPC1 overexpressed	biomass	Seo et al., 2018
	PtPTP overexpressed	biomass and lipids	Seo et al., 2020
	phosphoenolpyruvate carboxykinase knockdown	lipids	Yang et al., 2016
	DGAT2D overexpressed	lipids	Dinamarca et al., 2017
	DGAT2 and GPAT overexpressed	lipids	Zou et al., 2018
	PtME and PtD5b overexpressed	neutral lipid	Zou et al., 2019
	DGAT2 overexpressed	neutral lipid and PUFAs	Niu et al. , 2013
	DGAT2B overexpressed	PUFAs	Haslam et al,. 2020
	MCAT and D5b coordinated expressed	PUFAs	Wang et al., 2017
	ptTES1 disruption	TAGs	Hao et al., 2018
	UGP2 disruption	TAGs	Daboussi et al., 2014
	ptTES1 disruption	TAGs	Hao et al., 2018
	PtWS/DGAT overexpressed	TAGs	Cui et al., 2018
	GPAT1 and LPAT1 overexpressed	TAGs	Wang et al., 2018
	GPAT2 overexpressed	TAGs	Wang et al., 2020
	AGPAT1 overexpressed	TAGs and EPA	Balamurugan et al., 2017
	ER<DELTA>6FAD overexpressed	EPA	Zhu et al., 2017
	LPAAT overexpressed	EPA	Li et al., 2020
	DGAT2B expressed	EPA and DHA	Sayanova et al., 2020
	OtElo5, AppA, and PhyA expressed	EPA, DHA and recombinant phytase	Pudney et al., 2019
	OtElo5 expressed	DHA	Hamilton et al., 2014

Mutagenesis and selection of mutants are effective approaches to optimize the performance of strains aimed at the industry. This approach has been widely used because it allows a random approach without requiring prior knowledge of the biosynthetic pathways. This randomness in the production of mutants results in an expressive number of phenotypes, which can become a problem for effective selection of transformed ones. In the context of microalga species, high-yield methodologies based on fluorescence and flow cytometry techniques have been applied for screening strains with superior carotenoid biosynthesis (Mendoza et al., 2008). This approach was applied to *P. tricornutum* cells to select photo-oxidative stress-driven mutated strains that produce more carotenoids (Yi et al., 2015; Yi et al., 2018). The authors induced chemically about one thousand mutants and using flow cytometry it was possible to select five strains that showed an increase in 33% of the carotenoid concentration compared to the wild type (Yi et al. 2018). The method of flow cytometry in microalga strains enabled the phenotypic analysis of a high number of mutant genotypes and proved to be effective in reducing time and increasing efficiency in selection. As this technique uses fluorescence, it has been currently applied in the investigations of compounds like carotenoids and other pigments in microalga species.

Genetic Engineering Technologies

To increase the production of compounds of interest, genetic engineering techniques might have adopted in a first approach. However, the diploid genome of *P. tricornutum* and the lack of information about its sexual life are limitations to classical genetic approaches based on random insertion and mutagenesis, because it is very unlikely to obtain the inactivation of the two alleles of a gene simultaneously (Huang and Daboussi, 2017, Daboussi et al., 2020). This fact, coupled with the need to link genes and physiological processes, has guided studies in diatoms to use reverse genetic engineering such as RNAi (RNA interference) and the targeted genome engineering technique DSB (double-strand DNA break), repair mechanisms with the site-specific nucleases MNs (meganucleases), TALENs (transcription activator-like effector nucleases), and CRISPR/Cas9 (clustered regularly interspaced palindromic sequences/Cas9) (Huang and Daboussi, 2017). Unlike traditional random mutation, targeted genome engineering techniques permit gene editing at known sites in the genome. In this field CRISPR-Cas9 is a technology based on the immune system of prokaryotes and has revolutionized the genetic engineering of several organisms, especially eukaryotes, for its simplicity and, since it allows simple engineering and site-specific activity when compared to conventional genetic tools (Chae et al., 2017).

Diatoms can be genetically transformed at the level of nucleus, chloroplasts, and mitochondria, since its genome is contained in these compartments. Despite the lack of a proper transformation toolbox, the species *P. tricornutum* already has a genetic toolbox with very diverse elements of promoters and terminators, selectable markers to identify transformed cells, and tools to ensure the stability of the inserted vectors.

Lhcf (light harvesting complex) promoters, also called *fcp* (fucoxanthin chlorophyll a/c binding protein) and NR (nitrate reductase) are among the promoters used for *P. tricornutum* engineering. *Lhcf* promoters are efficiently used in *P. tricornutum* but are light dependent, which becomes an obstacle when it is necessary to express genes in dark conditions. NR promoter functions as an inducible promoter that can switch on and off transgene expressions and have been widely used in diatoms (Huang and Daboussi, 2017). Several other promoters have been designed for *P. tricornutum* engineering, e.g., *EF2* (elongation factor 2) (Seo et al., 2015), PtAPase (alkaline phosphatase) (Lin et al.,2017), V-ATPase C (vacuolar ATP synthase 16-kDa proteolipid subunit) (Watanabe et al., 2018), Pt211 (Zou et al., 2018), *HASP1* (highly abundant secreted protein 1) (Erdene-Ochir et al., 2019), and the iron responsive promoters *Isi1* (iron starvation induced protein1), *FBP1* (ferrichrome binding protein1 and *Fld* (flavodoxin) (Yoshinaga et al., 2014).

Selectable markers to identify *P. tricornutum* transformed strains include zeocin antibiotic resistance (Apt et al., 1996), nourseothricin (Zaslavskaya et al., 2001), and blasticidin-S (Buck et al., 2018). Methods for gene delivery in *P. tricornutum* include particle bombardment or biolistic (Apt et al., 1996, Falciatore et al., 1999), conjugation (Karas et al., 2015), and electroporation (Niu et al., 2012, Zhang and Hu, 2014, Xie et al., 2014).

RNAi has been used in *P. tricornutum* species for, e.g., knockdown expression and activity of NR (nitrate reductase) in order to obtain enhanced lipid biosynthesis (Levitan et al., 2015), identification of uridine-5'-monophosphate synthase gene (Sakaguchi et al., 2011), knockdown of phosphoenolpyruvate carboxykinase for lipid biosynthesis (Yang e t al., 2016), identification and localization of carbonic anhydrase in thylakoids (Kikutani et al., 2016), knockdown of a gene encoding lipid droplet associated protein to examine its function (Wang et al., 2017), and overexpression and knockdown of PGRL1 (proton gradient regulation 5-like photosynthetic phenotype 1) for investigation of its role in photosynthesis (Zhou et al., 2020), among others. In their turn, MNs, TALENs, and CRISPR/Cas9 have been used in *P. tricornutum* engineering to enhance TAG (triacylglycerol) production (Daboussi et al., 2014, Hao et al., 2018), photosynthesis mechanisms (Mann et al., 2017, Buck et al., 2019), multiple simultaneously gene knock-outs (Serif et al., 2018), and CRISPR/Cas9 application and optimization for this species (Nymark et al., 2016, Stukenberg et al., 2018, Russo et al., 2018, Slattery et al., 2018, Sharma et al., 2018, Madhuri et al., 2019).

Importantly, for engineered strains of microalgae it is relevant to highlight that a series of precautions must be taken not only to guarantee food security, for instance, but also the strain must be proven environmentally safe. Besides, a series of precautions must be taken to prevent these strains from leaking into the environment. Similarly, there are considerable bottlenecks in designing engineered strains as, for instance, antibiotic resistance, a widely used selection marker, should be substituted as soon as possible in the protocols to avoid possible consequences of antibiotic resistance.

Genome-Scale Metabolic Reconstruction and Flux Balance Analysis (FBA)

A genome-scale metabolic network reconstruction is a collection of genomes encoded metabolic functions based on the genome annotation of the target organism and biochemical databases considering hypothetical data and experimental evidence (Thiele and Palsson, 2010). Enzyme commission (E.C.) numbers, KEGG, and BRENDA databases are important sources to the identification of gene products for the metabolic models.

Genomic engineering through genome mapping and gene expression, silencing and deletion, enabled the development and application of metabolic engineering shedding light on the metabolism of organisms. The nuclear genome of diatoms contains genes of bacterial origins and genes originated of the secondary endosymbiosis (Allen et al., 2011). The genome sequencing of *P. tricornutum* (Bowler et al., 2008) made it possible to develop and make use of various genome-scale computational models and genetic engineering tools bringing up several characteristics about the metabolism of this species, some putative and others not. For example, the presence of an ornithine-urea cycle in diatoms (Allen et al., 2011), the prokaryotic Entner-Doudoroff glycolytic and phosphoketolase pathways (Fabris et al., 2012), and C4-like photosynthesis (Kroth et al., 2008).

A genome-scale metabolic network reconstruction is, however, subjected to missing, wrong or incomplete annotations and needs to be manually refined along its development (Thiele and Palsson, 2010). Enzymes listed in the biochemical databases are mostly not specific to an organism and may not be present in the target organism (Thiele and Palsson, 2010). Compartmentalization may also differ between organisms and transport reactions are needed for metabolite transport between membranes. For example, *Chlamydomonas reinhardtii*, a green alga, stores carbohydrates in the plastids while *P. tricornutum* accumulate them in vacuoles (Gruber and Kroth, 2017). Membrane transport systems are more complex systems because they involve different transport mechanisms requiring further studies, and therefore less experimental data is available. In this sense, the annotation of the genome and consequently the genome-scale metabolic model of a given species assumes a dynamic process, being refined as long as new data and tools are obtained for the species. To be more precise, genome-scale metabolic models need to constantly integrate missing metabolic functions, signaling, transport and network, and update incorrect annotations. Novel annotations for *P. tricornutum* genome were made over the last years (Yang et al., 2018, Rastogi et al., 2018), providing novel information for this species like revised genes and identification of novel genes and proteins.

Fluxomics is an important area in metabolic engineering as one aims at obtaining biotechnological products, since it is capable of making predictions of the flow of metabolites. An important *in silico* investigation tool obtained due to genome-scale metabolic network reconstruction is FBA (flux balance analysis) which avows simulations and optimizations of desired phenotypes. This tool allows the analysis of the flow of metabolites once it contains all known metabolic reactions and genes that encode each enzyme for an organism, thus making it possible to predict the growth rate of an organism or the rate of production of a certain metabolite of biotechnological interest (Orth et al., 2010). Constraint-based algorithms are constructed to predict intracellular fluxes in wild-types and genetically modified strains by gene knockout and amplification. MOMA (minimization of metabolic adjustment) and OptKnock are two popular algorithms to identify gene knockout targets and they have been widely used to enhance production of various chemicals in microorganism strains (Chae et al., 2017).

Two important genome-scale metabolic models for *P. tricornutum* were published and made available (Kim et al., 2016, Levering et al., 2016). Among other features, these models predicted intracellular fluxes under autotrophic, mixotrophic, and heterotrophic growth conditions (Kim et al., 2016), as well as a full range of metabolic reactions for the lipid metabolism and GPR (gene protein reaction) associations (Levering et al., 2016). As wild-type strains of *P. tricornutum* are not capable of growing on organic carbon sources, mixotrophic and heterotrophic modeling are especially useful for transformed lines phenotype simulations. Information provided about gene-protein-reaction (GPR) associations enables phenotype simulations at gene and enzyme levels.

Biotechnological Potential for High Value Compounds Production and Biorefinery Process

Potential biotechnological compounds naturally produced by *P. tricornutum* strains are fatty acids like PUFAs (EPA, DPA, DHA), pigments (carotenoids), its siliceous cell walls known as frustules, proteins, carbohydrates, and sterols (brassicasterol and campesterol) (Fabris et al., 2014, Jaramillo-Madrid et al., 2020). MUFAs and PUFAs gained attention due to their potential application to biofuels and source of omega-3-rich oil for human health, respectively (Cui et al., 2019). Among the pigments produced by this microalga, fucoxanthin has gained prominence because its important biological functions reported in countless works, e.g., antitumor (Hosokawa et al., 1999), anti-diabetic and anti-obesity (Maeda et al., 2007), anti-inflammatory and antioxidant (Shiratori et al., 2005, Pangestuti et al., 2011, D'Orazio et al., 2012, Xia et al., 2013). Defatted biomass of *P. tricornutum* is also a potential source of high-value protein (Cui et al., 2019), which can constitute a valuable residue in the biorefinery concept. Diatom cell walls, namely the frustules, have potential applications in nanotechnology due to the high quality of its structure that has not yet been achieved by our technology and can be useful as nanometric filters among other potential applications on a nanometric scale (Saade and Bowler, 2009). In addition, other studies have shown the possibilities of producing other important biotechnological products by modified strains of *P. tricornutum*, like the monoterpenoid geraniol (Fabris et al., 2020, George et al., 2020) and the triterpenoids lupeol and betulin (D'adamo et al., 2019).

As, at first, microalgae proved to be good producers of omega-3 fatty acids and carotenoids, metabolic engineering tools have been developed for the synthesis of these metabolites in *P. tricornutum*, in addition to microalgae biofuels which have been in fact driven the studies concerning biorefinery processes applied to microalgae. In the case of *P. tricornutum*, the great majority of works applying metabolic engineering has been focused on PUFAs for their high economic value, as well as on neutral lipids for biofuel production.

Due to the accumulation of PUFAs with high commercial value, most studies to date seek to increase the synthesis of lipids by overexpressing key lipogenic genes (Table 1) in *P. tricornutum* strains. The most studied genes to date are glycerol-3-phosphate acyltransferase (GPAT) and diacylglycerol acyltransferase (DGAT), which are involved in the triacylglycerol biosynthesis. DGATs genes encode a key transmembrane protein that has an enzymatic function. This protein is essential for lipid metabolism, as it catalyzes the terminal and unique step involved in the synthesis of triacylglycerol using diacylglycerol and fatty acyl-CoA as substrates. The GPAT gene esterifies the acyl group of acyl-ACP to the sn-1 position of glycerol-3-phosphate, an essential step in glycerolipid biosynthesis (UniProt Consortium, 2019). These two overexpressed genes increase the lipid content without affecting cell growth and photosynthetic efficiency, orchestrating even the expression of other important photosynthetic and lipogenic genes. In a biorefinery context, thinking about increasing simultaneously lipids and carotenoids sounds as a promising strategy (Zou et al., 2018; Wang et al., 2018).

There are countless other possibilities to produce valuable compounds from microalgae given that, in some aspects, not only marine organisms are more recently studied in relation to terrestrial organisms, but also the field of metabolic engineering still has a lot to be explored. *P. tricornutum* has been used as a platform for recombinant protein studying. For instance, phytase is a phosphatase enzyme that acts in the hydrolysis of phytic acid, an undigested plant phosphorus source for monogastric animals and it has been widely used as an additive in animal feed. Pudney et al., 2019 engineered *P. tricornutum* strains to co-express Δ5-elongase from *Ostreococcus tauri* and phytase from *Escherichia coli* and *Aspergillus niger* to obtain co-production of omega-3 LC-PUFAs and recombinant phytase. According those authors, besides the enhanced omega-3 LC-PUFAs production it was also possible to achieve phytase activity sufficient to treat up to 20 Kg of Atlantic salmon feed by the same engineered strain. Interestingly, *P. tricornutum* has also been engineered to secret monoclonal IgG antibodies against Marburg virus, which is relative to Ebola virus (Hempel, et al., 2017).

Biorefinery processes can be designed in many ways. Downstream processes and the final destination of the compounds obtained at the end of the production line are crucial to define a project. For example, high-value compounds as fucoxanthin and proteins for pharmaceutical purposes are best extracted first to avoid contamination. In addition to the possibility of producing and obtaining pigments and lipids in the biorefinery process, other configurations are proposed. Branco-Vieira et al., 2020 proposed three different biorefinery processes: *i)* a structure to produce biofuels from lipids (biodiesel), *ii)* carbohydrates fermentation (bioethanol), and *iii)* residual biomass (biomethane). Another configuration considers multiple high-value compounds productions (proteins, fucoxanthin, and silica) and a third option considers the production of fucoxanthin as the main product, and proteins, lipids and carbohydrates as co-products from *P. tricornutum* cell cultures. Zhang et al., (2018) proposed an integrated process for fucoxanthin, EPA (eicosapentaenoic acid) and chrysolaminarin production and extraction from *P. tricornutum* biomass. Interestingly, microalgal carbohydrates are more suitable for biofuels production because the absence of lignin associated with their chemical structures facilitates fermentation.

More complexity is added to metabolic engineering due to the fact that microorganisms can produce not one, but several important compounds with high economic value simultaneously, e.g., carotenoids and PUFAs. For evolutionary reasons, the increase in the production of one compound not rarely occurs at the expense of another or at the expense of cellular growth. This is especially true in the case of TAG accumulation under nitrogen starvation. Many studies have shown that *P. tricornutum* suffers cycle arrest while accumulating TAG under nitrogen starvation (Breuer et al., 2012, Kim et al., 2017, Remmers et al., 2018) a phenotype possibly related to carbon reallocation by downregulation of Calvin cycle, decreased biosynthesis of chrysolaminarin, and upregulation of TCA cycle and pyruvate metabolism (Levitan et al., 2015, Alipanah et al., 2015, Longworth et al., 2016, Bai et al., 2016, Kim et al., 2016, Butler et al., 2020). Indeed, this is one of the bottlenecks for microalgae biofuels industrial scale-up. Guerra et al., 2013, studying the metabolite pool in *P. tricornutum* nitrogen starved cells, suggests that a key point controlling carbon allocation under nitrogen starvation resides in GS-GOGAT (glutamine synthetase-glutamate synthase) and GDH (glutamate dehydrogenase) pathways, in a way that the ratio AKG/GL(N/U)(α-ketoglutarate/glutamine/glutamate) reports the nitrogen availability to the cell triggering gene responses which together with a high NADPH/NADP+ ratio increase lipid biosynthesis.

CO_2 also plays a key role when dealing with photoautotrophic microorganisms, not only because of its importance in central carbon metabolism, but also because of its high costs for microalgae cultivation on industrial scale. Under CO_2 starvation, *P. tricornutum* cells increased carbon flux toward PEP (phosphoenolpyruvate) and/or pyruvate biosynthesis (Heydarizadeh et al., 2017 and Heydarizadeh et al., 2019). Under low-light conditions, PEP and/or pyruvate were directed to provide substrates for respiration, as under high-light exposure microalgal cells directed toward lipids and proteins (Heydarizadeh et al., 2017) showing that *P. tricornutum* drives its metabolism toward lipid biosynthesis under carbon starvation as well.

There are successful studies in increasing the production of target compounds without affecting normal cell growth and development. Wang et al., 2017 applied a metabolic engineering strategy to *P. tricornutum* to increase PUFA production, expressing two key enzymes, i.e, malonyl CoA-acyl carrier protein transacylase and delta 5 fatty acid desaturase simultaneously, and were successful without affecting growth and photosynthetic yield of the microalgae. D'adamo et al., 2019, utilized metabolic pathway engineering to produce lupeol and betulin, two plant triterpenoids in *P. tricornutum*, expressing plant enzymes, and found no significant differences between wild-types and transformed strains for growth performance. Wu et al., 2019 engineered *P. tricornutum* for overexpression of glucose-6-phosphate dehydrogenase (G6PDH) which resulted in increased lipid content and growth. Besides, the transcript abundance and activity of G6PDH increased even more under high CO2 cultivation leading to a greater biosynthesis of EPA. Also, regarding nitrogen metabolism, Trentacoste et al., 2013 were able to increase lipid yields in *Thalassiosira pseudonana* by the knockdown of a multifunctional gene for lipase, phospholipase and lysophosphatidic acyltransferase activities without impairing growth.

Most likely the penalty on cell growth occurs in cases where the accumulation of lipids occurs due to a lack of certain limiting nutrients like nitrogen, which are important for the biosynthesis of primary metabolism compounds. Metabolic engineering makes it possible to intervene in key points of metabolism so that they are might be diverted to the production of compounds of interest without affecting the normal growth of cells. Consequently, the accumulation of biomass is not penalized. Thus, a challenge encountered in the biorefinery process is to find a balance in the simultaneous production of various compounds in order to maintain the process with the highest possible yield of targeted compounds and not affecting the normal cellular growth. Metabolic engineering plays a key role to understand in which situations metabolism sacrifices the production of certain compounds for the increasing synthesis of others.

Transporters are of key importance in intra-membrane and intracellular transport, but they have been scarcely studied in diatoms and also put complexity to the systems. Regarding intracellular transport, diatoms have the peculiarity of having plastids with four membranes being the external one continuous with the endoplasmic reticulum (ER), which means that proteins nuclear encoded to be imported into plastids have firstly to pass ER (Gruber and Kroth, 2017). As an example, we can cite the overexpression of a hypothetical gene for a pyruvate transporter. Pyruvate is a key metabolite for many important compounds like amino acids, terpenoids, and fatty acids. Seo et al., 2020 overexpressed a putative plastidial pyruvate transporter in *P. tricornutum* using particle bombardment to compartmentalize pyruvate in plastids and were able to enhance biomass, lipids, and growth up to 21.9%, 30%, and 8% respectively.

Case Study: Generating and Applying Metabolic Engineering Strategies to *Phaeodactylum tricornutum*

To create metabolic engineering strategies for the optimization of biorefinery processes two studies were performed using *P. tricornutum* as a model species. The first one applied systems analysis using metabolomics and bioinformatics tools to elucidate potential pathways to be inhibited or elicited, directing the metabolic flow to the synthesis of target compounds, i.e., eicosapentaenoic acid and fucoxanthin. The second study improved two models on a genomic scale seeking to predict ways based on restrictions and flow balance analysis to increase fucoxanthin synthesis.

Metabolomics

To create strategies for metabolic engineering using metabolomics focused on systems biology, biochemical data mining tools were used aiming at suppressing certain metabolic pathways for increasing the biosynthesis of eicosapentaenoic acid and fucoxanthin. For this purpose, supervised analyzes were performed, such as liquid and gas chromatography to quantify the compounds of interest, as well as unsupervised ones, e.g., vibrational infrared spectroscopy (FTIR) and nuclear magnetic resonance (NMR). The analytical platform adopted intended to metabolically profile *P. tricornutum* cells during cultivation to detect eventual data patterns that provide metabolic information that will assist in the optimization of the microalgae biorefinery process.

Supervised analyzes provided an overview of the contents of metabolites during the cell cultivation, showing stability in the fucoxanthin amounts and an oscillation in EPA concentration. In its turn, unsupervised analysis using FTIR spectroscopy gave rise to fingerprint regions in the spectra associated to the compounds of interest. Besides, it was noted that if the cell culture protocol aims at the maximum EPA production, it would indicate to interrupt the cultivation by the end of the logarithmic phase of cell growth. Regarding fucoxanthin, its contents were higher and somewhat stable from the end of the logarithmic phase until the end of the stationary phase. The stability of this oxycarotenoind when in relation to the cell biomass implies that this compound has a lot of potential to be worked in biorefinery processes. In their turn, protein contents showed augmented in the logarithmic phase of cell growth and further decreased over cultivation. Finally, at the end of the cultivation higher amounts of carbohydrates were detected to accumulate. This information is important to understand the oscillation of primary and secondary metabolites during cell cultivation, eventually subsidizing their analysis in large-scale industrial processes, focusing on the cell growth phases of higher amounts of target compounds as previously *in vitro* identified.

In a follow-up study to shed lights on metabolic engineering of *P. tricornutum* cells, NMR spectroscopy was applied to investigate the microalgal endometabolome. Such an approach allowed identifying a set of cell metabolites with the aid of the Bayesil software (Ravanbakhsh et al., 2015). However, the identification of the metabolite can also be carried out by other platforms or even manually with programs specific to the NMR spectra, such as Chenomx (Wishart et al., 2017). Importantly, a database containing reference NMR spectra for the metabolites found makes easier the annotation and identification of compounds, especially whether the NMR spectra have been recorded under similar experimental conditions to the samples in study. In the case of diatoms, the database DiatomCyc (Fabris et al., 2012) and KEGG (Kanehisa et al., 2007) are extremely relevant for this purpose.

After the identification of the compounds in the biochemical matrix of *P. tricornutum*, it was possible to investigate the main metabolic pathways and perform analysis of pathway topology and over-representation through the MetaboAnalyst platform (Chong, Wishart, and Xia, 2015). The main results from the NMR analysis indicated that the metabolic models developed led to the greatest impact on the carbohydrate and amino acid metabolism of *P. tricornutum*. Currently, with the support of computational platforms and databases that assist in the study of the metabolic pathways with high impact on cell metabolism, one can highlight which of them present the greatest potential for the optimization of a biorefinery process through metabolic engineering. However, this systems biology approach is generalist, disregarding the target compound(s) and requiring prior knowledge to select metabolic pathways that could afford higher amounts of a given compound of interest, without disturbing the whole cell's metabolism. This selection eventually might not be possible depending on the algorithms adopted, since many consider only the microalgal metabolism as a whole, hindering their usage under a biorefinery process analysis. This is, of course, a disadvantage of applying systems biology in the context of biorefineries, being inevitable the selection of pathways by researchers, or even by developing and validating machine learning models that perform this selection.

The metabolomics analysis performed by ^{1}H-NMR spectroscopy revealed that *P. tricornutum*'s biosynthetic pathways with the greatest impact and potential for application in metabolic engineering are concerned to alanine, glyoxylate, and pyruvate. These are important metabolites regarding the cell's thermodynamic state, also related to the synthesis of acetyl-CoA, a well-known building block of fat acids. Thus, it is speculated that higher amounts of these molecules available to the cell's metabolism might foster the biosynthesis of EPA and fucoxanthin. On the other hand, metabolic engineering strategies to inhibit the distribution of these molecules to other metabolic pathways require care, since they may negatively interfere on the cell metabolism in a way leading to the collapse of cell's functions and death.

These strategies generated from systems biology generate subsidies for metabolic engineering and can be applied to changes in synthetic levels, altering or synthetically blocking pathways with changes in the genetic level, or even more simplistically, using enzyme inhibitors specific to pathways that are not part of the synthesis of the compounds of interest, at determined times of cultivation.

The systems biology studies have advanced a lot, in confluence with the development of computational platforms and databases, thus facilitating the processing and understanding of omics data sets. Metabolomics allows one to overview the cell's metabolism and also the understanding as a whole of the biochemical machinery of cells in its dynamics, being a relevant way of rationally validate *in silico* metabolic models following a biorefinery concept of production.

Fluxomics: FBA Applied to Fucoxanthin Metabolism

Fluxomics is one of the most accurate and predictable tools for controlling metabolic flow and for genetic engineering purposes it is considered a much more reliable process of analysis. This type of application requires prior knowledge of cell metabolism with the use of genome-scale metabolic network reconstruction, allowing simulations and optimizations of desired phenotypes *in silico*. Strategies that can be carried out *in silico* are the most recommended actually, because they increase the predictability and are more ecological by reducing the use of reagents.

Two genome-scale metabolic models for the species *P. tricornutum* (Kim et al., 2016 and Levering et al., 2016) were utilized by Bauer et al., (2019) to predict ways to enhance fucoxanthin production using FBA. For that, firstly the metabolic model proposed by Kim et al., (2016) lacking reactions for carotenogenesis was completed to reach fucoxanthin biosynthesis, following its importation along with the metabolic model of Levering et al., (2016) into OptFlux software (Rocha et al., 2010) for FBA analysis. *In silico* simulations with these models predicted increased biomass flux up to 67% and 77%, by doubling the supply of inorganic carbon (CO_2 and HCO_3^-) and urea, respectively. For optimizations, FBA on OptFlux were carried out under MOMA (minimization of metabolic adjustment) algorithm and good optimization solutions for fucoxanthin biosynthesis pointed out to the knockout of genes/enzymes involved in the metabolism of important precursors for the non-mevalonate methylerythritol pathway - MEP (mostly pyruvate and glyceraldehyde-3-phosphate - G3P.

Fucoxanthin metabolism has yet to be fully elucidated, however some important points of this path are potentially established. In addition to the importance of the availability of precursors of the MEP pathway like pyruvate and G3P, 1-deoxy-D-xylulose 5-phosphate synthase (DXS) (Eilers et al., 2016), phytoene synthase (PSY) (Kaur and Spillane, 2015, Eilers et al., 2016), and a violaxanthin de-epoxidase–like (VDL) (Dautermann et al., 2020, Gaidarenko et al., 2020), have been shown to be important bottlenecks in fucoxanthin biosynthesis. DXS catalyzes the initial step of the MEP pathway and PSY guides the metabolic flux into carotenogenesis. VDL is closely related to the formation of allenic carotenoids diverting carotenoids function towards light harvesting in the evolution of marine algae including diatoms (Dautermann et al., 2020). According to Dautermann et al., (2020) VDL is related to the enzyme violaxanthin de-epoxidase-like which in turn catalyzes the formation of neoxanthin leading to the formation of diadinoxanthin and fucoxanthin.

Violaxanthin de-epoxidase (VDE) and VDE-related (VDR) which convert violaxanthin into zeaxanthin or diadinoxanthin into diatoxanthin, as well as zeaxanthin epoxidase 3 (ZEP3), responsible for the reversed reactions, also seem to be important in carotenoid formation. Transgenic strains of *P. tricornutum* designed to simultaneously overexpress these three genes increased carotenoid amounts up to 4-fold in fucoxanthin, 3-fold in diatoxanthin and up to 2-fold in β-carotene (Manfellotto et al., 2020), as silencing VDE reduced diatoxanthin synthesis (Lavaud et al., 2012).

Regarding carotenogenesis pathways, other branches besides the formation of carotenoids, such as the synthesis of hormones, exist and thus these models can still be completed in these pathways for better optimization results. Otherwise, computational models have to be constantly updated with novel information about *P. tricornutum* metabolome.

CONCLUSION

The development of metabolic engineering applied to microalgae has been growing in the last decade with the genome sequencing of some model species that facilitate studies in systems biology and thus, enable the understanding of algal metabolism as a whole. Systems biology analysis (e.g., genomic scale models and omics analysis), together with the increased availability of databases and open-access software, allow more precisely the discovery of bottlenecks in microalgal metabolism that can serve as targets for metabolic engineering.

P. tricornutum is considered the model species when looking at to produce more than one compound of high commercial value in a biorefinery process. With the publication of its genome in 2008, work in the field of metabolic engineering grew at an accelerated pace. The species is considered a model for the biorefinery because it produces a high content of lipids with a fatty acid profile rich in PUFAs of commercial interest (e.g., EPA and linoleic acid), highly antioxidant pigments (i.e., fucoxanthin), and carbohydrates such as chrysolaminarin. In addition to these compounds of biotechnological interest, the insertion of recombinant genes was effective in the production of other metabolites to enrich the cell-matrix of the culture, directing it to its application. Besides, the size of the *P. tricornutum*'s genetic material makes it an attractive system for metabolic engineering and enables the creation of successful transgenic strains.

Among the metabolic engineering strategies in *P. tricornutum*, those that use the knowledge of systems biology and apply genetic engineering are the most promising. Nonetheless, regarding the biorefinery concept applied to this species, it is crucial to understand the points of control and regulation of biosynthetic pathways of multiple compounds of biotechnological interest. In this sense, omics science technologies are truly useful because they aim to understand the systems as a whole.

The case studies herein reported where omics sciences were applied to understand the metabolism of *P. tricornutum* in connection to generate strategies aiming at an improved biorefinery process have shown promising targets. In the first case, metabolomics analysis allowed to have an overview of the oscillation of the metabolic profiles during cell cultivation. This approach intended to gain insights of the microalgal metabolism having more than one target class of compounds and facilitate mainly biotechnological processes that are not aimed at genetic engineering, but rather the optimization of an industrial cultivation process. The second case study applied fluxomics analysis to the *P. tricornutum*'s cell metabolism and gave important information to guide further *in vitro* experiments that seek the production of desired compounds and the understanding of their metabolism. Indeed, as such, fluxomics analysis coupled to biochemical investigations is thought as a two-way street that allows the constant improvement of methodological strategies to optimize the biosynthesis and accumulation of biotechnologically relevant molecules, reducing the need for practical experiments.

Finally, it is evident the need to develop a lot the biorefinery concept applied to *P. tricornutum* cell cultures in industrial scale. In the last decade, there has been an increase in metabolic engineering work that aims to increase the synthesis of compounds by *P. tricornutum*, however, there are few works addressing the biorefinery concept in a robust manner. Thus, in the future studies should focus on, for instance, multiple and simultaneous knock-outs genes, which makes metabolic modification complex but more promising, and understanding the regulatory networks of the microalga metabolism. The main challenge that must be overcome is to be able to balance the production of more than one compound without penalizing the cell growth of biomass.

Even though the idea of biorefinery is highly attractive, numerous advances and adjustments are still needed to make microalgae biorefinery economically feasible. The concept itself involves not only a multiplicity of products, but also protocols that range from the genetic engineering of organisms to upstream and downstream processes until obtaining the final products with the desired quality. Advances have occurred in all areas of the biorefinery process and need to be refined and evaluated on an industrial scale to become biotechnologically relevant.

REFERENCES

Alipanah, Leila, Rohloff, Jens, Winge, Per, Bones, Atle M. and Tore Brembu 2015. Whole-cell response to nitrogen deprivation in the diatom *Phaeodactylum tricornutum*. *Journal of Experimental Botany*, 66(20), 6281-6296. Doi: 10.1093/jxb/erv340.

Allen, Andrew E., Dupont, Christopher L., Oborník, Miroslav, Horák, Ales, Nunes-Nesi, Adriano, McCrow, John P., Zheng, Hong, Johnson, Daniel A., Hu, Hanhua, Fernie, Alisdair R. and Chris Bowler 2011. Evolution and metabolic significance of the urea cycle in photosynthetic diatoms. *Nature*, 473(7346), 203-207. Doi: 10.1038/nature10074.

Apt, Kirk E., Grossman, Peter G., and Arthur R. Kroth-Pancic 1996. Stable nuclear transformation of the diatom *Phaeodactylum tricornutum*. *Molecular and General Genetics MGG*, 252(5), 572-579. Doi: 10.1007/BF02172403.

Bai, Xiaocui, Song, Hao, Lavoie, Michel, Zhu, Kun, Su, Yiyuan, Ye, Hanqi, Chen, Si, Fu, Zhengwei and Haifen Qian 2016. Proteomic analyses bring new insights into the effect of a dark stress on lipid biosynthesis in *Phaeodactylum tricornutum*. *Scientific Reports*, 6(1), 1-10. Doi: 10.1038/srep25494.

Balamurugan, Srinivasan, Wang, Xiang, Wang, Hong-Lei, An, Chun-Jing, Li, Hui, Li, Da-Wei, Yang, Wei-Dong, Liu, Jie-Sheng and Hong-Ye Li 2017. Occurrence of plastidial triacylglycerol synthesis and the potential regulatory role of AGPAT in the model diatom

Phaeodactylum tricornutum. *Biotechnology for Biofuels*, 10(1), 97-111. Doi: 10.1186/s13068-017-0786-0.

Bauer, Cláudia M., Vilaça, Paulo, Ramlov, Fernanda, de Oliveira, Eva R., Cabral, Débora Q., Schmitz, Caroline, Gordo, Rafaela, Rocha, Miguel and Marcelo Maraschin 2018. In silico predictions for fucoxanthin production by the diatom *Phaeodactylum tricornutum*. In: *International Conference on Practical Applications of Computational Biology & Bioinformatics*. *Springer*, Cham. 12, 139-148. Doi: 10.1007/978-3-319-98702-6_17.

Bellou, Stamatia, Baeshen, Mohammed N., Elazzazy, Ahmed M., Aggeli, Dimitra, Sayegh, Fotoon and George Aggelis 2014. Microalgal lipids biochemistry and biotechnological perspectives. *Biotechnology Advances*, 32(8), 1476-1493. Doi: 10.1016/j.biotechadv.2014.10.003.

Benavides, Ana M. S., Torzillo, Giuseppe, Kopecký, Jirka, and Jiri Masojídek 2013. Productivity and biochemical composition of *Phaeodactylum tricornutum* (Bacillariophyceae) cultures grown outdoors in tubular photobioreactors and open ponds. *Biomass and Bioenergy*, 54, 115-122. Doi: 10.1016/j.biombioe.2013.03.016.

Bowler, Chris, Allen, Andrew E., Badger, Jonathan H., Grimwood, Jane, Jabbari, Kamel, Kuo, Alan, Maheswari, Uma, Martens, Cindy, Maumus, Florian, Robert P., Otillar, Rayko, Edda, Salamov, Asaf, Klaas, Vandepoele, Bank, Beszteri, Ansgar, Gruber, Marc, Heijde, Katinka, Michael, Mock, Thomas, Valentin, Klaus, Verret, Fréderic, John A., Berges, Brownlee, Colin, Cadoret, Jean-Paul, Chiovitti, Anthony, Choi, Chang J., Coesel, Sacha, De Martino, Alessandra, Detter, J. Chris, Durkin, Colleen, Falciatore, Angela, Fournet, Jérome, Haruta, Miyoshi, Huysman, Marie J. J., Jenkins, Bethany D., Jiroutova, Katerina, Jorgensen, Richard E., Joubert, Yolaine, Kaplan, Aaron, Kröger, Nils, Kroth, Peter G., La Roche, Julie, Lindquist, Erica, Lommer, Markus, Martin–Jézéquel, Véronique, J. Lopez, Pascal, Lucas, Susan, Mangogna, Manuela, McGinnis, Karen, Medlin, Linda K., Montsant, Anton, Secq, Marie-Pierre O., Napoli, Carolyn, Obornik, Miroslav, Parker, Micaela S., Petit, Jean-Louis, Porcel, Betina M., Poulsen, Nicole, Robison, Matthew, Rychlewski,

Leszek, Rynearson, Tatiana A., Schmutz, Shapiro, Jeremy H., Siaut, Magali, Stanley, Michele, Sussman, Michael R., Taylor, Alison R., Vardi, von Dassow, Assaf P., Vyverman, Wim, Willis, Anusuya, Wyrwicz, Lucjan S., Rokhsar, Daniel S., Weissenbach, Jean, Armbrust, E. Virginia, Green, Beverley R., Peer, Yves V., and Igor Grigoriev 2008. The Phaeodactylum genome reveals the evolutionary history of diatom genomes. *Nature*, 456, 239-244. Doi: 10.1038/nature07410.

Branco-Vieira, Monique, San Martin, Sergio, Agurto, Cristian, Freitas, Marcos A. V., Martins, Antonio A., Mata, Teresa M. and Nidia S. Caetano 2020. Biotechnological potential of *Phaeodactylum tricornutum* for biorefinery processes. *Fuel*, 268, 117357. Doi: 10.1016/j.fuel.2020.117357.

Breuer, Guido, Lamers, Packo P., Martens, Dirk E., Draaisma, René B. and René H. Wijffels 2012. The impact of nitrogen starvation on the dynamics of triacylglycerol accumulation in nine microalgae strains. *Bioresource Technology*, 124, 217-226. Doi: 10.1016/j.biortech. 2012.08.003.

Buck, Jochen M., Bártulos, Carolina R., Gruber, Asgar and Peter G. Kroth 2018. Blasticidin-S deaminase, a new selection marker for genetic transformation of the diatom *Phaeodactylum tricornutum. Peer J*, 6, e5884. Doi: 10.7717/peerj.5884.

Buck, Jochen M., Sherman, Jonathan, Bártulos, Carolina R., Serif, Manuel, Halder, Marc, Henkel, Jan, Falciatore, Angela, Lavaud, Johann, Gorbunov, Maxim Y., Kroth, Peter G., Falkowski, Paul and Bernard Lepetit 2019. Lhcx proteins provide photoprotection via thermal dissipation of absorbed light in the diatom *Phaeodactylum tricornutum*. *Nature Communications*, 10(1), 1-12. Doi: 10.1038/s41467-019-12043-6.

Butler, Thomas, Kapoore, Rahul V. and Seetharaman Vaidyanathan 2020. *Phaeodactylum tricornutum*: a diatom cell factory. *Trends in Biotechnology*, 38(6), 606-622. Doi: 10.1016/j.tibtech.2019.12.023.

Carrijo, Rodolfo D. S., Silva, Valeria C. F., dos Santos, Anne C. M., Costa, Mayra F., and Taciano P. Ferreira 2015. Uso de microalgas para a produção de biodiesel: vantagens e limitações. *Revista Eletrônica de Energia,* 5(1), 23-33. [Use of microalgae for biodiesel production: advantages and limitations. *Electronic Energy Magazine*]

Chae, Tong U., Choi, So Y., Kim, Je W., Ko, Yoo-Sung and Sang Y. Lee 2017. Recent advances in systems metabolic engineering tools and strategies, *Current Opinion in Biotechnology*, 47, 67-82. Doi:10.1016/j.copbio.2017.06.007.

Chew, Kit W., Yap, Jing Y., Show, Pau L., Suan, Ng H., Juan, Joon C., Ling, Tau C., Lee, Duu-Jong, Jo-Shu Chang 2017. Microalgae biorefinery: high value products perspectives. *Bioresource Technology*, 229, 53-62. Doi: 10.1016/j.biortech.2017.01.006.

Chong, Jasmine, Wishart, David S. and Jianguo Xia 2019. Using MetaboAnalyst 4.0 for comprehensive and integrative metabolomics data analysis. *Current Protocols in Bioinformatics*, 68, e86. Doi: 10.1002/cpbi.86.

Cui, Yulin, Zhao, Jialin, Wang, Yinchu, Qin, Son and Yandu Lu 2018. Characterization and engineering of a dual-function diacylglycerol acyltransferase in the oleaginous marine diatom *Phaeodactylum tricornutum*. *Biotechnology for Biofuels*, 11(1), 32. Doi: 10.1186/s13068-018-1029-8.

Cui, Yi, Thomas-Hall, Skye R. and Peer M.Schenk 2019. *Phaeodactylum tricornutum* microalgae as a rich source of omega-3 oil: Progress in lipid induction techniques towards industry adoption. *Food Chemistry*, 297, 124937. Doi: 10.1016/j.foodchem.2019.06.004.

Daboussi, Fayza, Leduc, Sophie, Maréchal, Alan, Dubois, Gwendoline, Guyot, Valerie., Perez-Michaut, Christophe, Amato, Alberto, Falciatore, Angela, Juillerat, Alexandre, Beurdeley, Marine, Voytas, Daniel F., Cavarec, Laurent and Philippe Duchateau 2014. Genome engineering empowers the diatom *Phaeodactylum tricornutum* for biotechnology. *Nature Communications*, 5(1), 1-7. Doi: 10.1038/ ncomms4831.

D'Adamo, Sarah, Schiano di Visconte, Gino, Lowe, Gavin, Szaub-Newton, Joanna, Beacham, Tracey, Landels, Andrew, Alle, Michael J. and Michiel Matthijs 2019. Engineering the unicellular alga

Phaeodactylum tricornutum for high-value plant triterpenoid production. *Plant Biotechnology Journal*, 17(1), 75-87. Doi: 10.1111/pbi.12948.

Dautermann, Oliver, Lyska, Dagmar, Andersen-Ranberg, Johan, Becker, Mirjana, Fröhlich-Nowoisky, Janine, Kramer, Li C., Mayr, Katharina, Heidi, Wipf, Niyogi, Krishna and Martin Lohr 2020. An algal enzyme required for biosynthesis of the most abundant marine carotenoids. *Science Advances*, 6(10), 9183. Doi: 10.1126/sciadv. aaw9183.

De Jong, Ed; Langeveld, H and Rene Van Ree Biorefinery. Biorefineries: giving value to sustainable biomass use. In: *The Biobased Economy.* Routledge, 132-151, 2013. Doi: https://doi.org/10.4324/978184 9774802

Desbois, Andrew P., Lebl, Tomas, Yan, Liming and Valerie J. Smith 2008. Isolation and structural characterisation of two antibacterial free fatty acids from the marine diatom, *Phaeodactylum tricornutum. Applied Microbiology and Biotechnology* 81(4), 755-764. Doi: https://doi.org/10.1007/s00253-008-1714-9.

Dinamarca, Jorge, Levitan, Orly, Kumaraswamy, G. Kenchappa, Lun, Desmond S. and Paul G. Falkowski 2017. Overexpression of a diacylglycerol acyltransferase gene in *Phaeodactylum tricornutum* directs carbon towards lipid biosynthesis. *Journal of Phycology*, 53(2), 405-414. Doi: 10.1111/jpy.12513.

D'Orazio, Nicolantonio, Gemello, Eugenio, Gammone, Maria A., De Girolamo, Massimo, Ficoneri, Cristiana, & Graziano Riccioni 2012. Fucoxantin: A treasure from the sea. *Marine Drugs*, 10(3), 604-616. Doi: 10.3390/md10030604.

Eilers, Ulrike, Bikoulis, Alexandros, Breitenbach, Jurgen, Büchel, Claudia Gerhard Sandmann1 2016. Limitations in the biosynthesis of fucoxanthin as targets for genetic engineering in *Phaeodactylum tricornutum. Journal of Applied Phycology*, *28*(1), 123-129. Doi: 10.1007/s10811-015-0583-8.

Erdene-Ochir, Erdenedolgor, Shin, Bok-Kyu, Kwon, Byeori, Jung, Choonkyun and Cheol-Ho Pan 2019. Identification and characterisation of the novel endogenous promoter HASP1 and its signal peptide from *Phaeodactylum tricornutum. Scientific Reports*, *9*(1), 1-10. Doi: 10.1038/s41598-019-45786-9.

Fabris, Michele, Matthijs, Michiel, Rombauts, Stephane, Vyverman, Wim, Goossens, Alain and Gino J.E. Baart 2012. The metabolic blueprint of *Phaeodactylum tricornutum* reveals a eukaryotic Entner–Doudoroff glycolytic pathway. *The Plant Journal*, 70(6), 1004-1014. Doi: 10.1111/j.1365-313X.2012.04941.x.

Fabris, Michele, Matthijs, Michiel, Carbonelle, Sophie, Moses, Tessa, Pollier, Jacob, Dasseville, Renaat, Baart, Gino J.E., Vyverman, Wim and Alan Goossens 2014. Tracking the sterol biosynthesis pathway of the diatom *Phaeodactylum tricornutum. New Phytologist*, 204, 521-535. Doi: 10.1111/nph.12917.

Fabris, Michele, George, Jestin, Kuzhiumparambil, Unnikrishnan, Lawson, Caitlin A., Jaramillo-Madrid, Ana C., Abbriano, Raffaela M., Vickers, Claudia E and Peter Ralph 2020. Extrachromosomal Genetic Engineering of the Marine Diatom *Phaeodactylum tricornutum* Enables the Heterologous Production of Monoterpenoids. *ACS Synthetic Biology*, 9, 598-612. Doi: 10.1021/acssynbio.9b00455.

Falciatore, Angela, Casotti, Rafaella, Leblanc, Catharine, Abrescia, Chiara and Chris Bowler 1999. Transformation of nonselectable reporter genes in marine diatoms. *Marine Biotechnology*, 1(3), 239-251.

Fu, Weiqi, Nelson, DR, Mystikou, Alexandra, Daakour, Sarah and Kourosh Salehi-Ashtiani 2019. Advances in microalgal research and engineering development. *Current Opinion in Biotechnology,* 59, 157-164. Doi: 10.1016/j.copbio.2019.05.013.

Gaidarenko, Olga, Mills, Daylan W., Vernet, Maria and Mark Hildebrand 2020. Overexpression of Thalassiosira pseudonana violaxanthin de-epoxidase-like 2 (VDL2) increases fucoxanthin while stoichiometrically reducing diadinoxanthin cycle pigment abundance. *BioRxiv*. 7, 1. Doi: 10.1101/2020.01.06.896316.

Gao, Wei, Wang, Yiqiao, Basavanagoud, B. and Muhammad K. Jamil 2017. Characteristics studies of molecular structures in drugs. *Saudi Pharmaceutical Journal*, *25*(4), 580-586. Doi: 10.1016/j.jsps.2017.04.027.

Ge, Feng, Huang, Weichao, Chen, Zhuo, Zhang, Chunye, Xiong, Qian, Bowler, Chris, Yang, Juan, Xu, Jin and Hanhua Hu 2014. Methylcrotonyl-CoA carboxylase regulates triacylglycerol accumulation in the model diatom *Phaeodactylum tricornutum. The Plant Cell,* 26(4), 1681-1697. Doi: 10.1105/tpc.114.124982.

George, Jestin, Kahlke, Tim, Abbriano, Raffaela M., Kuzhiumparambil, Unnikrishnan, Ralph, Peter J. and Michele Fabris M. 2020. Metabolic engineering strategies in diatoms reveal unique phenotypes and genetic configurations with implications for algal genetics and synthetic biology. *Frontiers in Bioengineering and Biotechnology*, 8, 513. Doi: 10.3389/fbioe.2020.00513.

Gruber, Ansgar and Peter Kroth 2017. Intracellular metabolic pathway distribution in diatoms and tools for genome-enabled experimental diatom research. *Philosophical Transactions of the Royal Society B: Biological Sciences*, 372(1728), 20160402. Doi: 10.1098/rstb.2016.0402.

Guerra, L. Tiago, Levitan, Orly, Frada, Miguel J., Sun, Jennifer S., Falkowski, Paul G. and G. Charles Dismukes 2013. Regulatory branch points affecting protein and lipid biosynthesis in the diatom *Phaeodactylum tricornutum. Biomass and Bioenergy*, 59, 306-315 Doi: 10.1016/j.biombioe.2013.10.007.

Hamilton, Mary L., Haslam, Richard P., Napier, Johnathan A. and Olga Sayanova 2014. Metabolic engineering of *Phaeodactylum tricornutum* for the enhanced accumulation of omega-3 long chain polyunsaturated fatty acids. *Metabolic Engineering*, 22, 3-9. Doi: 10.1016/j.ymben.2013.12.003.

Hao, Xiahui, Luo, Ling, Jouhet, Juliette, Rébeillé, Fabrice, Maréchal, Eric, Hu, Hanhua, Pan, Yufang, Tan, Xiaoming, Chen, Zhuo, You, Lingjie, Chen, Hong, Wei Fang and Yangmin Gong 2018. Enhanced triacylglycerol production in the diatom *Phaeodactylum tricornutum* by inactivation of a Hotdog-fold thioesterase gene using TALEN-based targeted mutagenesis. *Biotechnology for Biofuels*, 11(1), 312. Doi: 10.1186/s13068-018-1309-3.

Haslam, Richard P., Hamilton, Mary L., Economou, Chloe K., Smith Richard, Hassall, Kirsty L., Napier, Johnathan A. and Olga Sayanova 2020. Overexpression of an endogenous type 2 diacylglycerol acyltransferase in the marine diatom *Phaeodactylum tricornutum* enhances lipid production and omega-3 long-chain polyunsaturated fatty acid content. *Biotechnology for Biofuels*, 13, 1-17. Doi: 10.1186/s13068-020-01726-8.

Hempel, Franziska, Maurer, Michel, Brockmann, Bjorn, Mayer, Christian, Biedenkopf, Nadine, Kelterbaum, Anne, Becker, Sthephen and Uwe G. Maier 2017. From hybridomas to a robust microalgal-based production platform: molecular design of a diatom secreting monoclonal antibodies directed against the Marburg virus nucleoprotein. *Microbial Cell Factories*, 16(1), 1-10. Doi: 10.1186/s12934-017-0745-2.

Heydarizadeh, Parisa, Boureba, Wafâa, Zahedi, Morteza, Huang, Bing, Moreau, Brigitte, Lukomska, Ewa, Couzinet-Mossion, Aurelie, Wielgosz-Collin, Gaëtane, Martin-Jézéquel, Véronique, Bougaran, Gaël, Marchand, Justine, and Benoît Schoefs 2017. Response of CO2-starved diatom *Phaeodactylum tricornutum* to light intensity transition. *Philosophical Transactions of the Royal Society B: Biological Sciences*, 372(1728), 20160396. Doi: 10.1098/rstb.2016.0396,

Heydarizadeh, Parisa, Veidl, Brigitte, Huang, Bing, Lukomska, Ewa, Wielgosz-Collin, Gaetene, Couzinet-Mossion, Aurelie, Bougaran, Gaël, Marchand, Justine, and Benoît Schoefs 2019. Carbon orientation in the diatom *Phaeodactylum tricornutum*: the effects of carbon limitation and photon flux density. *Frontiers in Plant Science*, 10, 471. Doi: 10.3389/fpls.2019.00471.

Hosokawa, Masashi., Wanezaki, Satoshi, Miyauchi, Kozue, Kurihara, Hideyuki, Kohno, Hiroyouki, Kawabata, Jun, Odashima, Shizuo and Koretaro Takahashi 1999. Apoptosis-inducing effect of fucoxanthin on human leukemia cell line HL-60. *Food Science and Technology Research*, *5*(3): 243-246. Doi: 10.3136/fstr.5.243.

Huang, Weichao and Fayza Daboussi 2017. Genetic and metabolic engineering in diatoms. *Philosophical Transactions of the Royal Society B: Biological Sciences,* 372(1728), 20160411. Doi: 10.1098/rstb.2016.0411.

Jaramillo-Madrid, Ana C., Abbriano, Raffaela, Ashworth, Justin, Fabris, Michele, Pernice, Mathieu and Peter R. Ralph 2020. Overexpression of key sterol pathway enzymes in two model marine diatoms alters sterol profiles in *Phaeodactylum tricornutum. Pharmaceuticals*, *13*(12), 481. Doi: 10.3390/ph13120481.

Kanehisa, Minoru, Araki, Michihiro, Goto, Susumo, Hattori, Masahiro, Mika, Hirakawa, Masumi, Itoh, Katayama, Toshiaki, Kawashima, Shuichi, Okuda, Shujiro, Tokimatsu, Toshiaki andYoshiro Yamanishi 2007. KEGG for linking genomes to life and the environment. *Nucleic Acids Research*, 36, D480-D484. Doi: 10.1093/nar/gkm882.

Karas, Bogumil J., Diner, Rachel E., Lefebvre, Stephane C., McQuaid, Jeff, Phillips, Alex P.R., Noddings, Chari M., Brunson, John K., Valas, Ruben E., Deerinck, Thomas J., Jablanovic, Jelena, Gillard, Jeroen T.F., Beeri, Karen, Ellisman, Mark H., Glass, John I., Hutchison III, Clyde A., Smith, Hamilton O., Venter, J. Craig, Allen, Andrew E., Dupont Christopher L. and Philip D. Weyman 2015. Designer diatom episomes delivered by bacterial conjugation. *Nature Communications*, *6*(1), 1-10.Doi: 10.1038/ncomms7925.

Kaur, Simrat and Charles Spillane 2015. Reduction in carotenoid levels in the marine diatom *Phaeodactylum tricornutum* by artificial microRNAs targeted against the endogenous phytoene synthase gene. *Marine Biotechnology*, 17(1), 1-7. Doi: 10.1007/s10126-014-9593-9.

Kikutani, Sae, Nakajima, Kensuke, Nagasato, Chikako, Tsuji, Yoshinori, Miyatake, Ai and Yusuke Matsuda 2016. Thylakoid luminal θ-carbonic anhydrase critical for growth and photosynthesis in the marine diatom *Phaeodactylum tricornutum. Proceedings of the National Academy of Sciences*, 113(35), 9828-9833. Doi: 10.1073/pnas.1603112113.

Kim, Joomi, Brown, Christopher M., Kim, Min K., Burrows, Elizabeth H., Bach, Stéphane, Lun, Desmond S. and Paul G. Falkowski 2017. Effect of cell cycle arrest on intermediate metabolism in the marine diatom *Phaeodactylum tricornutum*. *Proceedings of the National Academy of Sciences*, 114(38), E8007-E8016. Doi: 10.1073/pnas.1711642114.

Kim, Joomi, Fabris, Michele, Baart, Gino, Kim, Min K., Goossens, Alain, Vyverman, Wim Falkowski, Paul G. and Desmond S. Lun 2016. Flux balance analysis of primary metabolism in the diatom *Phaeodactylum tricornutum*. *The Plant Journal*, 85(1), 161-176. Doi: 10.1111/tpj.13081.

Kroth, Peter G., Chiovitti, Anthony, Gruber, Ansgar, Martin-Jezequel, Veronique, Mock, Thomas, Parker, Micaela S., Stanley, Michele S., Kaplan, Aaron, Caron, Lise, Weber, Till, Maheswari, Uma, Armbrust, E.Virginia and Chris Bowler 2008. A model for carbohydrate metabolism in the diatom *Phaeodactylum tricornutum* deduced from comparative whole genome analysis. *PloS One*, 3(1), e1426. Doi: 10.1371/journal.pone.0001426.

Lavaud, Johann, Materna, Arnie C., Sturm, Sabine, Vugrinec, Sascha ans Peter G. Kroth 2012. Silencing of the violaxanthin de-epoxidase gene in the diatom *Phaeodactylum tricornutum* reduces diatoxanthin synthesis and non-photochemical quenching. *PLoS One*, 7(5), e36806. Doi: 10.1371/journal.pone.0036806.

Levering, Jennifer, Broddrick, Jared, Dupont, Christopher L., Peers, Graham, Beeri, Karen, Mayers, Joshua, Gallina, Alessandra A., Allen, Andrew E., Palsson, Bernhard O and Kasten Zengler 2016. Genome-scale model reveals metabolic basis of biomass partitioning in a model diatom. *PloS One*, 11(5), e0155038. Doi: 10.1371/journal.pone.0155038.

Levering, Jennifer, Dupont, Christopher L., Allen, Andrew E., Palsson, Bernhard O and Kasten Zengler 2017. Integrated regulatory and metabolic networks of the marine diatom *Phaeodactylum tricornutum* predict the response to rising CO2 levels. *Msystems*, 2(1), Doi: 10.1128/mSystems.00142-16.

Levitan, Orly, Dinamarca, Jorge, Zelzion, Ehud, Lun, Desmond S., Guerra, L. Tiago, Kim, Min Kyung, Kim, Joomi, Van Mooy, Benjamin A. S., Bhattacharya, Debashish and Paul G. Falkowski 2015. Remodeling of intermediate metabolism in the diatom *Phaeodactylum tricornutum* under nitrogen stress. *Proceedings of the National Academy of Sciences*, 112(2), 412-417. Doi: 10.1073/pnas. 1419818112.

Li, Shuangfei, Chen, Yufei, Wang, Yue, Yang, Xuewei, Wang, Jing and Huirong Chen 2020. Over-expression of LPAAT gene in *Phaeodactylum tricornutum* enhances fatty acid accumulation and increases fatty acid chain length. *Indian Journal of Pharmaceutical Sciences,* 82(2), 11-15. Doi: 10.36468/pharmaceutical-sciences.spl.76.

Hung-Yun, Lin, Yen, Shao-Chieh, Kuo, Po-Chih, Chung, Chih-Yu, Yeh, Kuan-Lin, Huang, Ching-Huei, Chang, Jeng and Han Lin 2017. Alkaline phosphatase promoter as an efficient driving element for exogenic recombinant in the marine diatom *Phaeodactylum tricornutum. Algal Research*, 23, 58-65. Doi: 10.1016/j.algal.2017. 01.007.

Longworth, Joseph, Wu, Danying, Huete-Ortega, María, Wright, Phillip C. and Seetharaman Vaidyanathan 2016. Proteome response of *Phaeodactylum tricornutum*, during lipid accumulation induced by nitrogen depletion. *Algal Research*, 18, 213-224. Doi: 10.1016/j.algal.2016.06.015.

Madhuri, Shvaita., Bártulos, Carolina. R., Serif, Manuel, Lepetit, Bernard and Peter G. Kroth 2019. A strategy to complement PtAUREO1a in TALEN knockout strains of *Phaeodactylum tricornutum. Algal Research*, 39, 101469. Doi: 10.1016/j.algal.2019.101469.

Maeda, Hayato, Hosokawa, Masashi, Sashima, Tokutake, Murakami-Funayama, Katsura and Kazuo Miyashita 2009. Anti-obesity and anti-diabetic effects of fucoxanthin on diet-induced obesity conditions in a murine model. *Molecular Medicine Reports*, 2(6), 897-902. Doi: 10.3892/mmr_00000189.

Manfellotto, Franscesco, Stella, Giulio R., Falciatore, Angela, Brunet, Chistophe and Maria I. Ferrante 2020. Engineering the unicellular alga *Phaeodactylum tricornutum* for enhancing carotenoid production. *Antioxidants*, 9(8), 757. Doi: 10.3390/antiox9080757.

Mann, Marcus, Serif, Manuel, Jakob, Torsten, Kroth, Peter G. and Christian Wilhelm 2017. PtAUREO1a and PtAUREO1b knockout mutants of the diatom *Phaeodactylum tricornutum* are blocked in photoacclimation to blue light. *Journal of Plant Physiology*, 217, 44-48. Doi: 10.1016/j.jplph.2017.05.020.

Mendoza, Héctor, De la Jara, Adelina, Freijanes, Karen, Carmona, Laura, Ramos, Ana A., de Sousa Duarte, Vanessa and João C. Serafim Varela 2008. Characterization of *Dunaliella salina* strains by flow cytometry: a new approach to select carotenoid hyperproducing strains. *Electronic Journal of Biotechnology*, 11(4), 5-6. Doi: 10.2225/vol11-issue4-fulltext-2.

Milićević, Dragan, Vranić, Danijela, Mašić, Zoran, Parunović, Nenad, Trbović, Dejana, Jelena, Nedeljković-Trailović and Zoran Petrović 2014. The role of total fats, saturated/unsaturated fatty acids and cholesterol content in chicken meat as cardiovascular risk factors. *Lipids in Health and Disease*, v. 13(1), 42, 2014. Doi: 10.1186/1476-511X-13-42.

Niu, Ying F., Yang, Zhi K., Zhang, Meng H., Zhu, Cong C., Yang, Wei D., Liu, Jie S. and Hong-Ye Li 2012. Transformation of diatom *Phaeodactylum tricornutum* by electroporation and establishment of inducible selection marker. *Biotechniques*, 52(6), 1-3. Doi: 10.2144/000113881.

Niu, Ying F., Zhang, Meng H., Li, Da W., Yang, Wei D., Liu, Jie S., Bai, Wei B. and Hong-Ye Li 2013. Improvement of neutral lipid and polyunsaturated fatty acid biosynthesis by overexpressing a type 2 diacylglycerol acyltransferase in marine diatom *Phaeodactylum tricornutum*. *Marine Drugs*, 11(11), 4558-4569. Doi: 10.3390/md 11114558.

Nymark, Marianne, Sharma, Amit K., Sparstad, Torfinn, Bones, Atle M. and Per Winge 2016. A CRISPR/Cas9 system adapted for gene editing in marine algae. *Scientific Reports*, 6(1), 1-6. Doi: doi.org/10.1038/srep24951.

Orth, Jeffrey D., Thiele, Ines and Bernhard Ø. Palsson 2010. What is flux balance analysis? *Nature Biotechnology*, 28(3), 245-248. Doi: 10.1038/nbt.1614.

Pangestuti, Ratih and Se-Kwon Kim 2011. Biological activities and health benefit effects of natural pigments derived from marine algae. *Journal of Functional Foods*, *3*(4), 255-266. Doi: 10.1016/j.jff.2011.07.001.

Peck, Philip, Bennett, Simon J., Bissett-Amess, Rachelle, Lenhart, Jennifer and Hamid Mozaffarian 2009. Examining understanding, acceptance, and support for the biorefinery concept among EU policy-makers. Biofuels, *Bioproducts and Biorefining*, 3(3), 361-383. Doi: 10.1002/bbb.154.

Pudney, Alex, Gandini, Chiara, Economou, Chloe K., Smith, Richard, Goddard, Paul, Napier, Johnathan A., Spicer, Andrew and Olga Sayanova 2019. Multifunctionalizing the marine diatom *Phaeodactylum tricornutum* for sustainable co-production of omega-3 long chain polyunsaturated fatty acids and recombinant phytase. *Scientific Reports*, 9(1), 1-10. Doi: 10.1038/s41598-019-47875-1.

Rastogi, Achal, Maheswari, Uma, Dorrell, Richard G., Vieira, Fabio R. J., Maumus, Florian, Kustka, Adam, McCarthy, James, Allen, Andy E., Kersey, Paul, Bowler, Chris and Leila Tirichine 2018. Integrative analysis of large-scale transcriptome data draws a comprehensive landscape of *Phaeodactylum tricornutum* genome and evolutionary origin of diatoms. *Scientific Reports*, 8(1), 1-14. Doi: 10.1038/s41598-018-23106-x.

Ravanbakhsh, Siamak, Liu, Philip, Bjordahl, Trent C., Mandal, Rupasri, Grant, Jason R., Wilson, Michel, Eisner, Roman, Sinelnikov, Igor, Hu, Xiaoyu, Luchinat, Claudio, Greiner, Russell and David S. Wishart 2015. Accurate, fully-automated NMR spectral profiling for metabolomics. *PloS One*, 10(5), e0124219. Doi: 10.1371/journal.pone.0124219.

Remmers, Ilse M., D'Adamo, Sarah, Martens, Dirk E., de Vos, Ric C., Mumm, Roland, America, Antoine H., Cordewener, Jan H.G., Bakker, Linda, Peters, Sander A., Wijffels, René H. and Packo P. Lamers 2018. Orchestration of transcriptome, proteome and metabolome in the diatom *Phaeodactylum tricornutum* during nitrogen limitation. *Algal Research,* 35, 33-49. Doi: 10.1016/j.algal. 2018.08.012.

Rocha, Isabel, Maia, Paulo, Evangelista, Pedro, Vilaça, Paulo, Soares, Simão, Pinto, José P., Nielsen, Jens, Patil, Kiran R., Eugenio C., Ferreira and Miguel Rocha 2010. OptFlux: an open-source software platform for in silico metabolic engineering. *BMC Systems Biology*, 4(1), 1-12. Doi: 10.1186/1752-0509-4-45.

Russo, Monia T., Cigliano, Riccardo A., Sanseverino, Walter and Maria I. Ferrante 2018. Assessment of genomic changes in a CRISPR/Cas9 *Phaeodactylum tricornutum* mutant through whole genome resequencing. *Peer J*, 6, 2-20.Doi: 10.7717/peerj.5507.

Saade, Anastasia and Chris Bowler 2009. Molecular tools for discovering the secrets of diatoms. *Bioscience*, 59(9), 757-765. Doi: 10.1525/bio. 2009.59.9.7.

Shimon Sakaguchi 2011. Regulatory T cells: history and perspective. *Regulatory T Cells*, 707, 3-17. Doi: 10.1007/978-1-61737-979-6_1.

Schiano di Visconte, Gino, Spicer, Aandrew, Chuck, Christopher J., Allen, and Michel J. Allen 2019. The microalgae biorefinery: A perspective on the current status and future opportunities using genetic modification. *Applied Science,* 9, 4793. Doi: 10.3390/ app9224793.

Seo, Seungbeom, Jeon, Hancheol, Hwang, Seongbin, Jin, EonSeon and Kwang S. Chang 2015. Development of a new constitutive expression system for the transformation of the diatom Phaeodactylum tricornutum. *Algal Research*, 11, 50-54. Doi: 10.1016/j. algal.2015.05.012.

Seo, Seungbeom, Jeon, Hancheol, Chang, Kwang S. and EonSeon Jin 2018. Enhanced biomass production by *Phaeodactylum tricornutum* overexpressing phosphoenolpyruvate carboxylase. *Algal Research*, 31: 489-496. Doi: 10.1016/j.algal.2017.08.017.

Seo, Seungbeom, Kim, Joon, Lee, Jun-Woo, Nam, Onyou, Chang, Kwang S. and EonSeon Jin 2020. Enhanced pyruvate metabolism in plastids by overexpression of putative plastidial pyruvate transporter in *Phaeodactylum tricornutum. Biotechnology for Biofuels*, 13(1), 1-11. Doi: 10.1186/s13068-020-01760-6.

Serif, Manuel, Dubois, Gwendoline, Finoux, Anne L., Teste, Marie A., Jallet, Denis and Fayza Daboussi 2018. One-step generation of multiple gene knock-outs in the diatom *Phaeodactylum tricornutum* by DNA-free genome editing. *Nature Communications*, 9(1), 1-10. Doi: 10.1038/s41467-018-06378-9.

Sharma, A. K., Nymark, M., Sparstad, T., Bones, A. M., and Winge, P. 2018. Transgene-free genome editing in marine algae by bacterial conjugation–comparison with biolistic CRISPR/Cas9 transformation. *Scientific Reports*, 8(1), 1-11. Doi: 10.1038/s41598-018-32342-0.

Shiratori, Kenji, Ohgami, Kazuhiro, Ilieva, Iliyana, Jin, Xue H., Koyama, Yoshikazu, Miyashita, Kazuo, Yoshida, Kazuhiko, Satoru, Kase and Shigeaki Ohno 2005. Effects of fucoxanthin on lipopolysaccharide-induced inflammation in vitro and in vivo. *Experimental Eye Research*, *81*(4), 422-428. Doi: 10.1016/j.exer.2005.03.002.

Slattery, Samuel S., Wang, Helen, Giguere, Daniel J., Kocsis, Csanad, Urquhart, Bradley L., Karas, Bogumil J. and David R. Edgell 2020. Plasmid-based complementation of large deletions in *Phaeodactylum tricornutum* biosynthetic genes generated by Cas9 editing. *Scientific Reports*, *10*(1), 1-12. Doi: 10.1038/s41598-020-70769-6.

Stukenberg, Daniel, Zauner, Stefan, Dell'Aquila, Gianluca and Uwe G. Maier 2018. Optimizing CRISPR/Cas9 for the diatom *Phaeodactylum tricornutum. Frontiers in Plant Science*, 9, 740. Doi: 10.3389/fpls.2018.00740.

Thiele, Ines and Bernhard Ø Palsson 2010. A protocol for generating a high-quality genome-scale metabolic reconstruction. *Nature Protocols*, *5*(1): 93. Doi: 10.1038/nprot.2009.203.

Tredici, M. R., Bassi, N., Prussi, M., Biondi, N., Rodolfi, L., Zittelli, G. C. and Sampietro, G. 2015. Energy balance of algal biomass production in a 1-ha "Green Wall Panel" plant: how to produce algal biomass in a closed reactor achieving a high net energy ratio. *Applied Energy*, 154, 1103-1111. Doi: 10.1016/j.apenergy.2015.01.086.

Trentacoste, Emily. M., Shrestha, Roshan P., Smith, Sarah R., Glé, Corine, Hartmann, Aaron C., Hildebrand, Mark and William H. Gerwick 2013. Metabolic engineering of lipid catabolism increases microalgal lipid accumulation without compromising growth. *Proceedings of the National Academy of Sciences*, 110(49), 19748-19753. Doi: 10.1073/pnas.1309299110.

Ubando, Aristotle T., Felix, Charles B. and Wei-Hsin Chen 2020. Biorefineries in circular bioeconomy: A comprehensive review. *Bioresource Technology*, 299, 122585. Doi: 10.1016/j.biortech.2019.122585.

UniProt Consortium 2019. UniProt: a worldwide hub of protein knowledge. *Nucleic Acids Research*, 47(D1), D506-D515. Doi: 10.1093/nar/gky1049.

Ventura, Sonia P. M., Nobre, Beatriz. P., Ertekin, Figen, Hayes, Maria, Garciá-Vaquero, Marco, Vieira, Flavia, Koç, Mehmet, Gouveia, Luisa, Barros, M. R. Aires and Antonia M. F. Palavra 2017. Extraction of value-added compounds from microalgae. In Microalgae-Based Biofuels and Bioproducts, *Woodhead Publishing*, 1, 461-483. Doi: 10.1016/B978-0-08-101023-5.00019-4.

Wang, Xiang, Liu, Yu-Hong, Wei, Wei, Zhou, Xia, Yuan, Wasiqi, Balamurugan, Srinivasan, Hao, Ting-Bin, Yang, Wei-Dong, Liu, Jie-Sheng and Hong-Ye Li 2017. Enrichment of long-chain polyunsaturated fatty acids by coordinated expression of multiple metabolic nodes in the oleaginous microalga *Phaeodactylum tricornutum*. *Journal of Agricultural and Food Chemistry*, *65*(35), 7713-7720. Doi: 10.1021/acs.jafc.7b02397.

Wang, Xiang, Dong, Hong P., Wei, Wei, Balamurugan, Srinivasan, Wei-Dong, Liu, Jie-Sheng and Hong-Ye Li 2018. Dual expression of plastidial GPAT1 and LPAT1 regulates triacylglycerol production and the fatty acid profile in *Phaeodactylum tricornutum. Biotechnology for Biofuels,* 11(1), 318. Doi: 10.1186/s13068-018-1317-3.

Wang, Xiang, Liu, Si F., Li, Ruo Y., Yang, Wei D., Liu, Jie S., Lin, Carol S. K., Balamurugan, Srinivasan and Hong-Ye Li 2020. TAG pathway engineering via GPAT2 concurrently potentiates abiotic stress tolerance and oleaginicity in *Phaeodactylum tricornutum. Biotechnology for Biofuels*, 13(1), 1-14. Doi: 10.1186/s13068-020-01799-5.

Wang, Xiang, Liu, Yu-Hong, Wei, Wei, Zhou, Xia, Yuan, Wasiqi, Balamurugan, Srinivasan, Hao, Ting-Bin, Yang, Wei-Dong, Liu, Jie-Sheng and Hong-Ye Li 2017. Enrichment of long-chain polyunsaturated fatty acids by coordinated expression of multiple metabolic nodes in the oleaginous microalga *Phaeodactylum tricornutum. Journal of Agricultural and Food Chemistry*, 65(35), 7713-7720. Doi: 10.1021/acs.jafc.7b02397.

Watanabe, Yumi, Kadono, Takashi, Kira, Nozomu, Suzuki, Kengo, Iwata, Osamu, Ohnishi, Kouhei, Yamaguchi, Haruo and Masao Adachi 2018. Development of endogenous promoters that drive high-level expression of introduced genes in the model diatom *Phaeodactylum tricornutum. Marine Genomics*, 42, 41-48. Doi: 10.1016/j.margen.2018.06.003.

Westerhoff, Hans V. and Bernhard O. Palsson 2004. The evolution of molecular biology into systems biology. *Nature Biotechnology, 22*(10), 1249-1252. Doi: 10.1038/nbt1020.

Wishart, David S., Greiner, Russel, Rosborough, Tim A., Lefebvre, Brent A., Epstein, Noah A., Newton, Jack B. and Warren R. Wong 2007. U.S. Patent No. 7,181,348. Washington, DC: U.S. *Patent and Trademark Office*. Accessed December 2020 at https://patents.google.com/patent/US7191069B2/en.

Wu, Songcui, Gu, Wenhui, Huang, Aiyou, Li, Yuanxiang, Kumar, Manoj, Lim, Phaik E., Huang, Li, Gao, Shan and Guangce Wang 2019. Elevated CO_2 improves both lipid accumulation and growth rate in the glucose-6-phosphate dehydrogenase engineered *Phaeodactylum tricornutum. Microbial Cell Factories*, 18(1), 1-16. Doi: 10.1186/s12934-019-1214-x.

Xia, Song, Wang, Ke, Wan, Linglin, Li, Aifen, Hu, Qiang and Chengwu Zhang 2013. Production, characterization, and antioxidant activity of fucoxanthin from the marine diatom *Odontella aurita. Marine Drugs*, 11(7), 2667-2681. Doi: 10.3390/md11072667.

Xie, Wei H., Zhu, Cong C., Zhang, Nai S., Li, Da W., Yang, Wei D., Liu, Jie S., Sathishkumar, Ramalingam and Hong-Ye Li 2014. Construction of novel chloroplast expression vector and development of an efficient transformation system for the diatom *Phaeodactylum tricornutum. Marine Biotechnology*, 16(5), 538-546. Doi: 10.1007/s10126-014-9570-3.

Yang, Juang, Pan, Yufang, Bowler, Chris, Zhang, Lingxiang and Hanhua Hu 2016. Knockdown of phosphoenolpyruvate carboxykinase increases carbon flux to lipid synthesis in *Phaeodactylum tricornutum. Algal Research,* 15, 50-58. Doi: 10.1016/j.algal.2016.02.004.

Yang, Mingkun, Lin, Xiaohuang, Liu, Xin, Zhang, Jia and Feng Ge 2018. Genome annotation of a model diatom *Phaeodactylum tricornutum* using an integrated proteogenomic pipeline. *Molecular Plant*, 11(10), 1292-1307. Doi: 10.1016/j.molp.2018.08.005.

Yi, Zhiqian, Su, Yixi, Xu, Maonian, Bergmann, Andreas, Ingthorsson, Saevar, Rolfsson, Ottar, Kourosh, Salehi-Ashtiani, Brynjolfsson, Sigurdur and Weiqi Fu 2018. Chemical mutagenesis and fluorescence-based high-throughput screening for enhanced accumulation of carotenoids in a model marine diatom *Phaeodactylum tricornutum. Marine Drugs*, 16(8), 272. Doi: 10.3390/md16080272.

Yi, Zhiqian, Xu, Maonian, Magnusdottir, Manuela, Zhang, Yuetuan, Brynjolfsson, Sigurdur and WeiqiFu 2015. Photo-oxidative stress-driven mutagenesis and adaptive evolution on the marine diatom *Phaeodactylum tricornutum* for enhanced carotenoid accumulation. *Marine Drugs*, 13(10), 6138-6151. Doi: 10.3390/md13106138.

Yoshinaga, Ryo, Niwa-Kubota, Megumi, Matsui, Hiroaki and Yusuke Matsuda 2014. Characterization of iron-responsive promoters in the marine diatom *Phaeodactylum tricornutum. Marine Genomics*, 16, 55-62. Doi: 10.1016/j.margen.2014.01.005.

Zaslavskaia, Lioudmila A., Lippmeier, J. Casey, Kroth, Peter G., Grossman, Arthur R. and Kirk E. Apt 2000. Transformation of the diatom *Phaeodactylum tricornutum* (Bacillariophyceae) with a variety of selectable marker and reporter genes. *Journal of Phycology*, 36(2), 379-386. Doi: 10.1046/j.1529-8817.2000.99164.x.

Zhang, Chunye and Hanhua Hu 2014. High-efficiency nuclear transformation of the diatom *Phaeodactylum tricornutum* by electroporation. *Marine Genomics*, 16, 63-66. Doi: 10.1016/j.margen.2013.10.003.

Zhang, Wenyuan, Wang, Feifei, Gao, Baoyan, Huang, Luodong and Chengwu Zhang 2018. An integrated biorefinery process: Stepwise extraction of fucoxanthin, eicosapentaenoic acid and chrysolaminarin from the same *Phaeodactylum tricornutum* biomass. *Algal Research*, 32, 193-200. Doi: 10.1016/j.algal.2018.04.002.

Zhou, Lu, Gao, Shan, Wu, Songcui, Han, Danxiang, Wang, Hui, Gu, Wenhui, Hu, Qiang, Wang, Jing and Guangce Wang 2020. PGRL1 overexpression in *Phaeodactylum tricornutum* inhibits growth and reduces apparent PSII activity. *The Plant Journal*, 103(5), 1850-1857. Doi: 10.1111/tpj.14872.

Zhu, Bao H., Tu, Chang C., Shi, Hong P., Yang, Guan P. and Ke H. Pan 2017. Overexpression of endogenous delta-6 fatty acid desaturase gene enhances eicosapentaenoic acid accumulation in *Phaeodactylum tricornutum. Process Biochemistry*, 57, 43-49. Doi: 10.1016/j.procbio.2017.03.013.

Zou, Li-Gong, Balamurugan, Srinivasan, Zhou, Tian-Bao, Chen, Jia-Wen, Li, Da-Wei, Yang, Wei-Dong, Liu, Jie-Sheng and Hong-Ye 2019. Potentiation of concurrent expression of lipogenic genes by novel strong promoters in the oleaginous microalga *Phaeodactylum tricornutum. Biotechnology and Bioengineering*, 116(11), 3006-3015. Doi: 10.1002/bit.27110.

Zou, Li-Gong, Chen, Jia-Wen, Zheng, Dan-Lin, Balamurugan, Srinivasan, Li, Da-Wei, Yang, Wei-Dong, Liu, Jie-Sheng and Hong-Ye 2018. High-efficiency promoter-driven coordinated regulation of multiple metabolic nodes elevates lipid accumulation in the model microalga *Phaeodactylum tricornutum. Microbial Cell Factories*, 17(1), 54. Doi: 10.1186/s12934-018-0906-y.

In: The Future of Metabolic Engineering ISBN: 978-1-68507-362-6
Editors: A. Sharma and D. Amin

Chapter 7

DECIPHERING THE METABOLIC ADJUSTMENTS OF ENGINEERED PLANTS USING GC-MS: A TYPICAL WORKFLOW

Maneesh Lingwan and Shyam Kumar Masakapalli*
BioX Center, School of Basic Sciences,
Indian Institute of Technology Mandi, Kamand,
Himachal Pradesh, India

ABSTRACT

Deciphering the metabolic adaptations of plants provides useful insights into their phenotypes often supporting the achievement of desired traits. The perturbations induced via stress or engineering often confer metabolic reprogramming in plant systems which can be profiled with precision using high throughput analytical platforms. However, the associated challenges are attributed to the subcellular compartmentation compounded with spatial and temporal regulations of metabolism. In addition, the diversity of the phytochemicals (primary and secondary metabolites) calls for adopting a robust workflow for targeted metabolic analysis. In this chapter, we focused on examining the experimental and analytical bottlenecks to monitor plant metabolites specifically using

---

* Corresponding Author's E-mail: shyam@iitmandi.ac.in.

Gas Chromatography-Mass Spectroscopy (GC-MS). We provide a robust workflow for extraction, separation, identification and comparative metabolic analysis of plant systems. In addition, we discussed a case study using our workflow that captured the metabolic variations in *Arabidopsis thaliana* genotypes which are either susceptible or tolerant to UV-B radiation stress. We finally discussed the scope of adopting multi-analytical platforms for defining precise plant metabolic phenotypes.

Keywords: engineered plants, GC-MS, metabolomics, profiling, extraction, analytical techniques

1. Introduction

Plants being sessile, encounter environmental changes and regulate their overall growth. Plant adaptation towards external stress or genetic engineering involves sensing, signaling and the activation of several genes and metabolites, which may be responsible for plant survival (Hirayama et al., 2010). Metabolites are small molecules and a set of all in the biological system at time and space are known as the metabolome, profiling of metabolites is of interest for agronomic, synthetic biology, metabolic engineering, health, and industrial applications (Hagel et al., 2008, Fraga et al., 2020). Metabolome in plant systems varies due to the influence of physiological, genetic, environmental and other factors (Fiehn et al., 2000). Plant metabolic networks are substantially more complex than other organisms as they maintain vast chemical repertoires with specialized pathways. The subcellular compartmentation along with profound tissue-specific and temporal physiology adds to the complexity (Pieterse et al., 2009, Fiehn et al., 2001). Capturing all the metabolites needs high throughput analytical platforms owing to the large range with different biochemical characteristics of structures, molecular weights, polarity etc. Metabolite profiling and Metabolomics is aimed at identifying and analysing all low molecular weight metabolites in a tissue at a given developmental stage (Fiehn et al., 2001).

In plants, sugars, amino acids, and organic acids play a significant role in central metabolism (Jorge et al., 2016). Often the central metabolism is robust and gets reprogrammed on perturbation due to stress or genetic engineering of targeted pathways. On the other hand, secondary metabolites (e.g., phenolics, flavonoids) and phytohormones deal with the plant defence system and specialised metabolism (Oksman et al., 2004, Pieterse et al., 2009). It is well established that in response to genetic engineering, plants regulate several metabolic adaptations which can be profiled. Qualitative and quantitative analysis of metabolites can improve our understanding of plant signalling and communication during perturbation and metabolic engineering. Primarily, it can reveal the effect of gene manipulation on the desired trait by high throughput profiling of differences and similarities between the metabolic phenotypes of wild type and engineered plants. The high throughput profiling of metabolites in a targeted or untargeted manner is achieved by a range of analytical platforms. It must be noted that these analytical platforms have different sensitivities and hence one needs to select based on the metabolites of interest as well the research emphasis one wants to understand using the metabolomics approach. Therefore, it's beneficial to explore techniques and combine various analytical platforms for metabolite profiling (Arbona et al., 2009, Shuman et al., 2011, Fiehn et al., 2002).

Mass spectrometry (MS) and Nuclear Magnetic Resonance (NMR) spectroscopy are widely used analytical technologies for plant metabolomics studies. In-plant metabolomics studies MS coupled with powerful chromatographic techniques are preferred over NMR due to its high sensitivity and broad coverage of the metabolome. MS can be coupled with either gas chromatography (GC) or liquid chromatography (LC). Liquid chromatography is a technique used for the identification of low abundant compounds in complex mixtures (Segers et al., 2020, Fiehn et al., 2002, Jorge et al., 2016). GC-MS permits the separation and characterization of the thermally stable compounds and provides in-depth coverage of metabolome data (Zhou et al., 2020, Shuman et al., 2011).

GC-MS is applied largely for the analysis of volatile compounds in complex samples (Aretz et al., 2016). The major advantage of GC-MS for metabolomics is the availability of both commercially and publicly available EI (electron ionisation) spectral libraries (e.g., NIST, FiehnLib). GC-MS based approach for metabolite profiling has gained relevance with great promise for plant metabolomics (Jorge et al., 2016, Shuman et al., 2011). GC-MS is ideal for the detection and analysis of volatile molecules as well as other classes of low volatile polar metabolites after their derivatization such as methoxyamination, silylation etc (Shuman et al., 2011, Lisec et al., 2006). A typical workflow of metabolite analysis in plant systems is presented in Figure 1. This workflow can be adopted to decipher the metabolic variations among various plant systems including the genetically engineered.

In this chapter, we have shown a typical workflow one can adopt to capture plant metabolic variations using GC-MS analysis. Also, a case study showcasing the variations in the levels of metabolites in genetically engineered *Arabidopsis thaliana (35S:BBX31)* vs the wild type (Col-0) under UV-B radiations using GC-MS is presented (Yadav et al., 2019a). *BBX31* is a B-box protein that is well known to regulate photomorphogenic responses in plants (Yadav et al., 2020) and hence we aimed to capture the effect of BBX31 overexpression on metabolic levels. The variations in the levels of selected metabolites in *BBX31* that may improve tolerance to UV-B as reported by Yadav et al., 2019 are further discussed.

2. Typical Workflow for Metabolic Analysis in Plant Systems Using GC-MS

The routine instruments, equipment, reagents required for metabolite profiling are listed in Table 1. The workflow is presented in the following sections.

Table 1. The routine instruments, equipment, reagents required for GC-MS based metabolite profiling

Instruments	• High-end GC-MS
Equipment	• Eppendorf tubes, 2.0 ml and 1.5 ml • Liquid nitrogen cane • Freezer (− 80 °C) for sample storage • Freeze-dryer for sample drying • Thermomixer C • Microcentrifuge • Speed Vac/vacuum concentrator • Ultrasonicator • Table-top vortex
Reagent for Extraction solvent	• GC-grade methanol, chloroform and water or other suitable solvent
Derivatisation reagent	• Methoxyamine hydrochloride • Pyridine • N-methyl-N-trimethylsilyltriflouroacetamide (MSTFA) • Ribitol (Internal standard)
GC-MS analytical parameters	• GC-ALS-MS 5977B (Agilent Technologies) • HP-5ms column (30 m x 250 µm x 1 µm) • Total run time : 50 min • Carrier gas: Helium • Mode : Split/Splitless • Flow rate : 0.6 ml min^{-1}
Data analysis	• Mass hunter , Metalign , library : NIST17 and Fiehn13
Statistical analysis	• Metaboanalyst or comparable software for multivariate analysis

2.1. Metabolite Extraction and Sample Preparation of Plant Systems

In GC-MS based metabolic profiling, sample preparation is the first crucial step that needs careful optimisation. The plant tissues (seedlings, leaf, shoot, root etc.) need to be quenched first using liquid nitrogen which will arrest their metabolic activity. The tissues need to be powdered fine, freeze-dried and stored in the freezer for long runs. For metabolite extractions one can take about ~20-30 mg lyophilised tissue for extraction using appropriate solvents (e.g., water, acetone, methanol, chloroform, ethyl ether or a mix of methanol, chloroform, and water in the ratio 3:1:1 v/v/v) and incubate (e.g., at 70ºC for 5-10 minutes) (Lisec et al., 2006). After that, centrifuge the extracts (at 10,000 g at room temperature for 15 min) to separate metabolites and the cell pellets. About 35-50 µl of sample aliquots from the supernatant can be taken and dried under vacuum and derivatised. In general, methoxamine trimethylsilane (MeOx-TMS) derivatization is performed to capture a broad range of metabolites in plants such as sugars, organic acids, amino acids etc. We present a typical workflow adopted for MeOx-TMS derivatisation and GC-MS analysis of *Arabidopsis thaliana* tissue extracts as outlined in Figure 2 and detailed elsewhere (Shree et al., 2019, Lingwan et al., 2020, Saini et al., 2020).

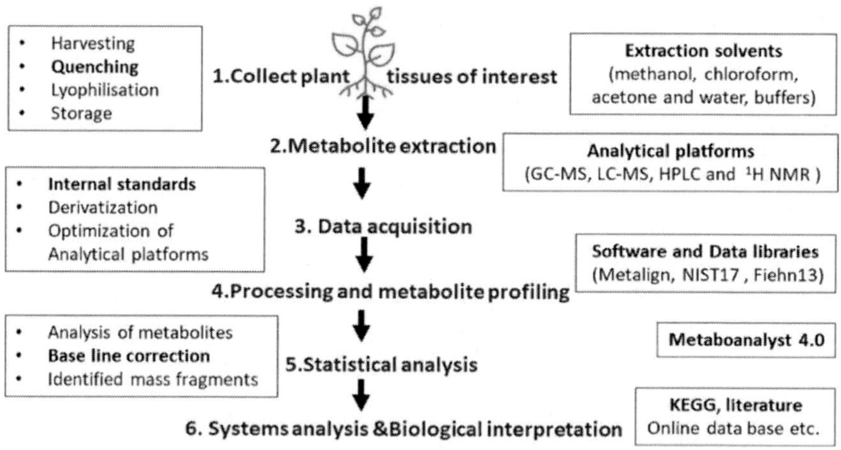

Figure 1. Typical workflow of metabolite profiling and metabolomics in plants. The steps involve 1. Choosing plant tissues of interest and quenching the samples using liquid nitrogen 2. Metabolite extraction using suitable solvents 3. Data acquisition using different analytical platforms 4. Data processing for metabolite profiling by suitable software tools 5. Multivariate statistical analysis for comprehensive metabolomics using Metaboanalyst (Xia et al., 2016) and 6. Biological interpretation.

2.2. GC-MS Data Acquisition

Acquisition of data in GC-MS should be aimed at appropriate separation and resolution of peaks. These are influenced by GC-MS columns and other parameters (temperature, carrier gas flow etc). GC columns such as DB-5MS, DB-2MS, DB-Wax and others are commercially available for different classes of metabolites (nonpolar, polar, lipophilic etc). In GC-MS, an automatic liquid sampler (ALS) holds vials and ensures injection (1-2 μl) of the sample from the front inlet to a fitted column. Different derivatised metabolites get volatilised and separated by temperature gradient and inert gas flow. We routinely use this temperature gradient for metabolite profiling in plants, oven temperature gradient set for 1 minute hold at 50°C, ramp up the temperature by 10°C min^{-1} to reach 200°C with a 4 min hold at 200°C, again ramp up by 10°C min^{-1} till 300°C and hold for 10 min. Helium is used as a carrier gas in a column with a constant flow rate of 0.6 to 1 ml/min. The output of data acquisition is represented as total ion chromatograms (TIC) with multiple peaks belonging to different metabolites (Masakapalli et al., 2014, Lisec et al., 2006).

2.3. GC-MS Data Pre-Treatment and Data Analysis

Data pre-treatment involves baseline correction of TIC spectra and it makes the spectra more comparable for data analysis. Although various tools are available, baseline correction using Metalign software is widely adopted (Lommen et al., 2012). The MassHunter or chemstation software (or similar that comes with GC-MS platform) is used for metabolite identification and profiling from each spectrum. The m/z fragments of each metabolite can be searched and assigned using suitable tools against commercially available spectral libraries e.g., National Institute of Standards and Technology library (NIST), Fiehn library (Kopka et al., 2005) or standards. The metabolic identities and abundances from each spectrum can be tabulated, which contains information for m/z fragments of metabolite, their retention time and probability of match against the available library. Further validation and precise quantification can be achieved from standard curves of commercially available metabolites. The tabulated data can be used for statistical analysis. The use of high-throughput multivariate statistical analysis tools and related databases (e.g., Metaboanalyst 5.0, Kyoto encyclopaedia of genes and genomes and PubChem) can be used for biological interpretation (Xia et al., 2016). The steps in GC-MS based metabolomics are depicted in Figure 2.

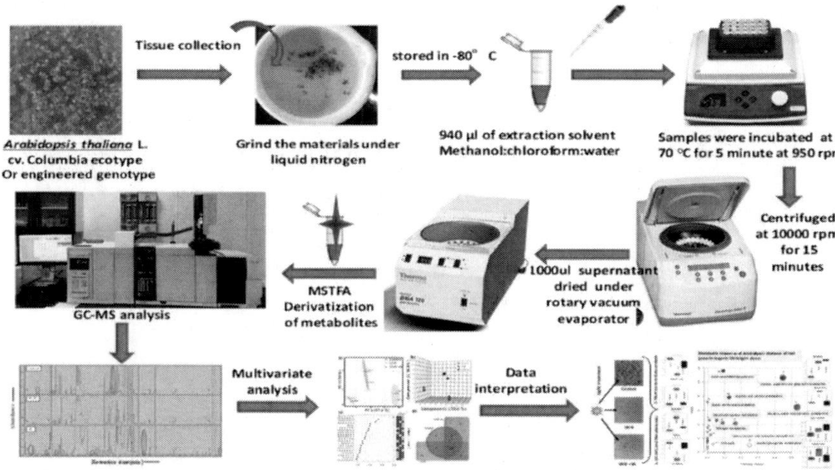

Figure 2. A typical workflow for sample preparation and analysis of plant extracts using GC-MS. Steps include grinding the tissue using liquid nitrogen, extraction with suitable solvents followed by derivatization. Data acquisition and processing of the Total ion chromatogram/spectra provides metabolite peaks. The fragments of each metabolite peak can be matched with the NIST library for identification. The peak abundances from multiple samples can be subjected to Metabolomics using Metaboanalyst 5.0 which can further support biological interpretation.

3. Case Study: GC-MS Analysis of *Arabidopsis thaliana* Wildtype (Col-0) and Engineered (*BBX31*) Genotype Under UV-B

BBX31 seedlings have more tolerance than wild-type under UV-B radiations (Yadav et al., 2019a, Yadav et al., 2019b). To investigate the basis of survival, comparative metabolic adaptation was captured in 13-days old seedlings of Col-0 and *35S: BBX31* treated with 24 hr of UV-B radiation and subjected to GC-MS analysis. For central metabolite profiling, 20 mg lyophilized samples were used for metabolite extraction. To each sample, 940 μl of the solvent mix (prepared in the ratio of 3:1:1 v/v/v of methanol, chloroform and water respectively) and 60 μl of an internal standard ribitol (0.2 mg ml^{-1}) was added. Further steps followed are similar as detailed in section 2.1. For profiling of phenols using GC-MS, the protocol from de Ascensao and Dubery 2003 was followed. The data acquisition, temperature gradient programme and metabolite analysis was followed as discussed in section 2.2 and 2.3.

GC-MS analysis of extracts from Arabidopsis Col-0 and *35S: BBX31* reported significant variations in the levels of phenols and other metabolites as shown in Yadav et al., 2019 (Figure 3A). UV-B triggered a higher accumulation of gallic acid, coumaric acid, vanillic acid and hydroxybenzoic acid belonging to the class of phenols in *35S:BBX31*. In the UV-B tolerant Arabidopsis BBX31, the abundance of sucrose and myoinositol also increased significantly (Figure 3B). Myoinositol is ubiquitous in plant systems and its enhanced levels can act as an early stress signal to plants (Casati et al., 2011). The enhanced levels of myoinositol, sucrose and phenolics indicate their protection role under UV-B stress. Further, Principal component analysis using Metaboanalyst revealed metabolic variations among Col-0 and *35S:BBX31* under UV-B radiation exposure (Figure 3B). Sucrose and other free sugars accumulation seem to play significant roles under various stresses including biotic (Benina et al., 2013). The enhanced metabolic levels of phenols in *35S:BBX31* may have a crucial role as UV-B absorbing molecules (Yadav et al., 2020). Enhanced levels of the key intermediate primary and secondary metabolites in *35S:BBX31* may confer additional UV-B protection (Yadav et al., 2019). In summary, GC-MS analysis has shown metabolic readjustment of primary and secondary metabolites in *35S:BBX31* seedlings under UV-B stress. For more details, one may refer to Yadav et al., 2019. The GC-MS based metabolite analysis workflow can be adapted to capture and compare metabolic variations conferred in plant systems due to genetic engineering.

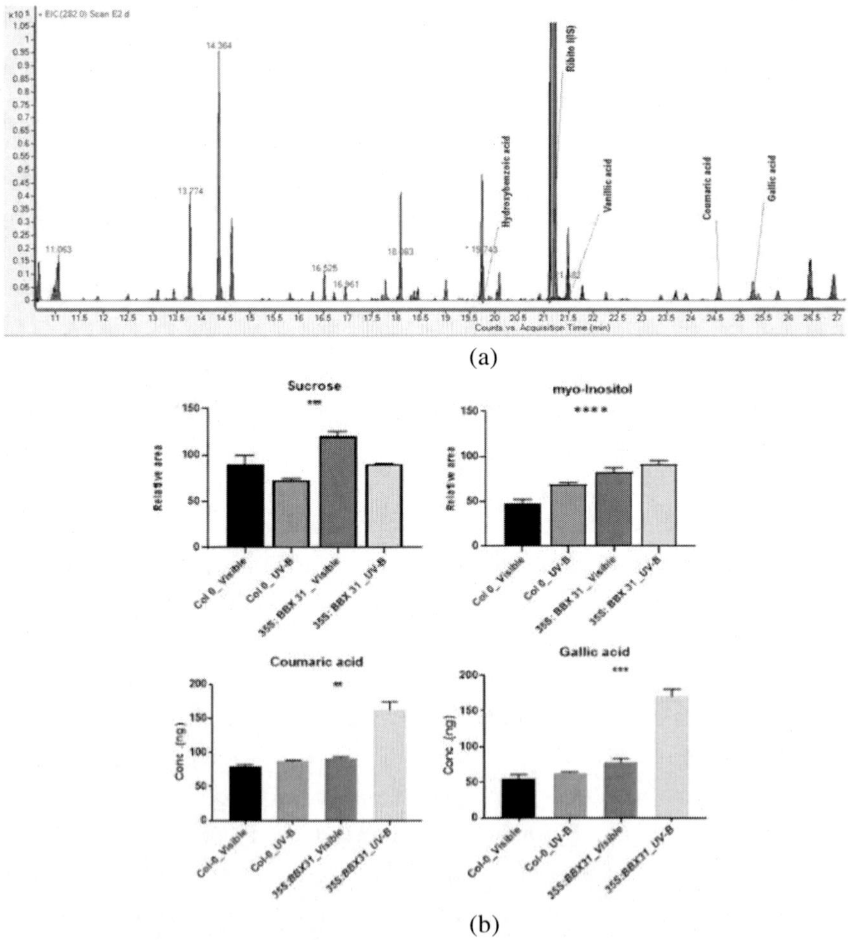

Figure 3. (Continued)

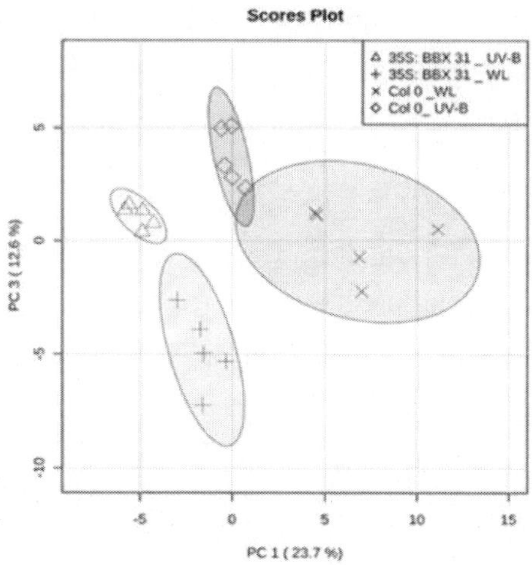

(c)

Figure 3. *35S: BBX31* reported significant variations in the metabolic levels of phenols and other metabolites under UV-B stress. A) Representation of GC-MS spectra showing phenols in *35S:BBX31* under UV-B. B) GC-MS captured quantitative metabolic readjustments in Col-0 and *35S:BBX31* under visible and UV-B exposure. Representative graphs of Sucrose, Myoinositol, Coumaric acid and gallic acid are shown. C) Principal component analysis using Metaboanalyst showed independent clusters which reveal variations among Arabidopsis genotypes Col-0 and BBX31 under visible (WL) and UV-B exposure. Error bars represent the standard deviation of 4 independent replicates. An unpaired t-test represent statistically significant differences of ***P < 0.001, **P < 0.01, *P < 0.05 in graphs.

4. ANALYTICAL PLATFORMS TO STUDY PLANT METABOLIC VARIATIONS

The total no. of distinct compounds in the plant kingdom is predicted to be approximately 200,000 which differs greatly in their chemical composition, structure and molecular weights. Profiling all the metabolites using a single analytical platform is a challenging task and is often recommended to adopt multi-analytical platforms. Although GC-MS has the advantages of having large spectral libraries, it doesn't capture thermally unstable as well as higher molecular weight metabolites. Further, a different class of metabolites may need different derivatisation which also are reported to introduce artefacts. For these reasons, GC-MS may provide limited coverage of metabolome (Aretz et al., 2016) thereby encouraging parallel use of other analytical techniques. Liquid chromatography-mass spectrometry (LC-MS) and UPLC-MS (Ultra Performance Liquid chromatography-mass spectrometry) are sensitive techniques with wide applications for the analysis of thermally unstable molecules in complex samples (Segers et al., 2020, Fiehn et al., 2002, Jorge et al., 2016). CE-MS (Capillary Electrophoresis) can capture and is limited for polar ionic metabolites. Other techniques are also gaining much relevance in metabolite analysis and can be adopted for plant systems. FT-ICR (Fourier transform ion cyclotron resonance) provides large detection of metabolites and is suitable for complex mixtures but confers high experimental costs. Near-Infrared Spectroscopy (NIR) and Raman Spectroscopy are highly specific for metabolic class. GALDI (Graphite Assisted Laser Desorption/Ionisation) and MALDI (Matrix-Assisted Laser Desorption Ionization) can also be used for detecting molecules and can be coupled with IMS (Imaging Mass Spectrometry) for detection of the spatial arrangement of metabolites in plants (Zhang et al., 2012, Hagel et al., 2008, Fraga et al., 2020, Dunn et al., 2005).

Focusing on GC-MS as an analytical platform for studying plant metabolism some examples can further be highlighted. GC-MS based analysis facilitated studying the a) effect of deoxynivalenol on the central carbon metabolism of wheat (Warth et al., 2015), b) nutraceutical potential of Cannabis seeds (Rashid et al., 2021), c) geographical location specific metabolites in tobacco leaves (Zhao et al., 2015) d) phytochemical composition of *Potentilla anserina* (Mari et al., 2013) and e) untargeted metabolomics to see the impact of azadirachtin in *Bactrocera dorsalis* (Zhou et al., 2020). In summary, metabolite profiling and metabolomics using GC-MS can comprehensively capture metabolic variations in engineered plant systems. With the recent advances in the sensitivity of analytical platforms and data analysis, GC-MS will continue playing a significant role in defining plant metabolic variations to perturbations as shown in the case study.

CONCLUSION

This chapter highlighted a typical GC-MS workflow that can be adopted for undertaking the metabolite profiling and data acquisition of typical plant systems. Also, the case study provided an illustrative example of how GC-MS based metabolite analysis could capture metabolic variations in response to perturbations - abiotic, biotic or genetic engineering. The metabolic analysis will continue to provide insights into plant systems that have significance to our Agriculture.

ACKNOWLEDGMENTS

Shyam Kumar Masakapalli acknowledges the Science and Engineering Research Board (SERB) Early career research funding (File No: ECR/2016/001176). Maneesh Lingwan acknowledges IIT Mandi and MHRD for financial support to carry relevant work. We thank Dr Sourav Datta from IISER Bhopal for access to plant material and support.

CONFLICT OF INTEREST

The authors declare no competing financial interests.

REFERENCES

[1] Hirayama, T., Shinozaki K. (2010) Research on plant abiotic stress response in the post-genome era: past, present and future. *Plant J.* 61, 1041-1052.

[2] Hagel, J. M., & Facchini, P. J. (2008). Plant metabolomics: analytical platforms and integration with functional genomics. *Phytochemistry reviews*, 7(3), 479-497.

[3] Fraga-Corral, M., Carpena, M., Garcia-Oliveira, P., Pereira, A. G., Prieto, M. A., & Simal-Gandara, J. (2020). Analytical metabolomics and applications in health, environmental and food science. *Critical Reviews in Analytical Chemistry,* 1-23.

[4] Fiehn, O., Kopka, J., Dörmann, P., Altmann, T., Trethewey, R. N., Willmitzer, L., Metabolite profiling for plant functional genomics. *Nature biotechnology,* 18(11), 115, 2000.

[5] Pieterse, C. M. J., Leon-Reyes A., Van der Ent S., Van Wees S. C. M. (2009) Networking by small-molecule hormones in plant immunity. *Nat. Chem. Biol.* 5, 308-316.

[6] Fiehn, O. (2001) Combining genomics, metabolome analysis, and biochemical modeling to understand metabolic networks. *Comp. Funct. Genomics,* 2, 155-168.

[7] Jorge, T. F., Mata, A. T., & António, C. (2016). Mass spectrometry as a quantitative tool in plant metabolomics. *Philosophical Transactions of the Royal Society A: Mathematical, Physical and Engineering Sciences,* 374(2079), 20150370.

[8] Oksman-Caldentey, K. M. and Inze, D. (2004) Plant cell factories in the postgenomic era: new ways to produce designer secondary metabolites. *Trends Plant Sci.*, 9, 433-440.

[9] Jorge, T. F., Rodrigues, J. A., Caldana, C., Schmidt, R., van Dongen, J. T., Thomas-Oates J., António C. (2016) Mass spectrometry-based plant metabolomics: metabolite responses to abiotic stress. *Mass Spectrom. Rev.* 35, 620-649.

[10] Fiehn, O. (2002) Metabolomics – the link between genotypes and phenotypes. *Plant Mol. Biol.,* 48, 155-171.

[11] Shuman, J. L., Cortes D. F., Armenta J. M., Pokrzywa R. M., Mendes P., Shulaev V. (2011) Plant metabolomics by GC-MS and differential analysis. *Methods Mol Biol.;* 678:229-46.

[12] Kim, H. K., Choi Y. H., Verpoorte R. (2011) NMR-based plant metabolomics: where do we stand, where do we go? *Trends Biotechnol.* 29, 267-275.

[13] Segers, K., Declerck, S., Mangelings, D., Heyden, Y. V., & Eeckhaut, A. V. (2019). Analytical techniques for metabolomic studies: a review. *Bioanalysis*, 11(24), 2297-2318.

[14] Dunn, W. B., & Ellis, D. I. (2005). Metabolomics: current analytical platforms and methodologies. *TrAC Trends in Analytical Chemistry,* 24(4), 285-294.

[15] Zhang, A., Sun, H., Wang, P., Han, Y., & Wang, X. (2012). Modern analytical techniques in metabolomics analysis. *Analyst,* 137(2), 293-300.

[16] Warth, B., Parich, A., Bueschl, C., Schoefbeck, D., Neumann, N. K. N., Kluger, B., ... & Schuhmacher, R. (2015). GC-MS based targeted metabolic profiling identifies changes in the wheat metabolome following deoxynivalenol treatment. *Metabolomics,* 11(3), 722-738.

[17] Rashid, A., Ali, V., Khajuria, M., Faiz, S., Gairola, S., & Vyas, D. (2021). GC-MS based metabolomic approach to understand nutraceutical potential of Cannabis seeds from two different environments. *Food chemistry,* 339, 128076.

[18] Zhao, Y., Zhao, J., Zhao, C., Zhou, H., Li, Y., Zhang, J., ... & Xu, G. (2015). A metabolomics study delineating geographical location-associated primary metabolic changes in the leaves of growing tobacco plants by GC-MS and CE-MS. *Scientific reports,* 5(1), 1-11.

[19] Mari, A., Lyon, D., Fragner, L., Montoro, P., Piacente, S., Wienkoop, S., ... & Weckwerth, W. (2013). Phytochemical composition of *Potentilla anserina* L. analyzed by an integrative GC-MS and LC-MS metabolomics platform. *Metabolomics,* 9(3), 599-607.

[20] Zhou, Y., Zhang, P. W., Chen, X. T., Liu, B. J., Cheng, D. M., & Zhang, Z. X. (2020). Integrated LC-MS and GC-MS-based untargeted metabolomics studies of the effect of azadirachtin on *Bactrocera dorsalis* larvae. *Scientific reports,* 10(1), 1-11.

[21] Lisec, J., Schauer, N., Kopka, J., Willmitzer, L., Fernie, A. R. Gas chromatography mass spectrometry-based metabolite profiling in plants. *Nature protocols*, 1(1), 387, 2006.

[22] Shree, M., Lingwan, M., & Masakapalli, S. K. (2019). Metabolite profiling and metabolomics of plant systems using 1H NMR and GC-MS. *OMICS-Based Approaches in Plant Biotechnology,* 129, 129.

[23] Lingwan, M., Shagun, S., Pant, Y., Kumari, B., Nanda, R., & Masakapalli, S. K. (2020). *Antiviral phytochemicals identified in Rhododendron arboreum petals exhibited strong binding to SARS-CoV-2 MPro and Human ACE2 receptor* (doi: https://10.20944/preprints202008.0530.v1).

[24] Saini, P., Bhatia, S., Mahajan, M., Kaushik, A., Sahu, S. K., Kumar, A., ... & Yadav, R. K. (2020). Elongated Hypocotyl5 negatively regulates Decrease Wax Biosynthesis to increase survival during UV-B stress. *Plant Physiology,* 184(4), 2091-2106.

[25] Masakapalli, S. K., Kruger, N. J., Ratcliffe, R. G. The metabolic flux phenotype of heterotrophic Arabidopsis cells reveals a complex response to changes in nitrogen supply. *The Plant Journal,* 74(4), 569-582, 2013.

[26] Lommen, A. MetAlign: interface-driven, versatile metabolomics tool for hyphenated full-scan mass spectrometry data preprocessing (2012). *Anal. Biochem*, 81(8), 3079-86, 2009.

[27] Kopka, J., Schauer, N., Krueger, S., Birkemeyer, C., Usadel, B., Bergmüller, E., Willmitzer, L. GMD@ CSB. DB: The Golm metabolome database. *Bioinformatics*, 21(8), 1635-1638, 2004.

[28] Xia, J., and Wishart, D. S. Using MetaboAnalyst 3.0 for comprehensive metabolomics data analysis. *Current protocols in bioinformatics*, 14-10, 2016.

[29] Yadav, A., Singh, D., Lingwan, M., Yadukrishnan, P., Masakapalli, S. K., & Datta, S. (2020). Light signaling and UV-B-mediated plant growth regulation. *Journal of integrative plant biology,* 62(9), 1270-1292.

[30] Yadav, A., Bakshi, S., Yadukrishnan, P., Lingwan, Dolde, U., Wenkel, S., Masakapalli, S. K. and Datta, S. (2019). The B-Box-containing microprotein miP1a/BBX31 regulates photomorphogenesis and UV-B protection. *Plant physiology,* 179(4), 1876-1892.

[31] Yadav, A., Lingwan, M., Yadukrishnan, P., Masakapalli, S. K., & Datta, S. (2019). BBX31 promotes hypocotyl growth, primary root elongation and UV-B tolerance in Arabidopsis. *Plant signaling & behavior,* 14(5), e1588672.

[32] Ascensao, A. R., Dubery I. A. (2003). Soluble and wall-bound phenolics and phenolic polymers in *Musa acuminata* roots exposed to elicitors from *Fusarium oxysporum* f.sp. *cubense. Phytochemistry*. Jul; 63(6):679-86.

[33] Aretz, I., & Meierhofer, D. (2016). Advantages and pitfalls of mass spectrometry-based metabolome profiling in systems biology. *International journal of molecular sciences*, 17(5), 632.

[34] Casati, P., Morrow, D. J., Fernandes, J., & Walbot, V. (2011). Rapid maize leaf and immature ear responses to UV-B radiation. *Frontiers in plant science*, 2, 33.

[35] Benina, M., Obata, T., Mehterov, N., Ivanov, I., Petrov, V., Toneva, V., & Gechev, T. S. (2013). Comparative metabolic profiling of *Haberlea rhodopensis*, *Thellungiella halophyla*, and *Arabidopsis thaliana* exposed to low temperature. *Frontiers in Plant Science*, 4, 499.

In: The Future of Metabolic Engineering ISBN: 978-1-68507-362-6
Editors: A. Sharma and D. Amin

Chapter 8

PLANTS AS A BIOREACTOR FOR EDIBLE VACCINES: EMERGING EXCELLENCE

***Ashraf Y. Z. Khalifa*[1,2]*, T. Savitha*[3]**
***and A. Sankaranarayanan*[4,5,*]**
[1]Department of Biological Sciences, College of Sciences,
King Faisal University, Al-Ahsa, Saudi Arabia
[2]Botany and Microbiology Department, Faculty of Sciences,
Beni-Suef University, Beni-Suef, Egypt
[3]Department of Microbiology, Tiruppur Kumaran College
for Women, Tiruppur, Tamil Nadu State, India
[4]C.G. Bhakta Institute of Biotechnology,
Uka Tarsadia University, Gopal Vidhya Nagar,
Gujarat State, India
[5]Department of Life Sciences, Sri Sathya Sai University
for Human Excellence, Kamalapur, Karnataka State, India

[*] Corresponding Author's E-mail: drsankarkamal@gmail.com.

ABSTRACT

Edible vaccines are subunit vaccinations in which the transgenic plant is induced to produce the encoded proteins once the selected genes are introduced in it. Plants have been identified as a valuable resource for developing promising expression systems for commercial vaccination antigen manufacturing. A number of plant-derived vaccine antigens have been shown to be safe and to elicit a sufficient immune response. As a result, transgenic plants, including edible plant components, are advocated as great options for vaccine production and inexpensive scale-up via culture. Designing methodologies for significant expression of vaccination antigens in plants has been motivated by improved understanding of plant molecular biology, and as a result, refinement of genetic engineering techniques. Potato, banana, lettuce, corn, soybean, rice, and legumes are examples of edible vaccination foods. They are simple to administer, preserve, and give to patients of various ages in a timely and cost-effective manner. Edible vaccinations have great potential for significantly decreasing diseases such as measles, hepatitis B, and cholera, particularly in lower-income nations. This chapter envisions to cover the important characteristics of plant as bio-factories for edible vaccines, including their manufacturing methods, plant species involved, applications, and future prospects.

Keywords: edible vaccine, transgenic plant, vaccine antigens, immunization, protein antigens

1. INTRODUCTION

Vaccines are biological measures that improve immunity to specific diseases and represent a substantial advancement in nineteenth-century. They have a protein moiety that resembles that of infective germs and is frequently found in fragile or robust outline of microorganisms. Vaccines act as a catalyst, causing the body's immune system to become familiar with the antigen. Currently, a new type of vaccination is being developed that will tout superiority to mask the risks associated with conservative immunizations. This type of immunization was developed using genetically modified plants. The gene encoding bacterial or viral agents can be recombined in plants without losing its immunogenic function, which is useful in the development of edible vaccines. The main mechanism of action of edible vaccinations is to activate the basic and mucosal insusceptibility reactions to foreign microorganisms. Consolidating transgene into the desired plant cell can result in edible vaccinations, which are now being developed for both veterinary and human usage (Kurup and Thomas 2020, 79).

Plants have been known to have a diverse antibody production framework capable of conveying a wide range of different defense antigens that can be isolated and delivered by infusion, with proof of oral delivery capability. Edible vaccines are made from transgenic plants and animals, or contain specialized molecules that stimulate an animal's defense response. To put it simply, edible vaccinations are medications made from plants or animals (Penney et al., 2011, 796). Hiatt et al. came up with the notion of developing a plant-based vaccination in 1989. Dr. Arntzen proposed the use of transgenic plants to manufacture and deliver subunit vaccinations in 1990. Arntzen revealed that the edible vaccines can overcome the limits of traditional vaccine development. The success is in the development of edible vaccine using surface antigen (*Streptococcus mutans*) (Mason et al. 1992; 11747) from hepatitis B in the tobacco plant (Saxena and Rawat 2013, 208). In the current circumstances, edible vaccines offer enticing potential for reducing the burden of diseases such as hepatitis B and diarrhea, where storing and directing antibody is frequently a problem. It is essential to choose plants, algae, yeast, insects, and lactic acid bacteria whose products are consumed as raw when making edible vaccinations or antibodies. Several efficient plant-determined immunization antigens were found to be expressed in various plant tissues in the last decade. According to a research, expression of antigen in seeds of a consumable vegetable such as nut or soybean could make it possible to obtain antigen sufficient for inoculating a large number of individuals from a single section of land crop.

In a few years, a slew of protein products made in "naturalized bioreactors" may be available on the market. Efforts to make these improvements should primarily focus on the approval and immateriality of plant-based normalized mucosal and edible vaccines, regulatory pharmacology, formulations, and the establishment of cost-effective good laboratory practice conventions.

2. Plant as a Bioreactor for Pharmaceutical Production

Plants are acquiring widespread acceptance as a general platform for the large-scale production of recombinant proteins because of their specific characteristics (Figure 1). The initial plant-derived recombinant pharmaceutical proteins are nearing the end of their clinical trials and a number of others are in the pipeline. There have been significant technological improvements in this booming area of applied biotechnology (Figure 2) during the last two years, as seen by the continued commercial development of innovative plant-based expression platforms. There has also been a significant progress in overcoming some of the limitations of plant bioreactors, such as poor yields and uneven product quality, which have hampered plant-derived pharmaceutical approval.

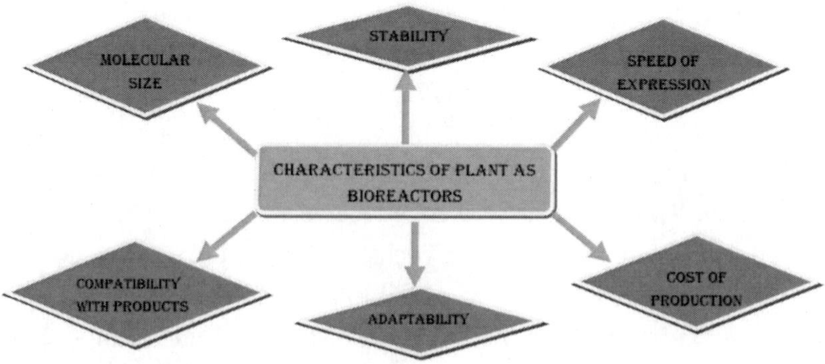

Figure 1. Depicts the characteristic features of plant as bioreactors.

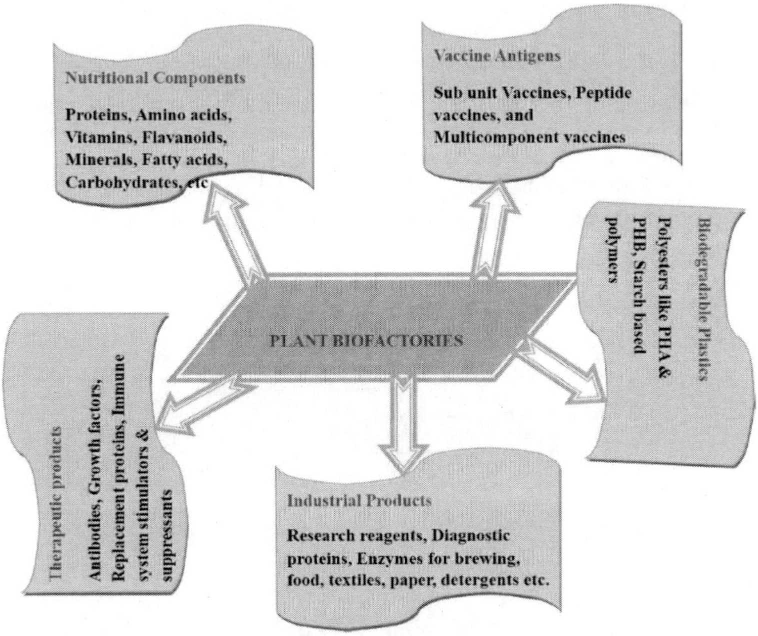

Figure 2. Emerging biotechnological arena as plant as bio factories.

2.1. Therapeutic Products

Biopharmaceuticals deals were expected to generate US$ 278.2 billion in revenue. Blood proteins, vaccine antigens, monoclonal antibodies, cytokines, therapeutic enzymes, growth factors, some fully functional mammalian proteins that transgenic plants can produce (Xu et al., 2011, 281; Davies 2010, 845; De Muynck et al., 2010, 530; Yusibov et al., 2016, 669). Plants can produce cost-effective, safer therapeutic proteins (Figure 3) than mammalian cells, and this would be the method of choice, for example, for the development of secretory antibodies (Paul and Ma 2011, 62). These recombinant therapeutic proteins are made from a variety of plant-based sources, including cells and field crops. The most well-known genus is tobacco. Plant-specific post translational modifications may pave the way for the development of new recombinant proteins with improved function and performance, as well as biosimilars

(Faye and Gomord, 2010, 528). Therapeutic targets are exclusively designed to supplement or produce new therapeutic proteins that amalgamate novel therapeutic targets in second-generation plant-made pharmaceutical proteins toward the progression of protein assembly, delivery, trafficking, stability, serum longevity, or protein solubility in either the production host or the target organism (Phoolcharoen et al., 2011, 695; Xu et al., 2007, 998; Xu et al., 2010, 850).

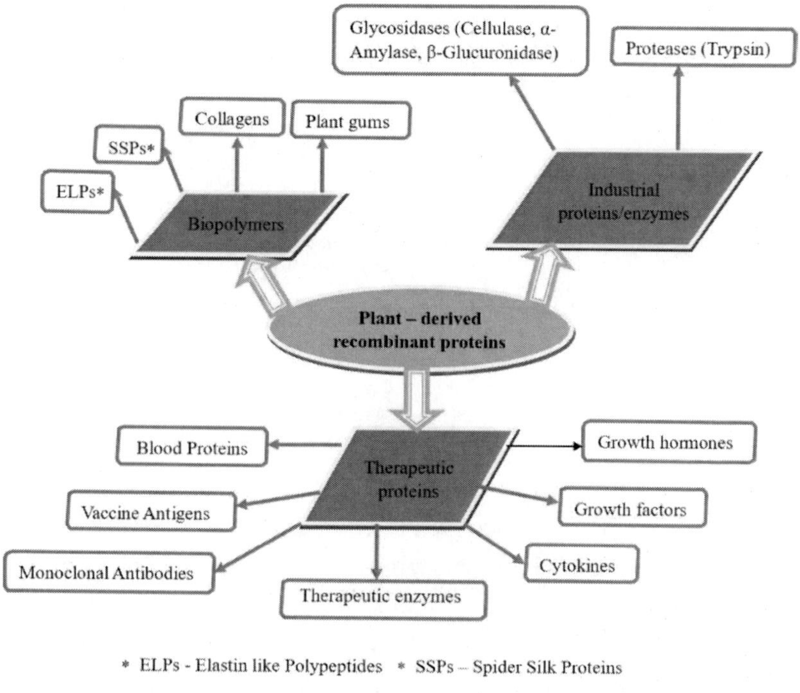

Figure 3. Plant – derived recombinant proteins.

2.2. Nutritional Components

Malnutrition is a widespread and deeply rooted socioeconomic problem that stems from a combination of factors including poverty, lack of access to food, inefficient food distribution, and an overreliance on subsistence mono-agriculture. Malnutrition is widespread in underdeveloped countries due to their reliance on staple foods that are low in several essential elements. Plant biotechnology has the potential to play a key role in the fight against starvation by creating nutritionally superior crops. Malnutrition can be addressed using three main techniques (Perez-Massot et al., 2013, 29-30), which are as follows:

- Increase the diversity of food intake
- To add nutrients to the diet artificially as supplements
- Bio fortification in which crops are modified or treated to accumulate additional nutrients

Dietary consumption of anthocyanins, a type of pigment produced by higher plants, has been linked to resistance to a variety of human diseases. However, the amount of anthocyanin in the majority of fruits and vegetables consumed may be insufficient to provide maximal benefit. Tomatoes are a promising option for transgenic flavonoid content increase. Flavonoids are a good source of hydrophilic dietary antioxidants, but lycopene, a lipophilic antioxidant, is the most abundant antioxidant in the tomato fruit (Butelli et al., 2008, 1301). In general, diets high in both soluble and membrane-associated antioxidants are considered to provide the best disease protection.

Phenylpropanoid amides (PAs) play an important role in plant defense as antimicrobial chemicals targeted at pathogens. They also represent a large family of nutraceuticals with powerful antioxidant and chemotherapeutic properties. Four genes involved in Pas synthesis have been cloned from various plant species including potatoes, tobacco, tomatoes, and peppers (Kang et al., 2009, 504). Other nutritional components including beta-carotene, fatty acids, vitamins, minerals, and carbohydrates can also be found in plants.

2.3. Vaccine Antigens

Surface protein A from *S. mutans* was embedded in tobacco as the first edible vaccination in 1990. Plant-derived vaccines in the form of seeds or fruits can be easily kept and moved from one location to another without being harmed or degraded. Cultivation in fields with comparably low inputs can successfully yield a large number of plant-derived vaccines. Vaccine antigens have been reported to be articulated in potato, tomato and carrot. Antigen genes that code for HBsAg, HIV *gag*, and rabies capsid proteins have been successfully expressed into tomato. *Escherichia coli* heat-labile enterotoxin B subunit in tobacco and potato, rabies virus G protein in tomato, hepatitis B surface antigen in tobacco and potato, Norwalk virus capsid protein in tobacco and potato, and cholera toxin B (CTB) subunit in potato are well studied. Antigens found in plants or plant products can be taken orally or injected intramuscularly or intravenously (Kumar et al., 2013, 6151).

3. Production of Edible Vaccine

The production of plant-based edible vaccine followed a series of specific concept. However, isolation and characterization of a specific antigen responsible for the disease is of primary interest, which shows a specific immune response (Plotkin, 2013, 32). If induced the immune response, the gene of interest cloned into vector and transformed into the plant of interest (Concha et al., 2017, 5). A basic schematic flow chart regarding the production of edible vaccine is shown in Figure 4. Both the direct and indirect gene delivery methods in the production of plant-based edible vaccines have their own advantages and disadvantages. In the direct method, the integration of a transgene into plant cell is done without the aid of any vector. In the indirect method, a gene can be combined with a vector. However, the expression of the gene in the plants is induced by stable transformation and transient transformation methods (Madhumita et al., 2014, 59; Clarke et al., 2017, 1615).

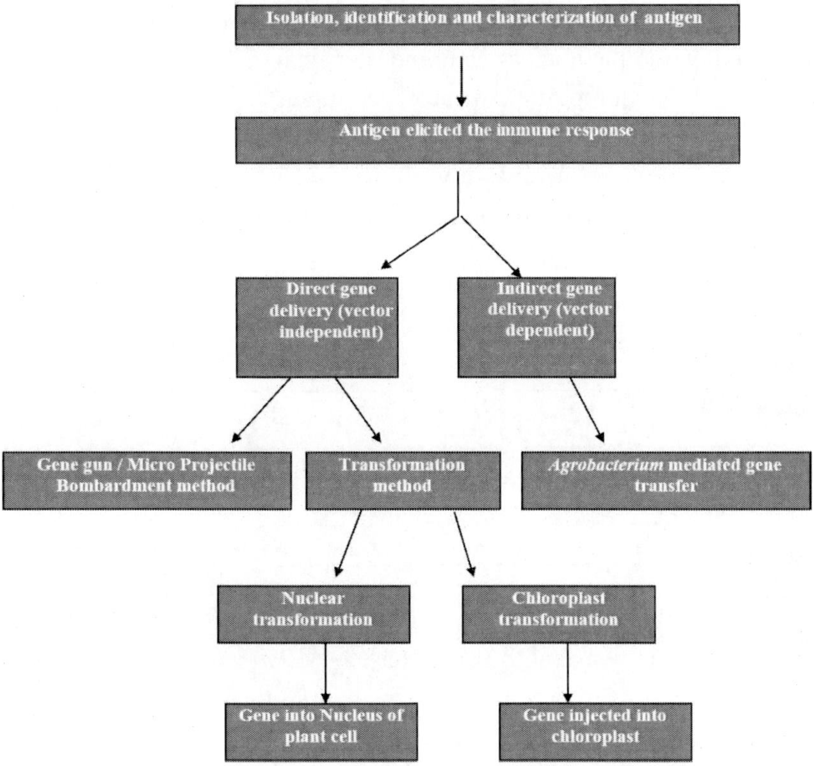

Figure 4. General concepts in the production of plant based edible vaccine.

3.1. Direct Gene Transfer

Direct gene transfer method is a simple method in which the nucleic material (DNA or RNA) is directly transferred into the plant cell. The other name of this method is biolistic method or micro-projectile bombardment method (Kurup and Thomas, 2019, 83). There are advantages and disadvantages of the direct gene transfer method, which are shown in Table 1.

In the direct gene transfer method, the precipitated DNA sequences are incorporated with metal nanoparticles and bombarded with the aid of a gene gun at an accelerated speed and introduced into the plant cell. The

DNA-containing micro particles are introduced inside the plant cell and integrated with the nuclear genome (Saxena and Rawat, 2014, 217). However, research reports have provided inputs regarding the chloroplast transformation method. Instead of non homologous recombination (nuclear transformation), injection of a gene into chloroplast is the most preferable method for the edible vaccine production (Shah et al. 2011, 3).

Table 1. Advantages and disadvantages of direct gene transfer method

S. No.	Advantages	Disadvantages
1.	Simple	Costly method
2.	Vector-independent method	Gene gun is required
3.	Gene transfer where *A. tumefaciens* mediated gene transfer is not possible	Harm the plant cells

As some of the economically important plant varieties (soy beans and most cereals) are resistant to vector-based (*Agrobacterium*) transmission, direct transfer method/bio-ballistic method is the most promising one (Concha et al., 2017, 6). Besides that, in the direct transfer method, using a gene gun, a higher number of particles make multiple hits, which leads to the transfer of the genes to many cells simultaneously. Further, according to Saxena and Rawat (2014, 217) the shape of the cell, its size and thickness of the cell wall not impart in its effectiveness. Chloroplast transformation method is an alternate and promising method to nuclear transformation method. The recombinant DNA plasmid is injected into the leaf by a gene gun. The plasmid carries a therapeutic gene, a marker gene, and a gene responsible for the enhancement of the translation of the therapeutic gene (Saxena and Rawat, 2014, 217). Apart from the gene gun, polyethylene glycol–mediated transformation and a galistan expansion femtosyringe microinjection techniques are also used for the delivery of genesin chloroplast (Saxena and Rawat, 2014, 218). The advantages of chloroplast transformation technology are shown in Figure 5.

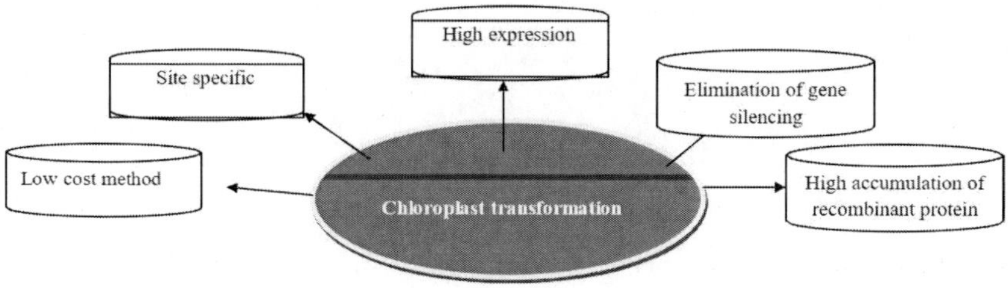

Figure 5. Advantages of Chloroplast transformation technique.

3.2. Indirect Gene Transfer

Indirect gene transfer method is nothing but vector-mediated gene transfer. *Agrobacterium* is a promising candidate in this method. Two promising *Agrobacterium* species are used in this method: *A. tumefaciens* and *A. rhizogenes.* The former contains tumor-inducing plasmid (Ti) and the latter carries the root-inducing (Ri) plasmid. The *Agrobacterium* species is usedin the production of plant-based vaccines against diarrhea, tuberculosis, dengue, avian flu virus, and Ebola (William, 2002, 59). A research report revealed *A. tumefaciens*-mediated gene transfer, Ti plasmid, insertion of sequence responsible for the production of protein (from pathogen), and a transgene integrated into plant genome. The schematic flow chart is shown in Figure 6.

The expression of protein in plant or fruit may be analyzed. Besides, it may be propagated by vegetative method or by seed (Tacket et al., 2000, 303). As this method is time consuming and requires sophisticated infrastructure, using this technique is difficult, especially in low-income countries. In chicks, for controlling the infectious Bursal disease virus (IBDV), the IBDV VP2 capsid, avian IgY (crystallized), and the transgene bombarded in yellow maize are designed as a prospective edible vaccine (Salem et al. 2020; 137). Besides that, recombinant virus particles infect the plant and the expression of viral gene along with the transgene during virus replication (Khan et al., 2020; 321).

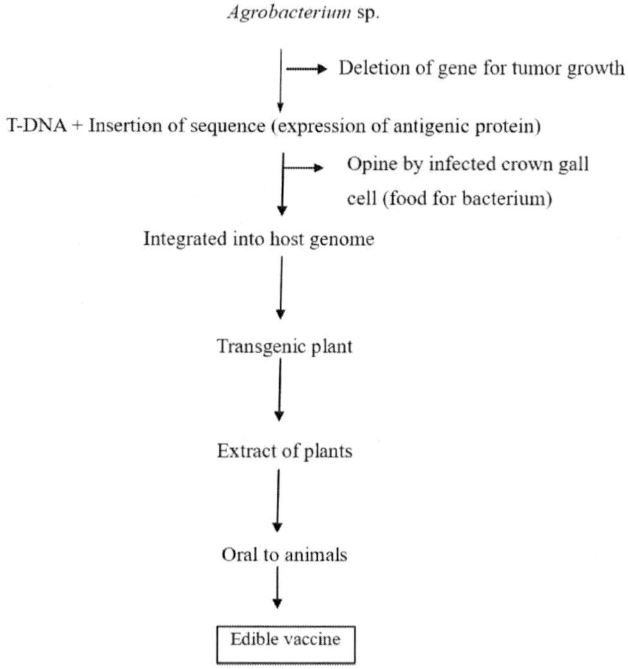

Figure 6. Brief flow chart of *Agrobacterium* mediated edible vaccine production.

4. Plant-Based Oral Vaccines

Plants are emerging as a new biopharmaceutical agent and due to the development of genetically engineered crops, the marketing value of the economically viable plant-based edible vaccines has increased (Hefferon, 2014, 650). There are many advantages of plant-based oral vaccine (Table 2).

A recent research report revealed the expression of pro *IGF-1* (insulin-like growth factor) gene in lettuce chloroplasts to facilitate oral delivery. This plant-based oral vaccine showed promising results in lab animals (diabetic mice) such as bone regeneration and increase in bone volume, density, and area (Park et al., 2020, 4).

Table 2. Comparison between injectable and plant based oral vaccines

S. No.	Injectable vaccine	Plant based Oral vaccine
1.	Fermentation facility is required for production	No fermentation facility
2.	Expensive protein purification technology is needed	No protein purification technology is required
S. No.	Injectable vaccine	Plant based Oral vaccine
3.	Cold storage facility is required	`No need of cold storage and stored in ambient temperature
4.	High-cost method	Less expensive method

In addition, the lyophilized plant cells stabilized and enhanced functionality upto 31 months when stored under ambient temperature (Park et al., 2020, 5). IBD, which is caused by virus, exerts deleterious effects on the life of chicks. The existing vaccines (live and attenuated) are not effective against the disease. The capsid of IBD virus with avian IgY in yellow maize was found to be effective in an *in vivo* study (Salem et al. 2020, 136). The list of various transgenic plants and expression of proteins are shown (Streatfield, 2006, 152). There are various approaches in the plant-based oral vaccines, which include delivery of purified antigen, delivery of processed plant materials, and delivery of unprocessed plant tissues. Among these, delivery of purified antigen is a costly method (Streatfield, 2006, 154).

5. Plant-Based Edible Vaccines and Their Advantages

In edible vaccines, different parts of the plant such as fruits, leaves, and plant-derived products are used as a vaccine, which stimulate the immune system in the host. The edible vaccines need not have pleasant odor, taste, or nutritious but they should be nontoxic to humans (Concha et al., 2017, 4). There are various advantages of plant-based edible vaccines (Figure 7).

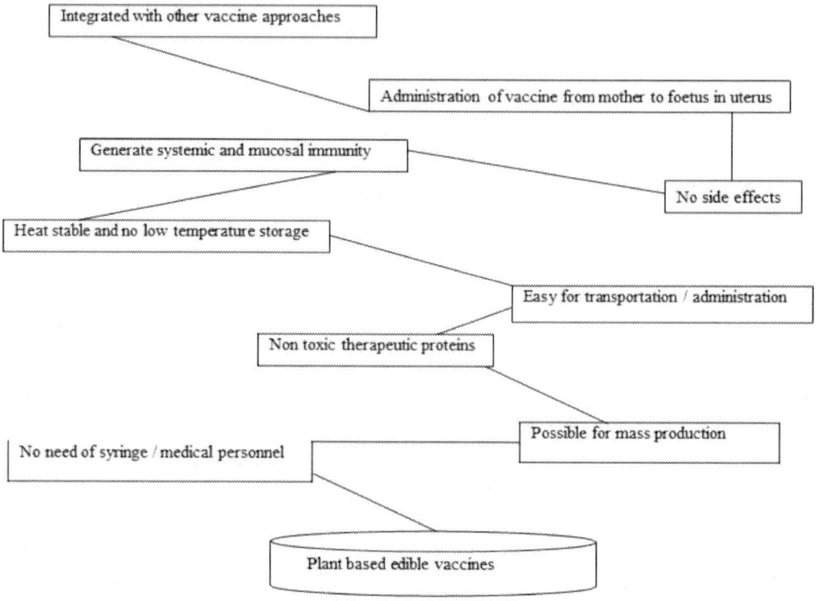

Figure 7. Advantages of plant based edible vaccines.

6. Significant Plant Species Used as Immunization Models

Vaccine production has been explored in a variety of plant species. If edible vaccines are developed from plants, it should be planned in such a way that it could be of raw consumption. If it is used for animal care, the plant material should be preferentially selected as a normal feed of the animal diet. Hence, the following section highlights the variety of plant species used as edible vaccines.

6.1. Potato

Potatoes may be used to make vaccines for tetanus, diphtheria, hepatitis B, and Norwalk virus infections. The initial challenge in developing a potato-based edible vaccine is enteritis, which is caused by a strain of *E. coli.* In humans, potato can serve as an oral strengthening agent for hepatitis B vaccines (Concha et al., 2017, 8). Potatoes were used to create a vaccine for mink enteritis virus. In addition, a potato edible vaccine was tested in wild rabbits as well as rabbit hemorrhagic virus. The most significant advantage of using potato to make edible vaccines is the ease with which they can be transformed and propagated. However, it requires refrigeration, and one of the biggest drawbacks is that cooking causes antigen to denature (Mason et al., 1996; 5335).

6.2. Rice

Rice is another possible plant crop that has been used to create edible vaccines. Its advantages over other plants include being widely used in baby food and having a high antigen expression level. However, it grows slowly and requires the use of a greenhouse. It is also self-contained and does not need refrigeration. A study conducted in 2007 showed that transgenic rice (*Oryza sativa)* induces a large number of antibodies against *E. coli*. In 2008, the expression of HBsAg in rice seeds was reported to be functional. Vaccines made from rice plants would have a huge impact on public health in areas where rice is the main source of food for the majority of population (Mason et al., 1996; 5335; Oszvald et al., 2007; 222).

6.3. Banana

Banana, a common fruit, is currently being studied as a potential vector for vaccines against both serious and widespread diseases. It is the most commonly used palatable fruit in the manufacture of edible vaccines. The major advantages include raw consumption, cost-effectiveness, different varieties, and easy availability compared with other plant species. Banana plants have a high capacity for expressing

HBsAg. The antigen is found in the banana leaf. The biggest drawback is that it takes 2–3 years for it to reach maturity, and it often spoils quickly after it is ripened (Qian et al., 2008; 621).

6.4. Tomato

Tomatoes are a standout competitor because they are simple to genetically alter and new crops can be grown in short time. To help delivery, the engineered tomatoes can be dehydrated or converted into paste. They are an important antidote against SARS, which is caused by corona virus. It is more effective against the Norwalk virus than vaccines made from potatoes. CT-B proteins from *Vibrio cholera* B toxin is expressed in the leaves, stem, fruits, and other tissues (Kumar et al., 2005, 485). This plant species also expresses HBsAg. A vaccine made from tomatoes alsoprotected against diseases such as pneumonia, septicemia and bubonic plagues. They have high vitamin A content, which improves immune response while also perishing quickly (Zhang et al., 2006, 419; Lou et al., 2007, 464).

6.5. Lettuce

Lettuce is an efficient sculpt system for preventing *E. coli*–related enteric diseases in both animals and humans. Lettuce expressing glycoprotein E2 of the classical swine hog pest virus has been recently developed. This plant is primarily eaten raw, and it has antiviral properties against hepatitis B virus (HBV). It may be plant material with the ability to be used as an edible vaccine (Srinivas et al., 2008, 1; Kim et al., 2007, 22).

6.6. Tobacco

Although tobacco is not an edible plant, it could serve as a model for developing an edible vaccine. In 1996, a plant vaccine for Norwalk virus, which causes gastroenteritis, was created. Also, transgenic tobacco that expresses VP1 protein was used to prepare vaccine against chicken having infectious anemia. Tobacco can concurrently express a polypeptide linked to hepatitis B, and it is also used to develop coccidiosis vaccine (Hahn et al., 2007, 85; Gomez et al., 2009, 337; Wigdorovitz et al., 1999, 85).

6.7. Alfalfa

Alfalfa is a plant that is used to make edible vaccines, mostly for veterinary purposes. Transgenic alfalfa with hog pest virus glycoprotein E2 was created in 2005. *Echinococcus granulosus* Eeg95–EgA31 was successfully developed and expressed using Alfalfa plants (Wigdorovitz et al., 1999, 86).

6.8. Carrots

Carrots are more nutritious and tastier, and can also be used to make edible vaccines. The transgenic carrot exerts potential effects against HIV, *E.coli*, and *Helicobacter pylori*. Consumption of this type of edible vaccine benefits people with weakened immune system (Karasev et al., 2005, 1875; Zhang et al., 2010, 127).

7. Applications of Edible Vaccines

Edible vaccines may be used for a number of therapeutic and diagnostic purposes, some of which are mentioned below:

7.1. Malaria

Countless efforts have been made and methods are used to develop a malaria vaccine. Currently, three antigens are being considered for the development of edible malaria vaccine. Merozoite surface protein (MSP) 4 and MSP 5 from *Plasmodium falciparum*, as well asMSP4/5 from *Plasmodium yoelii* are identified. In addition to CTB, oral administration of recombinant MSP4, MSP4/5, and MSP1 inducedantibody responses against the blood-stage parasites in mice. To express a minimum amount of antigen, a large quantity of antigen should be integrated into the plants (Mishra et al., 2008, 288; Chen and Lai 2015, 1–2). According to conventional wisdom, the mechanism responsible for malaria protection would require a multiple of 10–15 antigen targets to provide adequate protection against a variety of stages of malarial infection. Because of antigenic rivalry, a significant number of targets will not be appropriate for single dose vaccination (Haller et al., 2007, 1452). An anti malarial edible vaccine in transgenic tomato plants has been designed as a unique way to treat it.

7.2. Measles

Every year, measles causes 8,000,000 deaths worldwide. The live-attenuated vaccine (LAV) for measles has no oral effectiveness if a cold chain is not maintained. The presence of maternal antibodies in the LAV reduces its efficacy (Huang et al., 2001, 2163). Three hemagglutinin (H) and fusion proteins are ruined with two surface proteins of wild-type measles virus. The results showed that IgA antibodies were present in the feces of animals who had been vaccinated with MV-H. The best alternative for measles vaccination is a transgenic carrot plant, according to research findings (Naderi and Fakheri, 2015, 275–276).

7.3. Hepatitis

According to WHO estimates, 2billion people worldwide have had or are currently infected with HBV virus. More than 360 million people have infection all the time, and more than 600000 people die each year from HBV-related diseases such a sliver cirrhosis or hepatocellular carcinoma. Hepatitis B surface antigen (HBsAg) is used in the manufacture of an edible vaccine. Potatoes are being the plant of choice for developing an edible hepatitis vaccine. Their roots show more HBsAg expression than other parts of the plants (Malik et al., 2011, 18).

7.4. Autoimmune Diseases

Self-antigens are very beneficial in autoimmune illnesses such as type 1 diabetes. The core reason behind diabetes is cell damage and insulin production failure. Insulin is the most commonly recommended drug for type 1 diabetes; however, it is unable to entirely cure the disease. In diabetic mice, potatoes containing insulin or glutamic acid decarboxylase, as well as the benign B subunit of *Vibrio cholerae* toxin, showed a significant improvement. It has the potential to suppress immunological responses while maintaining insulin levels (Richter et al., 2000, 1167).

7.5. Diarrheal Diseases

Diarrhea is the third leading cause of death among Indian children. The most common cause of diarrheal illnesses is infection of the gastrointestinal system, which can be caused by bacteria, viruses, or parasitic organisms. Only a few mucosal vaccinations against pathogens have been licensed, despite the fact that several oral vaccinations have been developed for the prevention of diarrhea. An oral vaccine must travel through the hostile environment of the stomach and intestine in order to be effective. This can be accomplished by developing edible vaccine against enterotoxigenic *E.coli* (ETEC), cholera, and norovirus. According to (Mason et al., 1998; 1339) the gene expressing LT-B was transmitted to potato and tobacco leaves using *Agrobacterium tumefaciens*, which were fed to mice with anti-LT-B blood IgG and mucosal IgG. Transgenic corn expressing LT-B is also being investigated as a potential edible vaccination against ETEC infection. Corn protein expression is constant, and the period required for corn development is shorter. Transgenic corn stimulated IgG and IgA responses in mice in preclinical investigations (Ma et al., 1995; 716).

Cholera is a bacterial/diarrheal disease that has many symptoms similar to ETEC. The primary pathogens are *V. cholerae* O1 and O139. Anti-cholera toxin provides vaccine-induced protection against *V. cholerae*. In mice, a transgenic potato containing CY-B was responsible for the development of anti-CT-B antibodies in the intestine and blood (Streatfield et al., 2001, 2742). Noroviruses are enteric viruses that belong to the Caliciviridae family and are the most common cause of gastroenteritis. Manson et al. (1996; 5337) created transgenic potato and tobacco expressing capsid protein of Norwalk virus. Fecal and serum antibody responses were seen in preclinical studies in mice with these transgenic plants (Arakawa et al., 1999; 161–162).

7.6. Anthrax

Protective antigen (PA) was expressed in transgenic tobacco, by biolistic transfer of the *pag* gene to the chloroplast genome. Its effective expression can be used to cure cattle anthrax illness. For future research, *pag* gene should also be injected into tomatoes and spinaches (Xi et al., 1990; 1580).

7.7. Rabies

Rabies is a zoonotic, viral infection that causes death. Antibodies against rabies could be induced in mice by tomato plants expressing rabies antigens. Tobacco mosaic virus can also be used as an alternative. CaMV-transformed tomato plants harboring the rabies virus' glycoprotein gene were found to be immunogenic in animals.

8. Challenges and Future Prospective of Plant-Based Edible Vaccines

Though the edible vaccines are advantages, a few challenges remain in the usage of edible vaccines. Edible vaccines are having scientific, socioeconomic, and legislative issues. Basically plant-based edible vaccines are produced from genetically engineered plants (Sala et al., 2003, 805). Hence, there are still a lot of concerns over the use of genetically engineered plant products. To make effective immune response, a higher concentration of antigen is required. It seems little difficult to quantify and concentrate on antigen required for human/animals (Miranda et al., 2020, 76). Low-income countries are unable to establish infrastructures to develop plant-based edible vaccines. Hence these countries depend on existing vaccination programs (Altindis et al., 2014, 218). The societal acceptance of edible vaccines and various negative factors such as risk, public opinion, and geographical location are also a major concern at this juncture (Kurup and Thomas, 2020, 84).

Plant-based edible vaccines are an economical tool to deliver large volume of antigen to humans to combat various deleterious diseases (Streatfield, 2006, 156). If the plant material is not palatable, chances are there to purify the antigen and use it in the form of tablet/capsules.

Reports are available on the plant-based viral vaccines showing excellent results in breast cancer (Alejandra et al., 2017, 254), non-Hodgkin's lymphoma (McComick et al., 1999, 705), colorectal cancer (Ahn et al., 2014, 5), and Epstein–Barr virus (EBV)–associated malignancies (Alejandra et al., 2017, 254). Hence the future scope of edible vaccines is combating from normal diseases to life-threatening disease. These findings could widen the prospects of edible vaccines-based research.

Conclusion

Plant-based vaccines are an innovative scientific tool to treat various human and animal diseases including multidrug-resistant pathogens. Besides, plant-based edible vaccines can also provide a solution to malnutrition and immune prophylaxis. A wide variety of commonly available plant products can be used as an edible vaccine, a valuable point presented here. However, there are various socioeconomic, scientific, and ethical issues that plant-based edible vaccines need to overcome by strenuous scientific efforts.

References

Ahn, J., Lee, K. J. and Ko, K. (2014). Optimization of ELISA conditions to quantify colorectal cancer antigen antibody complex protein (GA733-FcK) expressed in transgenic plant. *Monoclon. Antib. Immunodiagn. Immunother.* 33(1): 1-7. https://doi.org/10.1089/mab.2013.0072.

Alejandra, W., Omar, G. and Sergio, R. (2017). Plant-made vaccines in the fight against cancer. *Trends in Biotechnology.* 35(3): 251-56.

Arakawa, T., Yu, J. and Langridge, W. H. (1999). Food plant-delivered cholera toxin B subunit for vaccination and immune tolerization. *Advances in Experimental Medicine and Biology.* 464: 161–178. https://doi.org/10.1007/978-1-4615-4729-7_13.

Altindis, E., Gulce, Iz. S., Ozgun Ozen, M., Nartop, P., Deliloglu Gurhan, I. and Gurel, A. (2014). Plant derived edible vaccines and therapeutics. In: *Frontiers in clinical drug research: anti-infective.* Atta-ur-Rahman Ed. Bentham Science Publishers; 2014. p. 200–36.

Butelli, E., Titta, L. and Giorgia, M., Mock, H., Matros, A., Peterek, S., Schijlen, E. G. W. M., Hall, R. D., Bovy, A. G., Luo, J. and Martin,C. (2008). Enrichment of tomato fruit with health – promoting anthocycanins by expression of select transcription factors. *Nature Biotechnology*. 26(11):1301–1308. https://doi.org/10.1038/nbt.1506.

Chen, Q. and Lai, H. (2015). Gene delivery into plant cells for recombinant protein production. *BioMed Research International.* 2015. https ://doi.org/10.1155/2015/932161.

Clarke, J. L., Paruch, L., Dobrica, M., Caras, L., Tucureanu, C., Onu, A., Ciulean,S., Stavaru, C., Eerde, A., Wang, Y., Steen, H., Haugslien,S., Petrareanu, C., Lazar, C., Popescu, C., Bock, R., Dubuisson, J. and Nichita, N. (2017). Lettuce-produced hepatitis C virus E1E2 heterodimer triggers immune responses in mice and antibody production after oral vaccination. *Plant Biotechnology Journal, 15*(12), 1611–1621.

Concha, C., Canas, R., Macuer, J., Torres, M. J., Herrada, A. A., Jamett, F. and Ibanez, C. (2017). Disease prevention: An opportunity to expand edible plant-based vaccines. *Vaccine (Basel).* 5(2): 14–23. https://doi.org/10.3390/vaccines5020014.

Davies, H. M. (2010). Review article: commercialization of whole-plant systems for bio manufacturing of protein products: evolution and prospects. *Plant Biotechnol J.* 8:845–861. doi: 10.1111/j.1467-7652.2010.00550.x.

De Muynck, B., Navarre, C. and Boutry, M. (2010). Production of antibodies in plants: status after twenty years. *Plant Biotechnol J.* 8:529–563. doi: 10.1111/j.1467-7652.2009.00494.x.

Faye, L. and Gomord, V. (2010). Success stories in molecular farming—a brief overview. *Plant Biotechnol J.* 8(5):525–528. https://doi.org/10.1111/j.1467-7652.2010.00521.x.

Gomez, E., Zoth, S. C., Asurmendi, S., Rovere, C. V. and Berinstein, A. (2009). Expression of hemagglutinin neuraminidase glycoprotein of Newcastle disease virus in agro infiltrated *Nicotiana benthamiana* plants. *Journal of Biotechnology*. 144: 337–340. doi:10.1016/ j.jbiotec.2009.09.015.

Hahn, B. S., Jeon, I. S., Jung, Y. J., Kim, J. B., Park, J. S., Ha, S. H., Kim, K. H., Kim, H. M., Yang, J. S. and Kim,Y. H. (2007). Expression of hemagglutinin-neuraminidase protein of Newcastle disease virus in transgenic tobacco. *Plant Biotechnology Reports*. 1: 85–92. doi: 10.1007/s11816-007-0012-9.

Haller, A. A., Lauer, G. M., King, T. H., Kemmler, C., Fiolkoski, V., Lu, Y., Bellgrau, D., Rodell, T. C., Apelian, D., Franzusoff, A. and Duke, R. C. (2007). Whole recombinant yeast-based immunotherapy induces potent T cell responses targeting HCV NS3 and core proteins. *Vaccine.* 25(8): 1452–1463. https://doi.org/10.1016/ j.vaccine.2006.10.035.

Hefferon, K. L. (2014). DNA virus vectors for vaccine production in plants: Spotlight on geminiviruses. *Vaccine*. 2(3): 642 – 653. https://doi.org/10.3390/vaccines2030642.

Huang, Z., Dry, I., Webster, D., Strugnell, R. and Wesselingh, S. (2001). Plant-derived measles virus hemagglutinin protein induces neutralizing antibodies in mice. *Vaccine*. 19(15–16): 2163–2171. https://doi.org/10.1016/s0264-410x(00)00390-x.

Kang, K., Lee, K., Sohn, A. O., Park, S., Lee, S, Kim, S. Y., Kim, Y. S. and Back, K. (2009). Ectopic expression of serotonin N-hydroxycinnamoyl transferase and different production of phenylpropanoid amides in transgenic tomato tissues. *Scientia Horticulture*. 120(4):504 – 510. doi:10.1016/j.scienta.2008.12.015.

Karasev, A. V., Foulke, S., Wellens, C., Rich, A., Shon, K. J., Zwierzynski, I., Hone, D., Koprowski, H. and Reitz, M. (2005). Plant based HIV-1 vaccine candidate: Tat protein produced in spinach. *Vaccine*. 23(15): 1875–1880. https://doi.org/10.1016/j.vaccine. 2004.11.021.

Khan, F., Szmigielski, R., Gul, A., Altay, V. and Ozturk M. (2020). Advancement in plant transgenomics approach for the bio pharmaceuticals and vaccine development. In: *Biodiversity and Biomedicine* (Eds. M. Ozturk, D. Egamberdieva, M. Pesic), Elsevier publications, 608 pp. doi.org/10.1016/B978-0-12-819541-3.00018-9.

Kim, T. G., Kim, M. Y., Kim, B. G., Kang, T. J., Kim, Y. S., Jang, Y. S., Arntzen, C. J. and Yang, M. S. (2007). Synthesis and assembly of *Escherichia coli* heat-labile enterotoxin B subunit in transgenic lettuce (*Lactuca sativa*). *Protein Expression and Purification*. 51(1): 22–27. https://doi.org/10.1016/j.pep.2006.05.024.

Kumar, B. V., Raja, T. K., Wani, M. R., Sheikh, S. A., Lone, M. A., Nabi, G., Azooz, M. M., Younis, M., Sarwat, M. and Ahmad, P. (2013).Transgenic plants as green factories for vaccine production. *African J biotechnol.* 12(43): 6147-6158. doi: 10.5897/ AJB2012.2925.

Kumar, G. B. S., Ganapathi, T. R., Revethi, C. B. J., Srinivas, L. and Bapat, V. A. (2005). Expression of hepatitis B surface antigen in transgenic banana plants. *Planta.* 222(3), 484–493. https://doi.org/ 10.1007/s00425-005-1556-y.

Kurup, V. M. and Thomas, J. (2020). Edible vaccines: promises and challenges. *Molecular Biotechnology*. 62: 79 – 90. https://doi.org/ 10.1007/s12033-019-00222-1

Lou, X. M., Yao, Q. H., Zhang, Z., Peng, Ri-He., Xiong, A. S. and Wang, H. K. (2007). Expression of the human hepatitis B virus large surface antigen gene in transgenic tomato plants. *Clinical and Vaccine Immunology*. 14(4): 464–469. https://doi.org/10.1128/cvi.00321-06.

Ma, J. K., Hiatt, A., Hein, M., Vine, N. D., Wang, F., Stabila, P., Dolleweerd, C. V., Mostov, K.and Lehner, T. (2006). Generation and assembly of secretary antibodies in plants. *Science.* 268: (5211), 716–719. doi: 10.1126/science.7732380.

Madhumita, N., Deepak, V. and Pallavi, U. (2014). Edible vaccines - A review. *Int J Pharmacother.* 58-61.

Malik, A., Vashisht, V. K., Rashid, R., Sharma, S. and Singh, J. (2011). Edible vaccine vegetables as alternative to needles. *International Journal of Current Research*. 33: 18–26.

Mason, H. S., Ball, J. M., Shi, J. J., Jiang, X., Estes, M. K. and Arntzen, C. J. (1996). Expression of Norwlak virus capsid protein in transgenic tobacco and potato and its oral immunogenicity in mice. *Proceedings of the National Academy of Sciences USA*. 93(11): 5335–5340. https://doi.org/10.1073/pnas.93.11.5335.

Mason H. S., Lam D. M., Arntzen, C. J. (1992). Expression of hepatitis B surface antigen in transgenic plants. *Proceedings of the National Academy of Sciences USA*. 89:11745-9.

Mason H. S., Haq, T. A., Clements, J. D., Arntzen, C. J (1998). Edible vaccine protects mice against *Escherichia coli* heat-labile enterotoxin (LT): Potatoes expressing a synthetic LT-B gene. *Vaccine* 16:1336-43.

McCormick, A. A. Kumagai, M. H., Hanley, K., Turpen, T. H., Hakim, I., Grill, L. K., Tuse, D., Levy, S. and Levy, R. (1999). Rapid production of specific vaccines for lymphoma by expression of the tumour-derived single- chain Fv epitopes in tobacco plants. *Proc. Natl. Acad. Sci.* U.S.A. 96(2):703–708. https://dx.doi.org/10.1073%2Fpnas.96.2.703.

Miranda, F. C., Ruiz-Cabello, V. C. and Hartado, M. C. (2020). Food biopharmaceuticals as a part of sustainable bioeconomy: Edible vaccines case study. *New Biotechnology*. 54: 74-79.

Mishra, N., Gupta, P. N., Khatri, K., Goyal, A. K. and Vyas, S. P. (2008). Edible vaccines: A new approach to oral immunization. *Indian Journal of Biotechnology*. 7: 283–294.

Naderi, S. and Fakheri, B. (2015). Overview of plant-based vaccines. *Research Journal of Fisheries and Hydrobiology*. 10: 275–289.

Oszvald, M., Kang, T. J., Tomoskozi, S., Tamas, C., Tamas, L., Kim, T. G. and Yang, M. S. (2007). Expression of a synthetic neutralizing epitope of porcine epidemic diarrhoea virus fused with synthetic b subunit of *Escherichia coli* heat labile enterotoxin in rice endosperm.

Molecular Biotechnology. 35, 215–223. https://doi.org/10.1007/BF02686007.

Park, J., Yan, G., Kwon, K. C., Liu, M., Gonnella, P. A., Yang, S. and Daniell, H. (2020). Oral delivery of novel human IGF-1 bio encapsulated in lettuce cell promotes musculoskeletal cell proliferation, differentiation and diabetic fracture healing. *Biomaterials*. i.odorg/10.1016/j.biomaterials.2019.119591.

Paul, M. and Ma, J. K. (2011). Plant-made pharmaceuticals: leading products and production platforms. *Biotechnol Appl Biochem*. 58:58–67. doi: 10.1002/bab.6.

Penney, C. A., Thomas, D. R., Deen, S. S. and Walmsley, A. M. (2011). Plant-made vaccines in support of the millennium development goals. *Plant Cell Reports*. 30: 789–798. https://doi.org/10.1007/s00299-010-0995-5.

Perez-Massot, E., Banakar, R., Gomez-Galer, S. et al. (2013). The contribution of transgenic plants to better health through improved nutrition: Opportunities and constraints. *Genes Nutr* 8: 29-41. https://doi.org/10.1007/s12263-012-0315-5.

Phoolcharoen, W., Dye, J. M., Kilbourne, J., Piensook, K., Pratt, W. D., Arntzen, C. J., Chen, Q., Mason, H. S. and Herbst-Kralovetz, M. M. (2011). A non-replicating subunit vaccine protects mice against lethal Ebola virus challenge. *Proc Natl Acad Sci*. 108(51):20695–20700. doi: 10.1073/pnas.1117715108.

Plotkin. (2013). *Vaccine Fact Book; Pharmaceutical Research and Manufacturers of America:* Washington, DC, USA; p. 97.

Qian, B., Shen, H., Liang, W., Guo, X., Zhang, C., Wang, Y., Li, G., Wu, A., Cao, K. and Zhang, D. (2008). Immunogenicity of recombinant hepatitis B virus surface antigen fused with preS1 epitope sex pressed in rice seeds. *Transgenic Research*. 17, 621–631. https://doi.org/10.1007/s11248-007-9135-6.

Richter, L. J., Thanavala, Y., Arntzen, C. J., and Mason, H. S. (2000). Production of hepatitis B surface antigen in transgenic plants for oral immunization. *Nature Biotechnology*. 18(11):1167 – 71. https://doi.org/10.1038/81153.

Sala, F., Manuela Rigano, M., Barbante, A., Basso, B., Walmsley, A.M. and Castiglione, S. (2003).Vaccine antigen production in transgenic plants: strategies, gene constructs and perspectives. *Vaccine* 21(7-8):803–8. 30.

Salem, R. Assem, S. K., Omar, O. A., Khalil, A. A., Basry, M .A., Waly, F. R., Samir, N. and El-Kholy, A. A. (2020). Expressing the immunodominant projection domain of infectious bursal disease virus fused to the fragment crystallizable of chicken IgY in yellow maize for a prospective edible vaccine. *Molecular Immunology.* 118 (2020): 132-141.

Saxena, J. and Rawat, S. (2014). Edible vaccines. In book: *Advances in biotechnology* (pp. 207–226). New Delhi: Springer. doi:10.1007/978-81-322-1554-7_12.

Shah, C. P., Trivedi, M. N., Vachhani, U. D. and Joshi, V. J. (2011). Edible vaccine: A better way for immunisation. *International Journal of Current Pharmaceutical Research. 3*(1): 1–4.

Srinivas, L., Kumar, G. B. S., Ganapathi, T. R., Revathi, C. J. and Bapat, V. A. (2008). Transient and stable expression of hepatitis B surface antigen in tomato (*Lycopersicon esculentum* L.). *Plant Biotechnology Reports.* 2 (1): 1–6. http://dx.doi.org/10.1007/s11816-008-0041-z.

Streatfield, S. J., Jilka, J. M., Hood, E. E., Turner, D. D., Bailey, M. R., Mayor, J. M., Woodard, S. L., Beifuss, K. K., Horn, M. E. and Delaney, D. E., et al., (2001). Plant-based vaccines: Unique advantages. *Vaccine.* 19(17–19): 2742–2748. https://doi.org/10.1016/s0264-410x(00)00512-0.

Streatfield, S. J. (2006). Mucosal immunization using recombinant plant-based oral vaccines. *Methods.* 38(2): 150-157.

Tacket, C. O., Mason, H. S., Losonsky, G., Estes, M. K., Levine, M. M., Arntzen, C. J. (2000). Human immune responses to a novel Norwalk virus vaccine delivered in transgenic potatoes. *J. Infect. Dis.* 182, 302–305.

Wigdorovitz, A., Perez Filgueira, D. M., Robertson, N., Carrillo, C., Sadir, A. M., Morris, T. J. and Borca, M. V. (1999). Protection of mice against challenge with foot and mouth disease virus (FMDV) by immunization with foliar extracts from plants infected with recombinant tobacco mosaic virus expressing the FMDV structural protein VP1. *Virology.* 264(1): 85–91. https://doi.org/10.1006/viro.1999.9923.

William, S. (2002). A review of the progression of transgenic plants used to produce plant bodies for human usage. *Journal of Young Investigators, 4*(2002), 56–61.

Xi, J. N., Graham, D. Y., Wang, K. N. and Estes, M. K. (1990). Norwalk virus genome cloning and characterization. *Science.* 250(4987): 1580–1583. https://doi.org/10.1126/science.2177224.

Xu, J., Ge, X. and Dolan, M. C. (2011). Towards high-yield production of pharmaceutical proteins with plant cell suspension cultures. *Biotechnol Adv.* 29:278–299. doi: 10.1016/j.biotechadv.2011.01.002.

Xu, J., Okada, S., Tan, L., Goodrum, K. J., Kopchick, J. J. and Kieliszewski, M. J. (2010). Human growth hormone expressed in tobacco cells as an arabinogalactan-protein fusion glycoprotein has a prolonged serum life. *Transgenic Res.* 19:849–867. doi: 10.1007/s11248-010-9367-8.

Xu, J., Tan, L., Goodrum, K. J. and Kieliszewski, M. J. (2007). High-yields and extended serum half-life of human interferon alpha 2b expressed in tobacco cells as Arabinogalactan-protein fusions. *Biotechnol Bioeng*. 97:997–1008.

Yusibov, V., Kushnir, N. and Streatfield, S. J. (2016). Antibody production in plants and green algae. *Annu Rev Plant Biol.* 67:669–701.doi: 10.1146/annurev-arplant-043015-111812.

Zhang, H., Liu, M., Li, Y., Zhao, Y., He, H., Yang, G. and Zheng, C. (2010). Oral immunogenicity and protective efficacy in mice of a carrot-derived vaccine candidate expressing UreB subunit against *Helicobacter pylori. Protein Expression and Purification.* 69 (2); 127–131. https://doi.org/10.1016/j.pep.2009.07.016.

Zhang, X., Buehner, N. A., Hutson, A. M., Estes, M. K. and Mason, H.S. (2006). Tomato is a highly effective vehicle for expression and oral immunization with Norwalk virus capsid protein. *Plant Biotechnology Journal.* 4(4): 419–432. https://doi.org/10.1111/j.1467-7652.2006.00191.x.

In: The Future of Metabolic Engineering
ISBN: 978-1-68507-362-6
Editors: A. Sharma and D. Amin

Chapter 9

RECENT DEVELOPMENT AND FUTURE PROSPECT OF PLANT-BASED VACCINES

Bairagi Ghanshamdas R. and Vipul P. Patel*, PhD
Sanjivani College of Pharmaceutical Education and Research,
Kopargaon, India

ABSTRACT

Plant based vaccines are the futuristic approach for the painless and tasteful consumption of fruits to not only get nourished but also to get the immunity against a specific pathogen. Being an advanced product of recombinant DNA technology and genetic engineering, the plant-based vaccines provide a bright future against fight with new diseases. This chapter describes about the various plant based vaccines developing for bacterial, viral, parasitic, allergenic antigens infections in human beings and animals.

* Corresponding Author's E-mail: v_pharmacy@yahoo.co.in.

1. INTRODUCTION

A plant-based vaccine is nothing but a type of recombinant or subunit vaccine. In this type of antigenic preparation, the antigens against particular pathogens are introduced into the selected plant. The plant-made vaccines are less expensive and show the minimum toxicity as compared to conventional vaccines. The characteristics features of plant-derived vaccines are less toxicity, heat stability, safety, easily scalable, high potential, cheap, and oral activity [1]. Plant-based vaccines are the arising type of vaccines and have a higher therapeutic effect in the treatment of many human and animal diseases and disorders. Most commonly the plant-based vaccine is used against various types of bacteria, parasites, and viruses, which causes illness in human beings [2].

There are some of the plant-based vaccines that have been produced biotechnologically and a few of them are currently under development stage and some of them are in the clinical trial phase [2]. Plant-based vaccines are produced or constructed mainly by two techniques such as plant virology and genetic engineering and these plant-derived vaccines are used in the treatment of cancer, autoimmune diseases, bacterial and viral infections in human beings as well as animals. The produced vaccines have greater tolerability, stability, immunogenicity, safety, and high potential against infectious bacteria and viruses analyzed in human clinical trials and animal models [3, 4].

It is observed that the strong immune response shown by the plant viruses is higher or similar to as compared to the mammalian viruses. Moreover, the plant viruses with introduced antigens are powerful vaccine components due to their strongly organized and repetitive spatial structure. The plant viruses are promising together with virus-like carriers, is a highly encouraging agent for antigen inhibition, stimulating the response of different immune cells in the host and bringing out the formation of specific antibodies against infectious and self-antigen targets [3].

Plant-based vaccines are also produced through some novel methods. These provide a needleless, convenient, and easy route of administration of vaccines. The plant-based vaccines are convenient to store at room temperature, easy to handle, not complicated for large-scale production, time-saving, and cost-effective. These vaccines are the alternative vaccines for the conventional vaccines that have high costs and are time-consuming. A broad variety of plants have been used for stable transient expression as well as for transgenic expression for the production of vaccines. Transgenic plants are used to produce mainly four types of vaccines such as bacterial vaccines, parasite vaccines, viral vaccines, and immunocontraceptive vaccines [2].

Different types of plants are used to produce vaccines such as tobacco, rice, maize, potato, alfalfa, lettuce, tomato, carrot, peanut, and soybean, and these plants are commonly called a bioreactor. The role of these plants as hosts, in which viral genes are introduced and this is carried out by the in vitro methods like protoplast or cell culture, and hairy root culture. The different species of these plants are utilized to produce recombinant vaccines [4, 5].

The vaccines produced from the plants are the emerging trend in the development of vaccines from the last century. The researchers developed plant-based vaccines by using bacteria, viruses and, pathogens as vectors against the various bacterial, viral, parasitic, allergenic antigens infections in human beings and animals. However, to date, only two products of plant-derived vaccines have been licensed and approved by the US Department of Agriculture (USDA) (a) plant made scFV mAB used in the production of a recombinant HBV vaccine in Cuba and (b) Newcastle disease virus (NDV) vaccine for poultry.

2. Edible Vaccine

The edible vaccines are genetically engineered preparation of subunits that contain antigens that generate or stimulate the immune response against the disease-causing pathogens. They are manufactured by using transient plants and they produced protein that has their therapeutic value and the prepared protein after consuming the human beings, they are digested and when the optimum amount of protein enters into blood circulation which generates the immune response against disease-causing pathogens. Transgenic plants and transient gene expression plants are mainly used as edible vaccine production systems. The preparation of edible vaccines required greenhouses, not the large stainless-steel tanks for cell culture and purification [6-8].

For the production of edible vaccines, the preferred antigen coding gene is sequestered or isolated from the bacteria and viruses. The isolated antigen is further proceeds for edible vaccine preparation in two ways. The first way is a transformation technique in which the viral vectors is loaded with the genes of interest and these vectors are introduced into the selected plant by using various techniques (Electroporation/ Electrotransfection and Gene Gun or Biolistic Method), which allowed to produce encoded proteins in altered plants called as transgenic plants and which is used for edible vaccines. The second way is a genetic engineering technique, recombinant virus is inserted into selected host plant and chimeric virions are produced in the plant which is extracted and purified from these plants and used for edible vaccines preparation [6-8].

The oral administration of edible vaccines does not require trained medical personnel, especially in children. The edible vaccines are safe and easy to scale up as compared to conventional vaccines. Along with that edible vaccines have good heat stability and genetic stability, are inexpensive, harmless, prevention against various disease viruses, and are stored at normal condition and nearby to the place of usage [9]. The syringes and needles are not required for edible vaccines which reduce the risk of several infections. Cold storage conditions are not required for edible vaccines. The important characteristic feature of edible vaccines is, it reduces risk and eliminates the risk of animal virus infections and contamination [10].

The hybrid plants are developed through transgenic technology therefore the edible vaccines have multicomponent efficiency. The edible is more effective than injected vaccines because the second-generation edible vaccine is used against multiple diseases as they allow the various antigens to prime M cell (microfold cell) simultaneously [11]. The insertion of selected genes or antigen or vector into selected host plants through various plants genetic engineering methods such as Electroporation method, Electrotransfection method, Gene Gun or Biolistic Micro projectile bombardment method, plasmid or vector system, Agrobacterium tumefaciens Plasmid vector carrier system.

The concept of edible vaccines is firstly applied in tobacco plants was used to express hepatitis B surface antigen to produce an immunological product, which is effective against hepatitis B and which is further used for immunotherapy in various diseases. The edible vaccines used against various diseases such as measles, cholera, dengue, hookworm, rabies, mouth diseases, and hepatitis B, hepatitis C, and hepatitis E are produced in plants such as tobacco, banana, potato, lettuce, rice, etc., [9-11].

The edible vaccines show immunity through the mucosal as well as systemic ways. The edible vaccines show a dual mechanism of action through the mucosa by protecting the human beings from the infection's pathogens e.g., Mycobacterium tuberculosis and carriers causing diarrhea, pneumonia. The edible vaccines also show the immunization effect in fetus and infant by fetus-in-utero by the transplacental movement of maternal antibodies and the infant through breastfeeding, respectively [10, 12].

There are several platforms used for high yield production plants required for recombinant protein generation and plant-based edible vaccines production as given below:

2.1. Molecular Farming

The concept of molecular farming are originated in 1990, initially, the principle of molecular farming was reported, and afterward production the human serum albumin and recombinant antibodies in transgenic plants and cell cultures or cell suspension. The molecular farming process was commercially started in tear 1996 which involved the production of protein-based technical reagents in maize seeds. In 2012, the approved recombinant pharmaceutical plant-derived products came into the market for human use [13]. Molecular farming is also defined as the use of plants for the production of biopharmaceutical products, recombinant protein products, and biologically active non-native proteins, where the aim is to recover and utilize the protein product rather than the plant itself [14-15].

Molecular farming is a biotechnological program that includes the genetic modification of agricultural products to produce proteins and chemicals for commercial and pharmaceutical purposes. A vast majority of developing countries cannot afford the high costs of medical treatments resulted from the existing methods. Hence, we need to produce not only the new drugs but also the cheaper versions of the present samples in the market. Molecular farming can offer efficient solutions for the current growing need for biomedicines [16].

The field of molecular farming is mainly considered for the production of affordable plant-derived pharmaceuticals products. In the past century, mammalian cell culture is used for the production of recombinant proteins or pharmaceutical products worldwide. Because of certain disadvantages of this method, plants are used for large-scale production of recombinant proteins with the advantage of the production of complex and multimeric proteins because plants can carry out proper folding and necessary posttranslational modifications needed for the functioning of such proteins. The main aim of molecular farming is to match the biochemical properties of plant-derived recombinant protein and glycosylation pattern and with the protein from the human origin [17].

The process of molecular farming used terrestrial plants like tobacco and cereals. Due to the biotechnological development, they also included other plant systems such as plant cell and tissue cultures, moss, algae, aquatic plants, and in vitro transcription and translation systems which are derived from plant cells. The protein required for the formation of recombinant proteins are obtained from the crude extract either by extraction and purification or at the end of the process the used plants are demolished and discarded and utilized as separate for other purposes [18].

The different plant-based systems or platforms are used for molecular farming has their advantages such as transgenic plants are cheaper and incredibly scalable on large production amount as compared to the CHO cells. The transient expression systems can produced or scaled up the plant-derived products more efficiently, easily, and quickly as compared to the fermenter-based platform. The plant cells can produce uniquely tailored glycan structures. These all the plant-derived platforms are animals free, which is increases the consumer acceptance of the product and inherently safe as compared to the pharmaceuticals products of mammalian cells because these plant-based platforms don't cause replication to viruses [18].

The production of biopharmaceuticals, biologically active non-native proteins, and recombinant proteins by the engineered plants is lower as compared to molecular farming. Examples of plant-made pharmaceuticals are human insulin, HIV-neutralizing antibodies, and human serum albumin (HSA) [18].

2.2. Transgenic Plants

The concept of transgenic plants is defined as these are the plants in which one or more genes (antigenic transgene) from another species of plants have been introduced into the plant nuclear genome using genetic engineering techniques or modifications of DNA by genetic engineering techniques. The main aim of the transgenic plants is to introduce new characteristics in the plant which does not occur naturally in all the species of the particular plant [19-21].

Genetic engineering techniques have been used to produce transgenic plant vaccines. In this, the specified gene is encoded for the appropriate antigen and gives a strong immune response in the body when modified after administered orally. This is one of the most widely used techniques for vaccine production in plants. In the transgenic plant, the genes are inserted artificially or by using different genetic engineering techniques or methods. The seven transgenes are genes that are introduced into the selected plant and they may be from another species of the plant or independently from other plants [19].

The production of transgenic plants is the simplest method to produce transgene cell lines and it is easy to scale up. The transgenic plant approach is mainly used for terrestrial crop species. Transgenic plants are produced by different methods like Agrobacterium-mediated transformation and the delivery of DNA-coated metal particles by microprojectile bombardment [29]. The plant-based vaccines are produced by transgene are safer to use because they do not cause a risk of microbiological contamination, pathogenicity as compared to the animal-derived vaccines [19].

Transgenic plants are produced by the technique called transformation. In this, the explants or callus tissue are transformed and afterward the selection of the whole plant by regeneration. In some of the cases of the plant species plants directly transformed without a tissue culture step, it is the characteristics feature of the particular plant species.

The transformation for transgenic plant production is mainly carried out by using A. tumefaciens and a direct transfer method (usually particle bombardment). Because transgene integration process is random and afterward the integrated transgenes may be subjected to the position effects that inhibit expression and many of the independent primary transformants are usually originated and those with the best expression levels are chosen for further propagation and the generated cells are ensure that the transgene is present in each cell.

In molecular farming, the different species of transgenic plants are used as a host for the production of biopharmaceuticals i.e., vaccines. There are two main platforms for transgenic plant products such as leafy crops and seed crops. The main platforms of molecular farming are leafy crops. The leafy crops mainly consist of tobacco, which produces massive amounts of biomass for the production of biopharmaceuticals. The leafy crops also consist of watery content i.e recombinant proteins, which are extracted after the harvesting to avoid degradation of recombinant proteins e.g., HIV-neutralizing human antibody 2G12 was produced in tobacco [13, 15].

The second platform is seed crops, in which in the dry environment the recombinant proteins were protected. In seed crops, the protected recombinant proteins are stable and give potential effect after storing. The seeds consist of vaccines and prophylactic antibodies and which are administered orally because seeds are edible. An example is a maize, rice, and barley, cereal seeds [13].

The mechanism of action of plant-based vaccines produced by the transgenic technique is stimulating the lymphoid structure in the intestine. After the oral administration of plant-derived vaccines, they stimulate mucosal immunity in the tissues of the respiratory system and this is helping to abolish the injection-related hazards. They maintain their antigenic property after degradation in the intestine. It is one of the characteristic features of transgenic plant vaccines [19]. The transgenic plants have the ability to large-scale production on the agricultural scale and the resulted plant has its potential and it is a permanent genetic resource for further use for pharmaceutical productions [13, 15].

Sometimes in genetic engineering methods introduced combination or more than one gene in the plant, is to make it advantageous, the high production yield for further use to plant-derived vaccines. The advantages of transgenic plants are higher yield, pest resistance, tolerance to heat, improved quality, cold and drought resistance, against a variety of biotic and abiotic stresses, improving shelf life. Transgenic plants have characteristic features such as industrial and pharmaceutical value because they express foreign and recombinant proteins. The vaccines or antibodies (Plantibodies) produced by transgenic plants are free from human infectious viruses and bacteria which reduces the cost of viruses and bacterial toxins screening [21].

Table 1. Subunit vaccine candidates expressed by transgenic plants or plant viruses [11, 22, 23]

Pathogen/host	Antigen	Production system	Efficacy and Application
Enterotoxigenic strains of *E. coli*/humans and farmed animals	Heat-labile toxin B subunit	Tobacco leaf	Immunogenic by oral delivery to mice
Vibrio cholerae/humans	Cholera toxin B subunit	Potato tuber	Immunogenic and protective by oral delivery to mice
Human immunodeficiency virus type 1/humans	V3 loop of gp120 protein	Alfalfa mosaic virus (tobacco leaf)	Immunogenic by Intraperitoneal delivery to mice
Pseudomonas aeruginosa/humans	Peptides of outer-membrane protein F	Cowpea mosaic virus (cowpea leaf)	Immunogenic and protective by subcutaneous delivery to mice
Staphylococcus aureus/humans	D2 peptide of fibronectin-binding protein FnBP	Cowpea mosaic virus (cowpea leaf)	Immunogenic by subcutaneous, nasal or oral delivery to mice or rats
Human rotavirus	HRV-VP7	Potato	Elicited serum IgG and mucosal IgA by oral delivery to mice
Measles virus hemagglutinin	MV-H	Tobacco	Indicative of humoral and mucosal immune responses by oral delivery
Clostridium tetani	TetC (tetanus vaccine antigen)	Tobacco chloroplast	Protective antibodies produced following intranasal immunization

Some of the disadvantages of the transgenic plant system as compared to the other plant platforms like a time-consuming i.e it required longer time for development because to carry out various processes (transformation, regeneration, analysis of transgenic, selection, and bulking up of the producer line), the variable effect of epigenetic events on transgene expression of the plants (post-transcriptional gene silencing, position effects) and the potential for transgene spread from some crops throughout crossing [20].

In 1983, the first transgenic plants were reported. The few important examples of transgenic plants are as follows tobacco, corn, tomato, potato, banana, alfalfa and, canola (all present the recombinant protein for vaccine production). Tobacco, potatoes, and banana plants were generally used for purpose of vaccine production for the human beings [21].

2.3. Transient Expression

The transient expression system is a technique in which the viral vectors or targeted proteins were checked for their expression and observed and finding the function of the recombinant protein. The transient expression is moved through recombinant proteins into a stable transgenic plant system and it is very easy to scale up at a large range. The transgenic plants used to either part or whole plants for vaccine production and it may be problematic and to avoid that transient expression system were suggested [24].

The transient expression is technique is performed by modification in plant viruses to add concerned genes on the viral vector for the production of target recombinant protein. Firstly, the host plant undergoes inoculation under suitable environmental conditions. The viruses which contain viral vector with additional genes are causes systemic infection to the host plant, which result in the production of an immense amount of recombinant proteins throughout the plant. The recombinant protein is produced until the host plants destroyed the virus or decreased the virus infection. The environmental risk is originated due to the use of infectious viruses. It is necessary to deconstruct the virus used for transient expression i.e removed the vital part of the virus machinery which causes systemic infection to the new host. After the destruction, the virus has the ability to local spreading and leading infection to the new host and because of this, a limited amount of recombinant proteins are produced or recovered [15].

The transient expression uses plant vector for gene transfer and the transient expression with plant viral vectors has the ability of high replication in the selected or host plants. By the use, transient expression system gives the high yield of vaccines. The plant viral expression vectors in the first-generation encode all the viral properties of genes and the protein or peptide of interest is encoded by the genes of vectors. These encoded proteins or peptides are subjected to viral polymerase, movement protein, and coat protein genes. The above technique is used to produce several plant-derived vaccines, such as the vaccine against norovirus by modifying tobacco mosaic virus (TMV) and vaccine against human papillomavirus and influenza virus, by modifying potato virus (PVX) or TMV. The vaccine production with the first-generation is not much safer because the viruses retain infectivity to plants, which affects the transgene properties. The second-generation viral vectors are safer as compared to the first-generation viral vectors because these vectors contain a minimum number of viral elements which is required for vector replication and based on an integrated system [5].

Agrobacterium tumefaciens is an important transient expression system, in this technique the infiltration of whole-plant tissue with a suspension of Agrobacterium and by this way, the T-DNA is transferred to a very high proportion of cells, where it can integrate or remain as an episome, and in either case, will express its payload. This technique produces very high levels of transient expression and it is important for the production of a high yield of transgenic plants. In the as early year, this technology was used to study plant-pathogen interactions and in the investigation of endogenous gene expression, and used to produce antibody molecules [4].

Table 2. Selected list of vaccine candidates expressed in plants against various diseases [14, 24-25]

Transformation Method	Expression System	Recombinant proteins or Antigen	Pathogen/Disease
Agrobacterium-mediated (Transient expression)	Tobacco (*Nicotiana benthamiana*)	H5N1 (avian influenza virus) & H1N1 (human) influenza strains	Haemagglutinin (HA)
		Bluetongue virus	Bluetongue virus-like particles
		Hepatitis B virus (HBV)	Hepatitis B virus surface antigen
		Norwalk virus	Norwalk virus capsid protein
		Infectious bursal disease virus (IBDV)	Structural proteinVP2
		Avian influenza A (H7N7)	Haemagglutinin (HA)
	Lettuce (*L. sativa*)	Ebola virus	6D8
	Tobacco (*Nicotiana benthamiana*)	Cancer (tumor marker)	scFvT84.66

The most used transient expression system is infiltration of tobacco leaves (Nicotiana benthamiana or Nicotiana tabacum) with A. tumefaciens and based on this transient expression various companies developed platforms for rapidly manufacturing vaccines such as influenza vaccines (virus-like particles (VLPs) produced by Medicago (Québec, Canada/Durham, NC, USA) and iBio/Caliber Therapeutics (Bryan, TX, USA) and ZMapp (Composed of three monoclonal antibodies) was produced by transient expression by Kentucky Bioprocessing and developed byLeaf Biopharmaceutical (the commercial arm of Mapp Biopharmaceutical, San Diego, CA, USA).

Based on the phenomenon of deconstructed tobacco mosaic virus, the different platforms have been developed like magnification system (Icon Genetics, Halle/Saale, Germany) and Launch Vector system (Fraunhofer CMB, Newark, DE, USA), used for high yield production of recombinant proteins [15].

The advantages of transient gene expressions system are widely accepted method for the production of an immense amount of recombinant proteins, is time-saving, and efficient. The detection of transient gene expressions within a short period means within 3 hr and 6 days, which reported that a transient gene expression is an effective tool.

2.4. Mechanism of Action of Edible Vaccines

Almost all the disease-causing pathogens cause infection to human beings through mucosal surfaces, mainly through the gastrointestinal tract, urogenital and respiratory tract, which is a major pathway for entry of pathogens into the host body. After the pathogen enters the body it is necessary to activate the mucosal immune system by edible vaccines because it is the first-line defense mechanism. The edible vaccine activates both systemic and mucosal immune responses and it shows the antibody-mediated and cell-mediated immune response in the host cells. The best way to administrate edible vaccines is by oral route because it is an efficient way of mucosal immunization or to induce mucosal immunity. The orally administered vaccine is passed through the gastrointestinal tract and the enzymatic degradation and hydrolyzation of edible vaccine by gastric enzymes are protected by bio-encapsulation i.e., the tough outer wall of transgenic plant cells and which are finally hydrolyzed (break up) and released in the intestines. The appropriate protein in the form of vaccines is released in the intestine. The released antigens from the vaccine are absorbed through the intestinal wall and these absorbed antigens are taken up by M cells in the intestinal lining that overlies peyerís patches [7, 10].

After that, it is further passed to the macrophages and B cells and moves towards the antigen-presenting cells (APC) and after that, they will be moved and present to T and B cells in the payer's patches. The antigens are also passed to the gut-associated lymphoid tissue (GALT), passed on to macrophages, antigen-presenting cells, and local lymphocyte populations, which induced serum IgG, IgE responses, The plasma cells are generated by the activation of immune cells and they move towards mesenteric lymph nodes for clonal amplification and maturation. After that, the distribution to other mucosal membranes surfaces secretes immunoglobulin A (IgA). When any pathogenic infection occurs then the local IgA response and memory helper T cells induces antibody production and secretion and the released antibody neutralizes the attack by disease-causing pathogens or infectious agent [6, 12, 24].

2.5. Advantages of Edible Vaccines

1) The edible vaccines have their ability and efficiency to stimulate immune response and initiative the immunization against disease-causing pathogens.
2) Edible vaccines are more effective than injected traditional vaccines.
3) The edible vaccines are usually in dried forms (in the form of seeds, fruits, and vegetables), less moisture content, therefore it does not require cold chain storage, and easy to produce and transport.
4) Edible vaccines are less expensive, harmless, and time-saving as compared to conventional vaccines.
5) Edible vaccines are administered by the oral route, decrease the use of syringes, and needles and do not require a trained medical person.
6) Transgenic plants are required to manufacture edible vaccines that do not require high-cost equipment and machinery.

7) The transgenic plant is easily grown in the soil, does not require high production cost, and special manufacturing premises like sterilization and cultivation.
8) Edible vaccines reduce the risk of contamination and do not contain heat-killed pathogens.
9) Edible vaccines are a source of new vaccines by combining the two or more antigens (second-generation vaccines).

2.6. Disadvantages of Edible Vaccines

1) The edible vaccines contain foreign antigens expressing by fruit and vegetables which may be developed an allergy in the people.
2) The consistency of edible vaccines form, changes from fruit to fruit, plant to plant, and generation to generation i.e efficiency changes.
3) The edible vaccines are produced in plant-like Beans, Potato, Rice, and Spinach cannot be used as raw, after the cooking protein in the plant is denatured and the efficacy of vaccines is lost.
4) The dosage frequency of edible vaccines is variant, therefore it is not suitable for infants.
5) Develop immune tolerance of the particular edible vaccine protein or peptide.

2.7. Significance of Plant-Based Edible Vaccine Technology

Plant-based edible vaccines are one of the most prominent and more recently proposed vaccine technologies. The use of the edible vaccine is increasing day by day because of its low cost, safe immunization and, the oral route of administration ad compared to conventional vaccines. The main objectives of plant-based edible vaccines are the insertion of antigen genes into the host plant by the process of transformation, then production of recombinant proteins in the transgenic plant, which administered by the oral route as food sources and generates antigen-specific immune responses against pathogens. The preservation of plant-based edible vaccines does not require the standard storage conditions; therefore it is more acceptable and developed in a foreign country without manufacturing premises. The storage conditions and manufacturing premises are removing the edible vaccine, which are approximately $300 million and which is used for further vaccines production in different fields.

The edible vaccines reduce the risk of contamination and eliminate poisonous compounds from human beings. Many edible vaccines are under the phases of clinical trials (Phase Ii & Phase III) and may be available in the future for human application. The different approaches are used to improve antigen expression in the transgenic plants like strong promoters, optimization of the amino acid codons in plants, use of vectors-based approach, and chloroplast transformation method. These approaches improve the high level of antigen genes protein in plants for representation of gene expression, which is later used for vaccine production [6, 27].

3. Applications

3.1. Cancer Therapy

The monoclonal antibodies are produced by the genetically engineered plants are used in cancer therapy. The monoclonal antibody (BR-96) is obtained by genetic modification of soybean, which is an effective agent that targets doxorubicin responsible for breast cancer, ovarian cancer, colon cancer, and lung tumors, Non-Hodgkinís lymphoma is also being examined [7, 10].

3.2. Birth Control

After administration of the Tobacco mosaic virus (TMV) fabricated the protein found in mouse zona pellucida (ZB3 protein). When TMV is injected it has the ability to preventing the fertilization of eggs in mice due to resulting antibodies [7, 10].

3.3. Hepatitis

The edible hepatitis B vaccine is produced by using a Hepatitis B surface antigen called HBsAg. The antigen HbsAG is introduced into the Potato plant for the development of an edible vaccine for hepatitis. The antigen HBsAg expression is seen more in roots than in other parts of the plant [28].

3.4. Anthrax

The pag gene is introduced into the tobacco plant by biolistic transfer, expressing PA antigen and it can be used in the treatment of anthrax disease of cattle. This Pag gene is also inserted or injected in spinach, tomatoes for future studies [28].

3.5. Chloroplast Transformation

The chloroplast genome has characteristics feature in its nature of maternal inheritance. Therefore there is a lack of protein in pollen and usually, the transgene cannot be transmitted through crops by cross-pollination. It may also result in the accumulation of remarkable higher quantities of transgenic protein [7, 10].

3.6. Role in Autoimmune Diseases

Plant-based edible vaccines are useful in the treatment of autoimmune diseases. The self-antigen production in plants is under the developmental stage for large-scale production and it is required a sufficient amount of self-antigen against autoimmune diseases. The number of diseases is under the clinical study of treatment of edible vaccines is multiple sclerosis, type 1 diabetes, rheumatoid arthritis, lupus, and transplant rejection. In one clinical study, the mice are fed with potatoes expressing insulin and a protein called GAD (glutamic acid decarboxylase), linked to CT-B subunit, which is previously susceptible to diabetes stain, It was observed that protein suppressing immune attack in diabetic mice and delayed the onset of high blood sugar level [7, 10, 28].

3.7. Recombinant Drugs/Proteins

Along with the vaccines and antibodies, the plants are genetically modified by engineered viral inoculations to produce enzymes and drugs like albumin, serum protease, and interferon. The examples are Interleukin-10 to treat Crohn's disease and glucocerebrosidase (hGC) production in tobacco plants for treating Gaucherís disease. The method of development of recombinant proteins is widely accepted and commercialized for large-scale production. The other example is hirudin, which is antithrombin, an antiviral protein that inhibits HIV in vitro, trichosanthin is a ribosome inactivator, and angiotensin-I (antihypertensive drug) [10, 6].

4. Future Prospective

The future of plant-based edible vaccine technology is very promising as this provides an effective, safe vaccine and eliminates the need to maintain and train a medical person for its delivery. An edible vaccine can be developed in large quantities at a low cost with less processing time during preparation. This vaccine can be developed even in a safe use environment. It is well known that traditional vaccines have played an important role in improving the health of the world's population, but their mass production is expensive and time-consuming, therefore, edible vaccine technology is a different way to solve these problems. Through this technology, an effective, efficient and safe vaccine can be given to the public, especially for a life-threatening disease that affects large numbers of people worldwide such as AIDS, dengue, malaria, heart disease, intestinal infections, and respiratory infections and disorders. In addition, in the future, a multidisciplinary vaccine such as an adjuvant may be developed by blending two generations of genetically modified plants that produce clinically important antigens [30].

The future of edible vaccines also depends on the WHO standard, especially in terms of cost-effectiveness, hygiene, and safety. Currently, microalgae are used as the most vital source of active molecules such as carotenoids, chlorophyll, enzymes, fatty acids, phycobiliproteins, and carotenoids. Microalgae can be used in the future with recombinant protein expression, cleansers, medications, antibodies, growth factors, hormones, and many other products, such as the anti-cancer agent i.e., Taxol, and can be used as an edible vaccine. They have many distinct advantages such as large-scale production and rapid transformation and rapid growth and stable expression levels with proper folding and inclusion of multiple antigens in a form of vaccines and effective delivery via the oral route. They can be used as a tablet for easy administration and immunogenic response [31-32].

There are several edible vaccines candidates are under the phase of clinical trials, they are given below:

Table 3. Vaccine candidate under clinical trials [5, 7, 8, 28, 29]

Pathogens	**Antigen**	**Host plants**	**Expression Systems**	**Clinical trial phase**	**Use**
Enterotoxigenic *E. coli*	LT-B	Potato	Transgenic	Phase 1	Diarrhoea
Enterotoxigenic *E. coli*	LT-B	Maize	Transgenic	Phase 1	Diarrhoea
Rabies Virus	Glycoprotein and nucleoprotein (fusion)	Spinach	Viral vector (transient)	Phase 1	Rabies
Hepatitis B virus	Viral major surface protein	Lettuce	Transgenic	Phase 1	Hepatitis
Norwalk Virus	Capsid protein	Potato	Transgenic	Phase,1	Diarrhoea
Hepatitis B virus	Viral major surface protein	potato	Transgenic	Phase 1	Hepatitis
Influenza virus	HA (VLP) (seasonal; quadrivalent)	*Nicotiana-bentham-iana*	Agrobacterial binary vector (transient)	Phase 1	Influenza virus Infection
HBV	HBV	*Sacchar-omyces cerevisiae*	-	Phase 2	Chronic HCV
HCV	HCV	*Saccharo-myces Cerevisiae*	-	Phase 2	Chronic HCV
Influenza virus (H5N1)	HA	*Nicotiana-bentham-iana*	Launch vector (transient)	Phase 1	
Vibrio cholerae	CTB	Rice	Transgenic	Phase 1	Cholera

References

[1] Kumar, Gaurav, Loganathan Karthik, and Kokati Venkata Bhaskara Rao. 2018. "Plant Vaccines: An Overview." *Microbial*

Bioprospecting for Sustainable Development, 249-263. doi:10.1007/ 978-981-13-0053-013.

[2] Laere, Erna, Anna Pick Kiong Ling, Ying Pei Wong, Rhun Yian Koh, Mohd Azmi Mohd Lila, and Sobri Hussein. 2016. "Plant-Based Vaccines: Production and Challenges." *Journal of Botany*, 1-12. http://dx.doi.org/10.1155/2016/4928637.

[3] Balke, Ina and Andris Zeltins. 2019. "Use of Plant Viruses and Virus-Like Particles for the Creation of Novel Vaccine." *Advanced Drug Delivery Reviews* 145: 119-129. doi:10. 1016/j.addr.2018.08. 007.

[4] Rybicki, Edward P. 2009. "Plant-Produced Vaccines: Promise and Reality". *Drug Discovery Today* 14 (1-2): 16-24. doi:10.1016/ j.drudis.2008.10.002.

[5] Takeyama, Natsumi, Horoshi Kiyono, and Yoshikazu Yuki. 2015. "Plant-Based Vaccines for Animals and Humans: Recent Advances in Technology and Clinical Trials." *Therapeutic Advances. In Vaccines*, 1-16.DOI: 10.1177/ 2051013615613272.

[6] Sohrab, Sayed Sartaj, Mohd. Suhail, Mohammad A. Kamal, Azamal Husen, and Esam I. Azhar. 2017. "Recent Development and Future Prospects of Plant-Based Vaccines." *Current Drug Metabolism* 18 (9): 831-841. doi:10.2174/1389200218666170711121810.

[7] Goyal, R, R Sharma, P Lal, and VG Ramachandran. 2007. "Edible Vaccines: Current Status and Future." *Indian Journal of Medical Microbiology* 25 (2): 93. doi:10.4103/0255-0857.32713.

[8] Munshi, Anjana, and Vandana Sharma. 2018. "Omics and Edible Vaccines." In Omics *Technologies And Bio-Engineering: Towards Improving Quality Of Life*, 129-141. Academic Press. DOI: https://doi.org/10.1016/B978-0-12-815870-8.00008-5.

[9] Khan, Arif, Asad Khan, Irfan Khan, Muhammad Aamir Shehzad, Waqar Ali, Atif Muhammad, and Muhammad Akif. 2019. "A Review on Natural Way of Vaccination: Plant Derived Edible Vaccines." *Journal of Vaccine and Immunology* 5 (1): 028-021. doi: https://dx.doi.org/10.17352/jvi.

[10] Jan, Naeema, Fouzia Shafi, Omar Bin Hameed, Khalid Muzaffar, Shuaib Mohammad Dar, Ishrat Majid, and GA Nayik. 2016. "An Overview on Edible Vaccines and Immunization." *Austin Journal of Nutrition and Food Sciences* 4 (2): 1078.

[11] Appaiahgari, Mohan Babu, Usha Kiran, Athar Ali, Sudhanshu Vrati, and Malik Zainul Abdin. 2017. "Plant-Based Edible Vaccines: Issues and Advantages." *In Biotechnology: Principles And Applications.* 329-366. Singapore: Springer. https://doi.org/10.1007/978-981-10-2961-5_13.

[12] Gunasekaran, B., and K.M. Gothandam. 2020. "A Review on Edible Vaccines and Their Prospects." *Brazilian Journal of Medical and Biological Research*. 53(2): e8749. http://dx.doi.org/10.1590/1414-431X20198749.

[13] Fischer, R., and S. Schillberg. 2017. "Molecular Farming" *Encyclopedia of Applied Plant Sciences* 2:77-82. http://dx.doi.org/10.1016/B978-0-12-394807-6.00159-3.

[14] Shanmugaraj, Balamurugan, Christine Joy I. Bulaon, and Waranyoo Phoolcharoen. 2020. "Plant Molecular Farming: A Viable Platform for Recombinant Biopharmaceutical Production." *Plants* 9 (7): 842. doi:10.3390/plants9070842.

[15] Fischer, Rainer, and Johannes F. Buyel. 2020. "Molecular Farming-The Slope of Enlightenment." *Biotechnology* Advances 40: 107519. doi:10.1016/j.biotechadv.2020.107519.

[16] Ahmad, K. 2014. "Molecular Farming: Strategies, Expression Systems, and Bio-Safety Considerations." *Czech Journal Of Genetics and Plant Breeding* 50 (No. 1): 1-10. doi:10.17221/187/2013-cjgpb.

[17] Tak, Himanshu, Sanjana Negi, T. R. Ganapathi, and V. A. Bapat. 2016. "Molecular Farming: Prospects and Limitation." *Banana: Genomics and Transgenic Approaches for Genetic Improvement*, 261-275. doi:10.1007/978-981-10-1585-4_18.

[18] Clark, Michael, and Maciej Maselko. 2020. "Transgene Biocontainment Strategies for Molecular Farming." *Frontiers in Plant Science* 11: 1-11. doi:10.3389/fpls.2020.00210.

[19] MM, Awale, and Mody SK. 2012. "Transgenic Plant Vaccine: A Breakthrough in Immunopharmacotherapeutics." *Journal of Vaccines & Vaccination* 03 (05): 147. doi:10.4172/2157-7560.1000147.

[20] Yusibov, Vidadi, and Shailaja Rabindran. 2008. "Recent Progress in the Development of Plant-Derived Vaccines". *Expert Review of Vaccines* 7 (8): 1173-1183. doi:10.1586/14760584.7.8.1173.

[21] Jhansi Rani, S., and R. Usha. 2021. "Transgenic Plants: Types, Benefits, Public Concerns, and Future." *Journal of pharmacy research* 6:879 e883. http://dx.doi.org/10.1016/j.jopr.2013.08.008.

[22] Streatfield, Stephen J, and John A Howard. 2003. "Plant-Based Vaccines." *International Journal for Parasitology* 33 (5-6): 479-493. doi:10.1016/s0020-7519(03)00052-3.

[23] HAN, Mei, Tao SU, Yuan-Gang ZU, and Zhi-Gang AN. 2006. "Research Advances on Transgenic Plant Vaccines." *Acta Genetica Sinica* 33 (4): 285-293. doi:10.1016/s0379-4172(06)60053-x.

[24] Moon, Ki-Beom, Ji-Sun Park, Youn-Il Park, In-Ja Song, Hyo-Jun Lee, Hye Sun Cho, Jae-Heung Jeon, and Hyun-Soon Kim. 2019. "Development of Systems for the Production of Plant-Derived Biopharmaceuticals." *Plants* 9 (1): 30. doi:10.3390/plants9010030.

[25] Sahai, Aastha, Anwar Shahzad, and Mohammad Shahid. 2013. "Plant Edible Vaccines: A Revolution in Vaccination." *In Recent Trends in Biotechnology and Therapeutic Applications of Medicinal Plants.* 225-252. https://doi.org/10.1007/978-94-007-6603-7_10.

[26] Bhatia, Saurabh, and Randhir Dahiya. 2015. "Edible Vaccines." *In Modern Applications of Plant Biotechnology in Pharmaceutical Sciences*, 333-343. Academia Press. http://dx.doi.org/10.1016/B978-0-12-802221-4.00009-1.

[27] Kim, Tae-Geum, and Moon-Sik Yang. 2010. "Current Trends in Edible Vaccine Development Using Transgenic Plants." *Biotechnology And Bioprocess Engineering* 15 (1): 61-65. doi:10.1007/s12257-009-3084-2.

[28] Kurup, Vrinda M, and Jaya Thomas. 2019. "Edible Vaccines: Promises and Challenges." *Molecular Biotechnology* 62 (2): 79-90. doi:10.1007/s12033-019-00222-1.

[29] Khan, Faria, Rafal Szmigielski, Alvina Gul, Volkan Altay, and Munir Ozturk. 2020. "Advancements In-Plant Transgenomics Approach for the Biopharmaceutics and Vaccines Production." *In Biodiversity and Biomedicine*, 317-333. Academia Press. DOI: https://doi.org/10.1016/B978-0-12-819541-3.00018-9.

[30] Sayed Sartaj Sohrab, Mohd. Suhail, Mohammad A. Kamal, Azamal Husen and Esam I. Azhar. 2017. "Edible Vaccine: Current Status and Future Perspectives". *Current Drug Metabolism*, 18; DOI: 10.2174/1389200218666170711121810.

[31] Specht, E. A.; Mayfield, S.P.2014. "Algae-based oral recombinant vaccines". *Front. Microbiol.* 17:5:60. DOI: https://doi.org/10.3389/fmicb.2014.00060.

[32] Yan, N.; Fan, C.; Chen, Y.; Hu, Z. 2016. The Potential for Microalgae as Bioreactors to Produce Pharmaceuticals. *Int. J. Mol. Sci.,* 17:17. DOI: 10.3390/ijms17060962.

In: The Future of Metabolic Engineering ISBN: 978-1-68507-362-6
Editors: A. Sharma and D. Amin

Chapter 10

METABOLIC ENGINEERING APPROACHES FOR PLANT SECONDARY METABOLITES BIOSYNTHESIS

Subir Kumar Bose[1,*], Ramesh Kumar[2,†] and Reshma Kapoor[2]

[1]Assistant Professor, Biotechnology and Microbiology Department (CCS Programme), Meerut Institute of Engineering and Technology (MIET), Meerut, Uttar Pradesh, India

[2]Assistant Professor, Riva University, Biochemistry Deptt, Kattigenahalli, Bangalore, India

ABSTRACT

A novel recombinant DNA technology (r-DNA technique) is introduced for the purpose of creating transgenic organism ultimately resulting in the production of desired product. Product produced from the help of r-DNA techniques may be animal and or plant based. The r-DNA technique is also known as pathway/or metabolic engineering because of it interferes or modulates some secondary metabolite

* Corresponding Author's E-mail: bbausubir01@gmail.com, subir.kumar@miet.ac.in.
† Corresponding Author's E-mail: kushwaha.rameshkumar@gmail.com.

pathways and results in production of desired metabolites. These transformations create transgenic/GMO organism whose genetic constituent has been modified using genetic engineering approach. In the modern era GMOs have a lot of applications in medical and biological, gene therapy (Drug SMA-type 1 disease), production of pharmaceutical products (Insulin, & Taxol), agriculture (Golden rice), Bt crop (Bt cotton, brinjal & maize) and also environmental safety like bioremediation etc. By using genetic engineering approaches several metabolic pathways in plants are altered to produce specific product that play important role in our routine life. In this book chapter we are focusing only plant-based product (secondary metabolites) which are easy to produced *via* transformation/mutation/genomic editing and the process known as genetic/pathway/or metabolic engineering. Many secondary metabolites of high demand are very expensive and very low in production when grow in natural/traditional conditions. In this book chapter, we summarize production strategies through transgenic/GMO approaches in natural or artificial environment that are succesfully applied in such cases. We also review various strategies that have been used to increase the yield of plant based compounds in these microbial models and also recent progress made in metabolic engineering and biotechnology/bioinformatics approaches (pathways engineering) and their implemention in microorganisms with particular focus on the model of *Escherichia coli* or *Saccharomyces cerevisiae* or in plant for improved and enhance plant-based novel secondary metabolites.

Keywords: secondary metabolites, metabolic engineering, MACs (medicinal and aromatic compounds), medicinal plant, recombinant DNA technology

ABBREVIATIONS

Abs	antibodies
Bt	*Bacillus thuringiensis*
GE	Genetically engineered
GMO	Genetically modified organism
MACs	medicinal and aromatic compounds
MAPs	medicinal and aromatic plants
r-DNA	Recombinant DNA

1. Introduction

Plant secondary metabolites are plant-based natural compounds mainly synthesized by medicinal and aromatic plants that indirectly participate in growth and development of plant. The plant kingdom manufactures more than 200,000 distinct chemical compounds, most of which arise from specialized metabolism (Connolly et al., 1991; Croteau et al., 2005; and Pyne et al., 2019). They are synthesized from different part of the plants *viz* leaves, stems, flowers, barks, wood, and sometimes in roots. It is always isolated, identified and purified natural products that are important for pharmaceuticals. Typical group of plants compounds (secondary metabolites) commonly used drugs are as terpenoids, alkaloids, polyketides, phenylpropanoids, and flavonoids. For examples, pain killer drug like morphine and codeine are obtained from *Papaver somniferum* L., anti-malarial drug artemisinin from *Artemisia annua* L., paclitaxel from *Taxus brevifolia* Nutt, genistein from *Glycine max* L. (Merr.), withanolide and withaferin are obtained from *Withania somniferra*, strong antioxidant and anti-inflammation natural compound curcumin are obtained from *Curcuma longa*, scopolamine from *Dubosia* species, campothecin from *Camptotheca acuminata* Decne, and podophyllotoxin from *Podophyllum* species.

According of World Health Organization (WHO) report more than 80% of the population in the world in less developed countries depend primarily on herbal medicine for basic healthcare needs and main reason being accessibility and minimal or no side effects of plant-based products (Vines 2004). The current herbal drug market has reached a level of US$62 billion, which is forecast to grow to US $5 trillion in 2050 (Joshi 2004). In the past 30 years, more than 25% of the new drug entities approved was based on a molecule of plant origin and about one third of the approximately 980 new pharmaceuticals originated from, or was inspired by, natural products (Newman 2007 and Terryn 2006).

While plants synthesize secondary metabolites for unique function towards in their own defense and protection from grazing animals, protecting from insect and from some disease infection as well as UV light; they also possess a number of medicinal properties. The plant having the capabilities to synthesize secondary metabolites that have medicinal properties are known as medicinal plants, for example *Withania somniferra, Mentha arvensis* and *Cinchona officinalis* etc., other properties include flavors (*M. arvensis* and *Cymbopogon martini*), fragrances (*Cymbopogon citratus* also known as lemongrass), repellents (Pyrethrum is obtained from *Chrysanthemum coccineum*), and nutrients (*Moringa oleifera* popularly known as drumstick plant) etc., (Bose et al., 2013; Sharma et al., 2013). Powerful anticancer drugs vinblastine and vincristine is obtained from the flowering plant *Catharanthus roseus.* Taxol is also an anticancer drug extracted from the bark of pacific yew tree (*Taxus brevifolia*) and is used in the treatment of breast, lung, and ovarian cancer, as well as Kaposi's sarcoma. The total cost of a typical weekly paclitaxel regimen is about $13,000, whereas the twice dose in week is $23,000 to $40,000, because of its very low production. For reducing the high cost of taxol, scientists resorted to increasing the production of taxol using a bacterial expression system. The taxol biosynthesis gene is inserted into the bacterial system, and thereby the production cost and time will be very low in the future. Microbial systems like *Escherichia coli*, *Saccharomyces cerevisiae*, and *Bacillus subtilis* have been successfully employed to produce taxadiene. In 2010, Ajikumar et al. reached a production of 1.0 g L^{-1} by following a two-module strategy in *E. coli.* They optimized a native upstream MEP pathway and a heterologous downstream taxadiene-forming pathway with the minimal accumulation of indole, which acts as an inhibitory compound.

In the modern scenario scientist are looking to increase yield and production of plant-based secondary metabolites. While with traditional cultivation it takes high cost and low yield as well as long time, scientists aim to tackle these drawbacks by adopting modern genetic engineering technique mainly r-DNA technology and applying it in plant system especially for the production of transgenic plant and microbial culture/tissue culture/hairy root culture etc. These techniques are frequently used for metabolic engineering purpose. The metabolic abilities of various microorganisms (including both yeast and bacterial system) for producing novel/desired compounds are widely exploited in the manufacturing of chemical and pharmaceutical industry. Furthermore, there is an increasing tendency to replace chemical drug because of their toxic harmful side effect, for which scientists are adopting biotechnological based production involving microbial fermentations. As microorganisms are typically evolved for survival and growth in their natural habitat, it is often compulsory to regain the genotype of the applied cell factory to obtain a desired phenotype. Through the use of directed manipulation of genetic constituent using pathway engineering (r-DNA technology), it is possible to change the metabolic abilities of many other microorganisms, an approach referred to as metabolic engineering and resulted increased production and yield of desired secondary metabolites in very short time period. In metabolic engineering the 'metabolic design problem' can be approached through the construction of a mathematical or *in silico* model with the help of bioinformatics approaches (Sliwoski et al., 2014; Van Drie 2007). The model is used to 'draw' an improved metabolic network by suggesting changes in the genotype of such model microorganism. Model molecular biology approaches are used to perform such hypothesis and *in silico* model into root level.

In light of the promising use of mathematical models of metabolic networks in metabolic engineering, these models are likely to find application in areas beyond industrial biotechnology because they represent a platform for computational analysis of genomic DNA (*in silico* biology). They will also play a role in life science applications; in medical science successfully developed many drugs for example norfloxacin, saquinavir (approved in 1995), ritonavir and indinavir (both approved in 1996) and tirofiban, a fibrinogen antagonist approved in 1998. The complex nature of cellular metabolism and regulation often poses difficulties in metabolic engineering and it is therefore often necessary to analyze the metabolism as a whole. Complete genome sequences of desired microorganism are openly available and provide a platform to build a metabolic models on a genomic scale. Furthermore, the progress in experimental biology has moved the focus of modern biology from a traditional approach to a 'global' holistic perspective of the cellular level. This has resulted in the establishment of huge experimental databases, or also called omics databases biology, and much information on different microorganisms is also available on such platform. First with the help of genome-level models predictions or hypotheses a model which is directly connected to the nature and level of abstraction in the model used, and constantly assess the importance of applying genome-scale information in pathway engineering. In this review, our study is based on exploring the biosynthesis of plant-based product through modified and altered genome scale into microbial models/system with the help of metabolic engineering and using r-DNA technology approaches. In this book chapter, we discuss the production strategies (short time production and maximum yields) of plant-based secondary metabolites in microbes and other organism and also plants (Transgenic plants) *via* metabolic engineering/r-DNA technology approach.

2. DISCUSSIONS

2.1. Biopharmaceutical Biosynthesis Using Genetic/Pathways Engineering

The use of plants or their extracts have been adopted in society since from ancient time for the treatment of various human diseases. Indeed, chemically synthesized drug are dominating over natural drug, however about one-fourth pharmaceutical drugs are obtained from plant origin (Mukherjee et al., 2017). However, modern biotechnology has led to modify metabolic pathway for interested compound to obtain from genetic engineering (GE) plant. Through GE methods, tremendous compound can be produced such as: a variety of proteins, including mammalian antibodies, antimicrobial peptide, edible vaccines, hormones, enzymes, interleukins, interferons, human serum albumin and others

(Khademi et al., 2019; Goldstein and Thomas, 2004). Genetically modified plants acting as bioreactor can proficiently produce pharmaceutical compound or recombinant protein in large quantity. The plant-based pharmaceutical compound/medicine/antibody is free from human disease and easily can be grown in field for large scale production at low cost (Yao et al., 2015). The first commercial biopharmaceutical i.e., recombinant human insulin was produced in 1982 from genetically modified bacteria. Later in 1984 genetically engineered plant was developed and created lot of hope for pharmaceutical product with genetic expression of human growth hormone, interferon, monoclonal antibodies, and serum albumin. Numerous biopharmaceutical products are under clinical trials and till date none of the commercially available products are currently produced in plants. The plant-based antibodies production named as "plantibodies" (Penney et al., 2011). Plant derived antibodies (Abs) have many applications like binding to pathogenic organisms, binding to tumour antigens, binding to a cellular receptor to up or down regulation of receptor function (Whaley et al., 2011). US-FDA approved several recombinant monoclonal Abs for production and there are many under the development process (Lu et al., 2020). Antibody can be full size with light and heavy chain or Fab region or single chain fragments or chimeric Abs obtained from GE plants (Ahmad et al., 2012). Plants like tobacco, potatoes, soybeans, alfalfa, rice and corn have been used Abs production (Lu et al., 2020). The gene designed such a way that a chimeric secretory IgG/IgA antibody has been raised in tobacco which proved effective towards a surface antigen of *Streptococcus* mutants. This antibody is applied topically to teeth and blocked colonization of the *S. mutans* which confirmed action against dental caries (Gidding, 2001). Another example is the humanized Abs against herpes simplex virus (HSV) is expressed in genetically modified (GM) soybeans and effective on HSV-2 transmission in mice. Similarly human Abs are produced from GM corn and reported up to a kg yield per acre and Abs performed action even after five years of storage at normal condition. GM rice and wheat expressed antibody against carcinoembryonic antigen that may be useful for in vivo tumor imaging

(Stoger et al., 2000). Even tumor specific vaccine is expressed in tobacco for the treatment of lymphoma cancer disease. Currently, f plant derived antibodies are raised for validation against the diagnosis of cancer, dental caries, herpes simplex virus, and respiratory syncytial virus.

In present era, scientists are looking for alternative of conventional vaccine i.e edible vaccine. Edible vaccines are plant based or animal based production of immune trigger agents for animal's immune response through transgenic technology (Kurup and Thomas, 2020). For edible vaccines production, as suitable plants, algae, yeast, insect cells and bacteria, are selected so their products can be consumed without loss of their efficacy. These types of edible vaccines are economic, easy to produce at location wise, needleless, convenient and safe to vaccine production (Sohrab et al., 2017). Many plant based vaccines have been reported and most of them are still at clinical trial phase only. For e.g., a Phase II clinical trial has lately been performed using a quadrivalent fu vaccine derived from plants and declared in recent times that a phase III clinical study has been undertaken (Pillet et al., 2019).

Recently, *Mycobacterium tuberculosis* early secretory antigenic protein (ESAT-6) was expressed in *Brassica oleracea* for edible vaccine production (Saba et al., 2020). Hepatitis B virus (HBV) chronically effect on liver all over the world, therefore hepatitis B surface antigen (HBsAg) is an effective protective antigen suitable for vaccine development. HBsAg genes were transformed in rice and banana for plant based vaccine production (Guan et al., 2010). Similarly, vaccine against tetanus, diphtheria, and Norwalk virus were developed in potato (Kurup and Thomas, 2020). Edible vaccines against the severe acute respiratory syndrome coronavirus (SARS-CoV) spike protein were expressed in tobacco plant (Li et al., 2006). HIV-1 vaccine has been rained the spinach plant (Karasev et al., 2005). Urease subunit B gene of *Helicobacter pylori*is is transformed in *Arachis hypogaea* for edible vaccine purpose (Yang et al., 2011). Some of the edible vaccines are at initial expression phase and some are at clinical trial phase. Of course edible vaccine and Ig production from plant source will be good fortune. However, till date, there is no edible vaccine are approved by US-FDA or Indian government at commercial scale production for public use.

2.2. Plant Based Secondary Metabolism Obtained in Microbial Systems

Microorganisms are very important source for the secondary metabolites and play an important role in medicine, agriculture, and industry. Many free living fungi and bacteria are reported for the production of therapeutic, herbicidal, growth promoting agents and tools for biochemistry (Singh et al., 2019). Metabolic pathways of eukaryotic and prokaryotic microorganism are different and lead to diversity of natural compound. However, metabolic pathway in eukaryotic fungi are close to higher plants, therefore some metabolites in fungi are shared with plant's metabolic pathway (Calvo et al., 2002). The secondary metabolites productions in free living microorganism differ from microorganism that reside within plant tissue called "endophyte" (Gouda et al., 2016). Plant endophytes can be defined as "the microorganisms inhabiting within plant tissue/organs for complete life cycle or part of their life without causing apparent harm to the host" (Ancheeva et al., 2020).

Endophytic fungi are recognized as an alternative source of plant secondary metabolite. Because endophyte produces similar to plant secondary metabolites such as alkaloids, coumarins, flavonoids, lignans, saponins, terpenes, quinones, and xanthones molecules, which is synthesized from very few key precursor molecules by little variations in basic metabolic pathways (Ancheeva et al., 2020). Alkaloids are organic compounds containing at least one nitrogen atom which are derived from aromatic amino acids and dimethlyallyl pyrophosphate through the shikimic acid and the mevalonate pathways. Several endophytic fungi have been isolated from respective host and are known to produce same alkaloid as in their host (Table 1). For example, a antitumor camptothecin is mainly sourced from medicinal plant *Nothapodytes nimmoniana*, and same camptothecin compound has been isolated from several endophytic fungi such as *Botryosphaeria parva, Diaporthe conorum, Fusarium oxysporum, Fusarium sacchari, Fusarium solani, Fusarium subglutinans, Fusarium verticillioides, Galactomyces sp., Irpex lacteus, Phomopsis sp.,* and *Fusarium sp* (Shweta et al., 2010). *Catharanthus roseus* is a main source of well-known anticancer alkaloids molecule called vinblastine and vincristine. However, same alkaloid (vinblastine and vincristine) compound has been reported in *C. roseus* endophytic fungi *Fusarium oxysporum, Talaromyces radicus,* and *Eutypella spp* (Palem et al., 2015).

Coumarins are lactone containing group of molecule which are synthesized from cinnamic acid through the shikimic acid pathway (Annunziata et al., 2020). Isofraxidin is a coumarin, known for anti-inflammatory activity, hyperpigmentation agent and cytotoxic effect on cancer cell. Initially, this compound was reported from *Acanthopanax senticosus* and *Apium graveolens*. Thereafter, same compound has been reported from endophytic fungi *Annulohypoxylon bovei var. microspora* of *Cinnamomum* sp host plant. Some other coumarins compound from endophytic fungi are mentioned in table 1. Flavonoids are synthesized through phenylpropanoid pathway from basic phenylalanine molecule *via* series of enzymatic reaction such as phenylalanine ammonia lyase (PAL), chalcone synthase, chalcone isomerase, and flavonol reductase (Sharma et al., 2019). Derivatization of basic structure diversifies the flavonoids molecule into flavones, isoflavones, flavonols, flavanones, catechins, and anthocyanins. *Curcuma* species are primary source of curcumin flavonoid, while in Indian tradition this plant is in use since from ancient time during various health conditions. Curcumin has been recognized due to its medicinal properties, strong anti-inflammatory and antioxidant activities. Interestingly, same compound has been also reported from endophytic fungi *Chaetomium globosum* (Wang et al., 2012). Some other flavonoids such as chalcone, kaempferol, rutin, rotenone, vitexin, silymarin and quercetin are reported from endophytic fungi which are shown in Table 1.

Table 1. Table show list of some secondary metabolite reported in fungi as same to host metabolite

Sr No.	Sec. metabolite	Class	Endophytic Fungi	Plant host	Reference
1.	Vinblastine, Vincristine	Alkaloid	*Talaromyces radicus, Fusarium oxysporum*	*Catharanthus roseus*	Palem et al.,2015, Kumar et al., 2013
2.	Camptothecin	Alkaloid	*Fusarium solani*	*Camptotheca acuminata*	Liu et al., 2010
3.	Cinchona	Alkaloid	*Arthrinium, Fomitopsis, Diaporthe, Penicillium, Phomopsis, Schizophyllum*	*Cinchona ledgeriana*	Maehara et al., 2013
4.	Piperine	Alkaloid	*Periconia sp. Colletotrichum gloeosporioides*	*Piper longum, Piper nigrum*	Verma et al., 2011, Chithra et al., 2014
5.	Capsaicin	Alkaloid	*Alternaria alternata*	*Capsicum annuum*	Devari et al., 2014
6.	Sanguinarine	Alkaloid	*Fusarium proliferatum*	*Macleaya cordata*	Wang et al.,2014
7.	Isofraxidin	Coumarins	*Annulohypoxylon bovei var. microspora*	*Cinnamomum sp*	Yamazaki et al., 2010, Cheng et al., 2011
8.	Mellein	Coumarins	*Nigrospora* sp.	*Moringa oleifera*	Zhao et al., 2012
9.	Scopoletin, Umbelliferone	Coumarins	*Penicillium* sp	*Avicennia*	Huang et al., 2012
10.	Curcumin	Flavonoids	*Chaetomium globosum*	*Curcuma wenyujin*	Wang et al., 2012
11.	Chalcone	Flavonoids	*Ceriporia lacerata*	*Cleistocalyx operculatus*	Wang et al., 2013
12.	Kaempferol	Flavonoids	*Fusarium chlamydosporum*	*Tylophora indica*	Chaturvedi et al., 2014
13.	Quercetin	Flavonoids	*Aspergillus nidulans, Aspergillus oryzae*	*Ginkgo biloba*	Qiu et al., 2010
14.	Rutin	Flavonoids	*Aspergillus flavus*	*Aegle marmelos*	Patil et al., 2015
15.	Rotenone	Flavonoids	*Penicillium sp.*	*Derris elliptica*	Hu et al.,2005
16.	Vitexin	Flavonoids	*Colletotrichum* sp.	*Ginkgo biloba*	Zhou et al., 2011

Table 1. (Continued)

Sr No.	Sec. metabolite	Class	Endophytic Fungi	Plant host	Reference
17.	Silymarin	Flavonoids	*Aspergillus iizukae*	*Silybum marianum*	El-Elimat et al., 2014
18.	Podophyllotoxin	Lignans	*Fusarium solani*	*Podophyllum hexandrum*	Nadeem et al., 2012
19.	Sesamin, Syringaresinol, Ketopinoresinol	Lignans	*Annulohypoxylon ilanense*	*Cinnamomum cassia*	Cheng et al., 2013
20.	Phillyrin	Lignans	*Colletotrichum gloeosporioides*	*Forsythia suspensa*	Zhang et al., 2012
21.	Diosgenin	Saponins	*Fusarium sp.*	*Dioscorea nipponica*	Ding et al.,2014
22.	Ginsenoside	Saponins	*Aspergillus* sp., *Fusarium* sp., *Verticillium* sp.	*Panax ginseng*	Wu et al., 2013
23	Gymnemagenin	Saponins	*Penicillium oxalicum*	*Gymnema sylvestre*	Parthasarathy et al., 2014
24	Azadirachtin	Terpenes	*Penicillium (Eupenicillium) parvum*	*Azadirachta indica*	Kusari et al., 2012
25	Asiaticoside	Terpenes	*Colletotrichum gloeosporioides*	*Centella asiatica*	Gupta et al., 2018
26	Camphor	Terpenes	*Nodulisporium* sp.	*Lagerstroemia loudoni*	Suwannarach et al., 2013
27.	Cineole	Terpenes	*Nodulisporium* sp	*Thelypteris angustifolia*	Hassan et al., 2013
28.	Ginkgolide	Terpenes	*Fusarium oxysporum*	*Ginkgo biloba*	Cui et al., 2012
29.	Taxol	Terpenes	*Penicillium raistrickii*	*Taxus brevifolia*	Stierle et al., 2000
30.	Withanolide	Terpenes	*Taleromyces pinophilus*	*Withania somnifera*	Sathiyabama and Parthasarathy,2018
31.	Xanthatin	Terpenes	*Paecilomyces* sp	*Panax ginseng*	Xu et al., 2009
32.	Anthraquinone	Quinones	*Coniothyrium* sp.	*Salsola oppostifolia*	Sun et al., 2013
33.	Questin	Quinones	*Eurotium cristatum*	*Sargassum thunbergii*	Du et al., 2014
34.	Questinol	Quinones	*Penicillium glabrum*	*Punica granatum*	Hammerschmidt et al., 2012
35.	1,7-dihydroxyxanthone	Xanthones	*Penicillium* sp.	*Avicennia*	Huang et al., 2012

Sr No.	Sec. metabolite	Class	Endophytic Fungi	Plant host	Reference
36.	Lapachol	Xanthones	*Aspergillus niger*	*Tabebuia argentea*	Channabasava et al., 2014
37.	Hypericin	Xanthones	*Chaetomium globosum*	*Hypericum perforatum*	Kusari et al., 2008
38.	Lawsone	Xanthones	*Gibberella moniliformis*	*Lawsonia inermis*	Sarang et al., 2017

Lignans are group of polyphenols which is synthesized by a series of enzymes pinoresinollariciresinol reductase, PAL, cinnamoyl CoA reductase, and cinnamyl alcohol dehydrogenase (Satake et al., 2017). Podophyllotoxin is an aryl tetralin lactone lignan that disorganizes microtubules acting as anti-tubulin agent. Several endophytic fungi *Diphylleia sinensis,* and *Dysosma veitchii* have been isolated from *Podophyllum hexandrum* and were producing same podophyllotoxin (Table 1). Other lignans like sesamin, syringaresinol, ketopinoresinol were isolated from *Annulohypoxylon ilanense*, an endophyte of *Cinnamomum cassia* (Cheng et al., 2013).

Terpenes is a largest group of compound consist of isoprene units which are synthesized through mevalonate (operated in cytoplasm) and/or methylerythritol 4-phosphate (MEP-operated in chloroplast) pathway by sequence of enzymatic reaction (Zhao et al., 2013). In plants, both pathways are involved for isoprene unit synthesis, however in case of fungi only mevalonate pathway operates for the synthesis of terpenoids molecule. Taxol is well known anti-cancer agent and their primary source is *Taxus brevifolia* medicinal tree. However, many endophyte has been isolated as alternative source of taxol production in laboratory condition (Singh et al., 2021). For example, *Penicillium raistrickii*, an endophytic fungus has been reported from *Taxus brevifolia* for taxol production (Stierle et al., 2000). Saponins consist of steroid or triterpenoid aglycone and one or more sugar moiety. Saponins are synthesized from intermediates molecules of the phytosterol pathway by involving several enzymes such as oxidosqualene cyclases, cytochromes P450, and UDP-glycosyltransferases (Moses et al., 2014). There are several plant derivate saponins reported from endophytic fungi. However, some saponins like diosgenin, ginsenoside and gymnemagenin were isolated from endophyte *Fusarium sp., Aspergillus sp.,* and *Penicillium oxalicum* respectively (Ding et al.,2014; Wu et al., 2013; Parthasarathy et al., 2014) (Table 1).

Phenolic compound quinones are synthesized through many pathways such as, isoprenoid quinones are synthesized through shikimate pathway by utilizing chorismite derived precursor molecules; terrequinone are synthesized by multimodular nonribosomal peptide synthetases enzyme from tryptophan; dopaquinone are synthesized from tyrosine by action of tyrosinase, and benzoquinone are synthesized from catechol by the presence of catechol oxidase/ polyketide synthase enzyme. Anthraquinone, questin and questinol quinones were isolated from *Coniothyrium sp. Eurotium cristatum*, and *Penicillium glabrum* respectively and their corresponding host were *Salsola oppostifolia, Sargassum thunbergii, Punica granatum* (Sun et al., 2013; Du et al., 2014; Hammerschmidt et al., 2012). Xanthones are oxygenated heterocyclic group which are synthesized via the polyacetate/polymalonate pathway through the internal cyclization of polyketide molecule. Several endophytic fungi were reported for alternative source of xanthone but some xanthone like 1,7-dihydroxyxanthone, lapachol, hypericin and laws one are isolated from *Penicillium sp., Aspergillus niger, Chaetomium globosum* and *Gibberella moniliformis* respectively (Table 1).

Microbes are very important sources for secondary metabolite and many microbes originated product are available in market for research and therapeutic purpose. Of course the exploration of endophytic fungi as alternative sources of plant metabolite is an emerging and interesting research.

2.3. Production of Plant-based Compound *via* Tissue Culture

Plants produce a variety of secondary metabolites that are useful in the form of pharmaceuticals, flavors and fragrances, neutraceuticals, dyes etc. These secondary metabolites are mostly found in minute quantities, usually <1% of the plant dry weight. In addition, a number of these compounds are produced by plants that are rare, endangered or grow in specific habitats, or in tree species that takes several years for the accumulation of these metabolites. Production of metabolites from cultivation of plants is further hindered by the lack of cultivation technologies for a number of species, environmental and seasonal influence on biosynthesis and yield of metabolites, requirement of particular developmental stage for the accumulation of compounds in higher levels etc. Obtaining useful metabolites from plants in large quantities required for their various applications for mankind is challenging. Plant biotechnology tools therefore are immensely useful for the large scale, round the year production uninfluenced by the environmental and seasonal variations; without disturbing or damaging the natural populations of rare and endangered species.

Biotechnological production of plant secondary metabolites is also referred to as plant biofactories or green factories (Marchev et al. 2020). A number of secondary metabolites of pharmaceutical significance are produced in callus and cell suspension cultures. Certain secondary metabolites biosynthesized in specialized tissues or organs of the plants cannot be produced in undifferentiated cell cultures. In such cases, organ cultures in the form of shoots, adventitious roots and hairy roots are employed. Further, advances in functional genomics have enabled enhanced production of secondary metabolites as well as production of novel compounds. Metabolic engineering for the production of secondary metabolites involves strategies like decreasing the catabolism of the target compound, over-expression/repression of genes or transcription factors in the pathway, preventing feedback inhibition, decreasing the flux of competitive pathways, conversion of existing product into new product (Marchev et al. 2020) etc.

Metabolically engineered transgenic hairy roots are an ideal tool for increasing the levels of valuable secondary metabolite which are naturally found in low levels in plants (Kowalczyk et al. 2020). Overexpression of pathway genes or transcription factors can be employed both in homologous and heterologous plant systems. Transgenic hairy roots of *Rauwolfia serpentina* overexpressing tryptophan decarboxylase from *Catharanthus roseus* (*Cr*tdc) showed increased alkaloid production (Mehrotra et al. 2013). Overexpression of Squalene synthase 1 gene from *Panaxginseng* (*Pg*SS1) in hairy roots of *Senna obtusifolia* resulted in higher betulinic acid accumulation (Kowalczyk et al. 2020). Overexpression of transcription factor *Sm*bHLH148 in hairy roots of *Salvia miltiorrhiza* significantly increased the accumulation of phenolic acids (Xing et al. 2018). Similarly, hairy root lines of *Lycium ruthenicum* Murr. Overexpressing *Lr*TCP4 accumulated higher kukoamine A (Chahel et al. 2019). In another study, *Scutellaria baicalensis* hairy roots expressing maize (*Zm*Lc) and *Arabidopis* (*At*PAP1) transcription factors increased production of root specific flavones (Park et al. 2021).

Transgenic cell lines are also used for secondary metabolite production. Cell cultures of *Silybum marianum* expressing *Vitis vinifera* stilbene synthase gene (*Vv*STS3) produced higher t-resveratrol (Hidalgo et al. 2017). In metabolically engineered *Vitis amurensis* Rupr cell culture, overexpression of *Picea jezoensis PjSTS1a*/*Pj*STS2/*Pj*STS3 enhanced total stilbene content (Suprun et al. 2020). Expression of *Cicer arietinum* chalcone synthase gene (*Ca*CHS) increased naringenin synthesis in *Silybum marianum* cell cultures (Villar et al. 2020). Alternatively, ectopic expression and overexpression of transcription factors is also successfully applied for alteration of metabolite productions in cell cultures. Ectopic expression of *Arabidopsis thaliana* transcription factor (*At*MYB12) in *Silybum marianum* cell cultures enhanced levels of chlorogenic acid derivatives (Villar et al. 2020) while overexpression of *Citrus sinensis* transcription factor (*Cs*MADS6) in its callus elevated caretenoid content (Lu et al. 2018).

Combing the overexpression of pathway gene and transcription factor is also employed for improving secondary metabolite production. In *Catharanthus roseus* hairy roots, co-overexpression of strictosidine glucosidase (SGD) and ORCA3 resulted in increased production of alkaloids (Sun and Peebles 2015).

Plant systems are also engineered for the synthesis of compounds that are not naturally produced in the species. The co-expression of tryptophan decarboxylase *Os*TDC and anthranilate synthase alpha-subunit mutant (OASA1D) in *Oryza sativa* calli resulted in synthesis of serotonin and its derivatives in rice (Dubouzet et al. 2013). The expression of bacterial HCHL (p-hydroxycinnamomyl-CoA hydratase/lyase from *Pseudomonasfluorescens*) gene in *Beta vulgaris* hairy roots resulted in the production of vanillin (Singh et al. 2015). Further, biotransformation strategy is also utilized for plant secondary metabolite production. Biotransformation of precursors like trans-cinnamic acid and cinnamyl alcohol in callus cells of *Rhodiola rosea* resulted in increased metabolite production (Mirmazloum et al. 2019). Cell and callus cultures of *Citrus aurantium, C. paradisi, C. sinensis* are reported for biotransformation yielding narirutin, naringin and limonin (Shakour et al. 2020).

Biosynthetic pathways can be also transferred to another plant species for production of the desired metabolites in cell or root cultures. Artimisinin production in *Nicotiana tabacum* was successfully achieved by the transfer of pathway genes from *Artimisia annua* (Malhotra et al. 2016). Similarly, genistein production in plants that do not produce it like *N. tabacum, Lactuca sativa* and *Petunia hybrida* was possible by introducing gene from *Glycine max* (Barone et al. 2020). The recent developments in gene editing using CRISPR technology offer new prospects in metabolic engineering and secondary metabolite production in plants. CRISPR based gene editing is reported in hairy roots of *Salvia miltiorrhiza, Brassica carinata* (Marchev 2020). Even though not in the scope of this chapter it is noteworthy that model plant systems like *Nitotiana* are used in molecular farming for the production of engineered proteins, antibodies and vaccines (Dhama et al. 2020).

2.4. Pathway Engineered Plant/GMO Crop

The emerging role of genetic engineering technology for the modification of plant metabolic pathways has played an important role in the improvement of the quality and quantity of field crops as well as medicinal plants, and these genetically modified plants are known as a genetically engineered (or modified/transgenic organisms) organisms. Through genetic engineering method, the beneficial foreign gene(s) have been introduced or suppressed the expression of desire gene(s) in plants (Khan et al., 2018).Genetically modified (GM) plants have been developed with several useful trait like, abiotic stress tolerance, biotic stress tolerance, secondary metabolite accumulation and nutritional improvement (Parmar et al., 2017). Till date, nearly 525 different gene alternations have been reported in 32 different crops for agriculture purpose from different parts of the world (Kumar et al., 2020). GM plants have been shown to increase crop yields, reduce pesticide and insecticide use, reduce CO_2 emissions, and decrease the cost of crop production (Kumar et al., 2020). In context with gene alteration, different approaches are adopted such cis genesis, intra genesis, and most recently genome editing.

2.4.1. r-DNA Technology/ or Pathways Engineering Approach: bt Cotton, bt Brinjal and Others

Bt Crops are transgenic crops that produce the same toxin as the bacterium *Bacillus thuringiensis* in the plant cell, thereby, protecting the crops from pest's damage. *Bacillus thuringiensis* is a gram-positive, spore-forming bacteria which is mainly found in the soil. The bacterium secretes specific proteins known as "cry proteins" that are toxic to insect gut. The Cry toxin becomes active by proteolytic enzymes in the alkaline gut juice (pH 8–10). The toxins released by the plant would paralyze the guts of the caterpillar, when they feed on the plants. A few of the *Bt* crops include cotton, brinjal, maize, etc. When an insect feeds on the transgenic plants, the toxic cry protein present in the plants crystallizes the digestive system of insects, and finally results in death. This Cry toxins are toxic to specific species of insects belongs to orders: Lepidoptera, Hymenoptera, Coleoptera, Diptera, and Nematoda. However, the main positive advantage is that, it has no harmful effects on the human digestive system. In 2016, the total world area cultivated with genetically modified crops (GM crops) reached about 185 million ha (Abbas et al., 2018). However genetically modified plant produced first time in the laboratory was tobacco in 1983 and as herbicide-resistant in France and the USA (Martineau 2001). The most widely used *Bt* vegetable crop is sweet corn (Shelton et al. 2013).

2.4.2. Genome Editing

The clustered regularly interspaced short palindromic repeats (CRISPR/Cas9 enzymes) system has demonstrated great potential in various crop species due to simplicity of use and versatility of the reagents (Jiang et al., 2013; Sun et al., 2015; Svitashev et al., 2016; Zhang et al., 2016; Shimatani et al., 2017; Soyk et al., 2017). Engineered CRISPR/Cas9 nucleases target DNA adjacent to the 5′-NGG-3′, protospacer adjacent motif (PAM), in a single guide RNA (sgRNA) specific manner (Jinek et al., 2012). CRISPR/Cas12a that has PAM requirement of “TTTN,” allowing targeting of AT rich regions, is emerging as equally effective alternative, implemented in various plants (Kim et al., 2018; Tang et al., 2017). For multiplexing, to target more than one gene at a time, Cas12a requires only a single RNA Pol III promoter to drive several crRNAs, whereas Cas9 requires relatively large constructs (Zetsche et al., 2017).

The three major methods of genetic transformation in plants are: *Agrobacterium*-mediated transformation, biolistic and protoplast transfection. By far the most commonly used method to introduce genome-editing reagents in potato is by *Agrobacterium*-mediated transformation. A binary T-DNA vector is used to deliver and express the reagents in plant cells. Once inside the nucleus, the T-DNA randomly integrates into the plant/host genome leading to stable transformation resulting in persistent activity of reagents. However, there is a possibility that it remains extra-chromosomal leading to transient gene expression.

The most common method is polyethylene glycol (PEG)-mediated protoplast transfection. Protoplasts facilitate direct delivery of DNA into cells with gene-editing reagents expressed as plasmid DNA for transient transformation. Protoplasts have greater transformation efficiency compared to other methods (Jiang et al., 2013; Dlugosz et al., 2016; Baltes et al., 2015). Rainbow papaya is the best example of genome editing, where *Carica papaya* L. is a genetically engineered (GE) cultivar with resistance to papaya ringspot virus (PRSV) disease. This Rainbow cultivar currently accounts for about 70% of Hawaii's papaya acreage. Other techniques like physical mutation and polyploidy etc are also helpful in the modification of metabolic pathways.

2.5. Expression of Plant Genes in Microbial Species

Plants are major source of secondary metabolites, approximately 200,000 chemical compounds with distinct properties and most of them are synthesized from specialized metabolism. These compounds provide defense against biotic stress, interspecific competition and exploited as source of medicines, fragrances, flavors, nutrients, repellants, and colorants (Isah 2019). Apart from the vast chemical diversity, natural compounds are present at very low concentrations within plant system. Structural complexity of natural metabolites limit the chemical synthesis of plant compounds and lead to alternatives like transgenic plants or transgenic microbial system (Srinivasan and Smolke, 2020). Microbial systems overwhelm many of the problems obstructing traditional chemical synthesis or genetic modification of plant, thus providing an alternative avenue for exploring plant specialized pathways. Two microbes, *E. coli* and *S. cerevisiae* have been exploited extensively in industrial biotechnology. Although native plants can be genetically modified to accumulate target plant metabolites but metabolic engineering is technically more challenging in plants than in microbes. The translational expression of plant genes in microbial system has technical problems, such as plant enzymes translocate to specific organelles like chloroplast and other cell organelles (Pyne et al., 2019). Most of terpene synthases enzymes are plastid localized and their N-terminal has plastid signal sequence. Another challenge is the functional expression of membrane bound cytochrome P450 (CYP) and cytochrome P450 reductase (CPR) enzymes in microbial system and this problem is overcome by using eukaryotic expression systems. These CYP enzymes have important role such as hydroxylation, epoxidations, and dealkylations in secondary metabolic pathways (Meunier et al., 2004). Cytochrome P450 oxidase linked with reduction of NADPH through shuttle system with cytochrome P450 reductase (CPR) enzymes to electrons transfer between NADPH and CYP. Therefore, truncation of N-terminal and function expression of CYP and CPR enzymes required attention toward eukaryotic microbial system over prokaryote, thus plant genes in microbial system may leads to poor result or unsuitability of the protein function (Williams et al., 1998).

Pathways genes of terpenoids, alkaloids, phenyl propanoids, flavonoids and others compounds are transformed into *S. cerevisiae* and *E.coli* for heterologous production (Ozber et al., 2020). Artemisinin is well known natural antimalarial drug produced by the plant *Artemisia annua*. A genes encoding artemisinic acid to artemisinin from *A. annua* are transfered to yeast for semi-synthetic artemisinin production in bioreactor from basic artemisinic acid precursor molecule (Kung et al., 2018). Ajikumar and his group (in 2010) used multivariate-modular pathway engineering in *E. coli* for over production of taxol from isoprenoid pathway. Genetically modified yeast produce monoterpenoid indole alkaloids such as nepetalactol (Campbell et al., 2016). Phenylalanine, Tyrosine, and their derivatives (e.g., phenethylamine, tyramine, and dopamine) are the source of enormous type of alkaloids, such as benzylisoquinoline alkaloids (BIAs) and the phenethylisoquinoline alkaloids. Recently, genetically modified *S. cerevisiae* has been developed to produce tropane alkaloids (hyoscyamine and scopolamine) by providing simple sugars and amino acids (Srinivasan and Smolke, 2020).

Cluster of ten genes were incorporated into *S. cerevisiae* for de novo noscapine and halogenated alkaloids production (Li et al., 2018). Thebaine is opiate alkaloid, two genes from plant were inserted in *E. coli* and significantly produced thebaine (2.1 mg/l) in culture media (Nakagawa et al., 2016).

Limonene is a monoterpene and for their heterologous production, *S. cerevisiae* was genetically modified with limonene synthase gene of *Perilla frutescens* and *Citrus limon* for limonene production in de novo condition (Jongedijk et al., 2015). Over expression of hexaprenyl pyrophosphate synthetase and mutated farnesyl pyrophosphate synthetase in *S. cerevisiae* results in significantly increased linalool, geraniol, α-terpineol, farnesol and nerolidol production (Camesasca et al., 2018). Flavonoids is another large group of secondary metabolite, plant genes were expressed in *E. coli* and yeast to produce dihydrochalcone phloretin, flavanon liquiritigenin, flavanols kaempferol, quercetin, and fisetin (Pyne et al., 2019). In 2017, Rodriguez and coworker developed genetically engineered *S. cerevisiae* with cluster of plant gene for de novo production of six different flavonoids naringenin, liquiritigenin, kaempferol, resokaempferol, quercetin, and fisetin directly from glucose, without supplementation of expensive intermediates (Rodriguez et al., 2017). Even, polyphenolic anthocyanin pigments are produced by genetic modification of *E. coli* and *S. cerevisiae* polycultures (Jones et al., 2017; Levisson et al., 2018). In the rise of environmental stresses and the demand of plant-based product, the microbial system can be prominent source for all kinds of plant secondary metabolites.

3. Future Aspects

In the future, pathway engineering study will be improved for the discovery of new phytomolecules towards food and pharmaceuticals industries.

Adopting metabolic engineering approach a number of expensive phytomolecules, which are naturally produced in plants over a prolonged period, will be produced in very short duration using heterologous expression system for e.g., bacterial or yeast system. Hopefully, in future physical vaccination program will be replaced by plant based edible vaccine by adopting this pathways engineering approach. In case of agriculture sector, especially in the plant biotechnology field, scientific knowledge will improve and in the future more new designer and biofortified crops will be obtained.

CONCLUSION

This book chapter briefly summarises production of plant secondary metabolites using transgenic approach *via* r-DNA technologyor pathway engineering. MAPs that possess secondary metabolites, which are the active ingredients, are found to play an important role in agriculture as well as pharmaceuticals. The function/properties of phyto-molecules obtained from MAPs is mainly depends on their purity (quality and quantity). Intake of these secondary metabolites in the form of the drug may cure the health-related issues and protect against chronic disorders, such as inflammation, diabetes, cancer, cardiovascular, and neurodegenerative diseases. When obtaining secondary metabolites from plants poses challenge in the form of low quantities, prolonged time or environmental influence, it can be obtained *via* pathway engineering or transgenic plants or heterologous production in microbial system. Such an approach can aid in fulfilling the increasing global demand for plant secondary metabolites for pharmaceutical and cosmetic industries as well as provide a sustainable solution. The present investigation based on the plant-based compound obtained from metabolic engineering approach can improve the basic knowledge on obtaining plant-based secondary metabolites (drug) and other important natural molecules.

ACKNOWLEDGMENTS

Author SKB is highly thankful to Chancellor, Vice Chancellor, and Head of Biotechnology and Microbiology Departments, MIET, Meerut, UP, India, for providing necessary infrastructure and facilities.

REFERENCES

[1] Abbas, M. S. T., 2018. "Genetically engineered (modified) crops (*Bacillus thuringiensis* crops) and the world controversy on their safety." *Egyptian Journal of Biological Pest Control* 28: https://doi.org/10.1186/s41938-018-0051-2.

[2] Ahmad, Z. A., Yeap, S. K., Ali, A. M., Ho, W. Y., Alitheen, N. B., and Hamid, M. 2012. "scFv antibody: principles and clinical application." *Clinical and Developmental Immunology* 2012:980250. doi: 10.1155/2012/980250.

[3] Ajikumar, P. K., Xiao, W. H., Tyo, K. E. J., Wang, Y., Stephanopoulos, F.G, 2010. "Isoprenoid pathway optimization for Taxol precursor overproduction in *Escherichia coli*." *Science* 330:70-74. doi: 10.1126/science.1191652.

[4] Ancheeva, E., Daletos, G., Proksch, P., 2020. "Bioactive Secondary Metabolites from Endophytic Fungi." *Current medicinal chemistry* 27(11):1836-1854. doi: 10.2174/092986732 6666190916144709.

[5] Annunziata, F., Pinna, C., Dallavalle, S., Tamborini, L., and Pinto, A. 2020. "An Overview of Coumarin as a Versatile and Readily Accessible Scaffold with Broad-Ranging Biological Activities." *International Journal of Molecular Science* 21(13):4618. doi:10.3390/ijms2113461.

[6] Baltes, N. J., Hummel, A. W., Konecna, E., Cegan, R., Bruns, A. N., Bisaro, D. M., et al. (2015). "Conferring resistance to geminiviruses with the CRISPR-Cas prokaryotic immune system." *Nature Plants* 1:15145. doi: 10.1038/nplants.2015.145.

[7] Barone, R. P., Knittel, D. K., Ooka, J. K., Porter, L. N., Smith, N. T., Owens, D. K., 2020. "The Production of Plant Natural Products Beneficial to Humanity by Metabolic Engineering." *Current Plant Biology* 24:100121. doi: 10.1016/j.cpb.2019.100121.

[8] Bose S. K., Yadav R. K., Mishra S., Sangwan R. S., Singh A. K., Mishra B., Srivastava A. K., Sangwan N. S. 2013. "Effect of Gibberellic acid and Calliterpenone on Plant Growth Attributes Trichomes, Essential oil Biosynthesis and Pathway Gene Expression in Differential Manner in *Mentha arvensis* L." *Plant Physiology and Biochemistry* 66:150-158. http://doi.org/10.1016/j.plaphy.2013.02.011.

[9] Calvo, A. M., Wilson, R. A., Bok, J. W., and Keller, N. P. 2002. "Relationship between Secondary Metabolism and Fungal Development." *Microbiology and Molecular Biology Reviews* 66(3):447-59. doi: 10.1128/MMBR.66.3.447-459.2002.

[10] Camesasca, L., Minteguiaga, M., Farina, L., Salzman, V., Aguilar, P. S., Gaggero, C., and Carrau, F. 2018. "Overproduction of Isoprenoids by *Saccharomyces cerevisiae* in a Synthetic Grape Juice Medium in the Absence of Plant genes." *International Journal of Food Microbiology* 282:42-48. doi.org/10.1016/j.ijfoodmicro.2018.05.025.

[11] Campbell, A., Bauchart, P., Gold, N. D., Zhu, Y., De Luca, V., and Martin, V. J., 2016. "Engineering of a Nepetalactol-Producing Platform Strain of *Saccharomyces cerevisiae* for the Production of Plant Seco-Iridoids." *ACS Synthetic Biol*ogy 5(5):405-14. doi: 10.1021/acssynbio.5b00289.

[12] Chahel, A. A., Zeng, S., Yousaf, Z., Liao, Y., Yang, Z., Wei, X., Ying, W., 2019. "Plant-Specific Transcription Factor LrTCP4 Enhances Secondary Metabolite Biosynthesis in *Lycium ruthenicum* Hairy Roots." *Plant Cell, Tissue and Organ Culture* 136:323-337. doi:10.1007/s11240-018-1518-2.

[13] Channabasava, R., and Govindappa, M., 2014. "First Report of Anticancer Agent, Lapachol Producing Endophyte, *Aspergillus niger* of *Tabebuia argentea* and its in vitro Cytotoxicity Assays."

Bangladesh Journal of Pharmacology 9:12-139. Doi" 10.3329/bjp.v9i1.15622.

[14] Chaturvedi, P., Gajbhiye, S., Roy, S., Dudhale, R., and Chowdhary, A., 2014. "Determination of Kaempferol in Extracts of *Fusarium chlamydosporum,* an endophytic Fungi of *Tylophora indica* (Asclepeadaceae) and Its Anti-microbial Activity." *Journal of Pharmacy and Biological Sciences* 9:51–55.

[15] Cheng, M. J., Wu, M. D., Chen, J. J., Hsieh, S. Y., Yuan, G. F., Chen, I. S., and Chang, C. W. 2013. "Secondary Metabolites from the Endophytic Fungus of *Annulohypoxylon ilanense." Chemestry of Natural Compound* 49:523-525. doi: https://doi.org/10.1007/s10600-013-0658-1.

[16] Cheng, M. J., Wu, M. D., Hsieh, S. Y., Su, Y. S., Chen, I. S., and Yuan, G. F. 2011. "Secondary metabolites from the endophytic fungus *Annulohypoxylon boveri* var. microspora BCRC 34012." *Chemistry of Natural Compound* 47:536-540. https://doi.org/10.1007/s10600-011-9990-5.

[17] Chithra, S., Jasim, B., Sachidanandan, P., Jyothis, M., and Radhakrishnan, E. K., 2014. "Piperine Production by Endophytic Fungus *Colletotrichum gloeosporioides* isolated from *Piper nigrum." Phytomedicine* 21:534-540. https://doi.org/10.1016/j.phymed.2013.10.020.

[18] Connolly J. D. and Hill R. A. (1991). *Dictionary of terpenoids.* Chapman and Hall: London, UK.

[19] Croteau R. B., Davis E. M., Ringer K. L., Wildung M. R. (2005). "(-) Menthol Biosynthesis and Molecular Genetics." *Natur Wissenschaften* 92:562-77. https://doi.org/10.1007/s00114-005-0055-0.

[20] Cui, Y., Yi, D., Bai, X., Sun, B., Zhao, Y., Zhang, Y., 2012. "Ginkgolide B produced Endophytic Fungus (*Fusarium oxysporum*) Isolated from *Ginkgo biloba." Fitoterapia* 83:913-920. https://doi.org/10.1016/j.fitote.2012.04.009.

[21] Devari, S., Jaglan, S., Kumar, M., Deshidi, R., Guru, S., Bhushan, S., Kushwaha, M., Gupta, A. P., Gandhi, S. G., Sharma, J. P., et al.

2014. "Capsaicin Production by *Alternaria alternata,* an Endophytic Fungus from *Capsicum annum,* LC-ESI-MS/MS analysis." *Phytochemistry* 98:183-189. https://doi.org/10.1016/j.phytochem.2013.12.001.

[22] Dhama, K., Natesan, S., Yatoo, M. Iqbal., Patel, S. K., Tiwari, R., Saxena, S. K., and Harapan, H., 2020. "Plant-based Vaccines and Antibodies to Combat COVID-19: Current Status and Prospects." *Human Vaccines and Immuno therapeutics* 16:2913-2920. doi:10.1080/21645515.2020.1842034.

[23] Ding, C. H., Du, X. W., Xu, Y., Xu, X. M., Mou, J. C., Yu, D., Wu, J. K., Meng, F. J., Liu, Y., Wang, W. L., et al. 2014. "Screening for Differentially Expressed Genes in Endophytic Fungus Strain 39 during co-Culture with Herbal Extract of Its Host *Dioscorea nipponica Makino." Current Microbiology* 69:517-524. https://doi.org/10.1007/s00284-014-0615-7.

[24] Dlugosz, E. M., Lenaghan, S. C., and Stewart, C. N. (2016). "A robotic platform for high-throughput protoplast isolation and transformation." *Journal of Visualized Experiments* 115:e54300. doi: 10.3791/54300.

[25] Du, F. Y., Li, X. M., Song, J. Y., Li, C. S., and Wang, B. G. 2014. "Anthraquinone Derivatives and an Orsellinic acid Ester From the Marine Alga-Derived Endophytic Fungus *Eurotium cristatum* EN-220." *Helvetica chimica acta* 97:973-978. https://doi.org/10.1002/hlca.201300358.

[26] Dubouzet, J. G., Matsuda, F., Ishihara, A., Miyagawa, H., and Wakasa, K., 2013. "Production of indole alkaloids by metabolic engineering of the tryptophan pathway in rice." *Plant Biotechnology Journal* 11:1103-1111. doi:10.1111/pbi.12105.

[27] El-Elimat, T., Raja, H. A., Graf, T. N., Faeth, S. H., Cech, N. B., and Oberlies, N. H. 2014. Flavonolignans from *Aspergillus iizukae,* a fungal endophyte of milk thistle (*Silybum marianum*). *Journal of Natural Product* 77:193-199. https://doi.org/10.1021/np400955q.

[28] Giddings, G., 2001. "Transgenic plants as protein factories." *Current Opinion in Biotechnology* 12:450-454. https://doi.org/10.1016/S0958-1669(00)00244-5.

[29] Goldstein, D. A., and Thomas, J. A., 2004. "Biopharmaceuticals derived from genetically modified plants." *Quarterly Journal of Medicine* 97(11):705-16. doi: 10.1093/qjmed/hch121.

[30] Gouda, S., Das, G., Sen, S. K., Shin, H. S., and Patra, J. K., 2016. "Endophytes: A Treasure House of Bioactive Compounds of Medicinal Importance." *Frontiers in Microbiology* 7:1538. doi: 10.3389/fmicb.2016.01538.

[31] Guan, Z. J., Guo, B., Huo, Y. L., Guan, Z. P., and Wei, Y. H., 2010. "Overview of expression of Hepatitis B Surface Antigen in Transgenic Plants." *Vaccine* 28(46):7351-62. doi: 10.1016/j.vaccine.2010.08.100.

[32] Gupta, S., Bhatt, P., and Chaturvedi, P. 2018. "Determination and Quantification of Asiaticoside in Endophytic Fungus from *Centella asiatica* (L.) Urban." *World Journal of Microbiology and Biotechnology* 34:111. https://doi.org/10.1007/s11274-018-2493-9.

[33] Hammerschmidt, L., Wray, V., Lin, W., Kamilova, E., Proksch, P., and Aly, A. H. 2012. "New Styrylpyrones from the Fungal Endophyte *Penicillium glabrum* Isolated From *Punica granatum.*" *Phytochemistry Letters* 5:600-603. https://doi.org/10.1016/j.phytol.2012.06.003.

[34] Hassan, S. R., Strobel, G. A., Geary, B., and Sears, J. 2013. "An endophytic *Nodulisporium sp.* from Central America Producing Volatile Organic Compounds with Both Biological and Fuel Potential." *Journal of Microbiology and Biotechnology* 23:29-35. doi 10.4014/jmb.1208.04062.

[35] Hidalgo, Diego, Ascensión Martinez-Marquez, Rosa Cusidó, Roque Bru-Martinez, Javier Palazon, and Purificacion Corchete. 2017. "*Silybummarianum* cell cultures stably transformed with Vitis vinifera stilbene synthase accumulate t-resveratrol in the extracellular medium after elicitation with methyl jasmonate or

methylated β-cyclodextrins." *Engineering in Life Sciences* 17:686-694. doi:10.1002/elsc.201600241.

[36] Hu, M. Y., Zhong, G. H., Sun, Z. T., Sh, G., Liu, H. M., and Liu, X. Q. 2005. "Insecticidal activities of secondary metabolites of endophytic *Pencillium sp.* in *Derris elliptica Benth." Journal of applied pharmaceutical science* 129:413-417. https://doi.org/10.1111/j.1439-0418.2005.00988.x.

[37] Huang, Z., Yang, J., Cai, X., She, Z., and Lin, Y. 2012. "A new furanocoumarin from the mangrove endophytic fungus *Penicillium* sp. (ZH16)." *Natural Product Research* 26:1291-1295. https://doi.org/10.1080/14786419.2011.569502.

[38] Isah T. 2019. "Stress and Defense Responses in Plant Secondary Metabolites Production." *Biological Research* 52(1):39. doi: 10.1186/s40659-019-0246-3.

[39] Jiang, W., Zhou, H., Bi, H., Fromm, M., Yang, B., and Weeks, D. P. (2013). "Demonstration of CRISPR/Cas9/sgRNA-mediated targeted gene modification in *Arabidopsis*, tobacco, sorghum and rice." *Nucleic Acids Research* 41:e188. doi: 10.1093/nar/gkt780.

[40] Jones, J. A., Vernacchio, V. R., Collins, S. M., Shirke, A. N., Xiu, Y., Englaender, J. A., Cress, B. F., McCutcheon, C. C., Linhardt, R. J., Gross, R. A., Koffas, M. A. G., 2017. "Complete biosynthesis of anthocyanins using *E.coli* polycultures mBio." *Journal of Clinical Microbiology* 8(3): e00621-27. https://doi.org/10.1128/mBio.00621-17.

[41] Jongedijk, E., Cankar, K., Ranzijn, J., van der Krol, S., Bouwmeester, H., and Beekwilder, J. 2015. "Capturing of the Monoterpene *Olefin Limonene* Produced in *Saccharomyces cerevisiae." Yeast* 32(1):159-71. doi: 10.1002/yea.3038.

[42] Joshi, K., Chavan, P., Warude, D., and Patwardhan, B. 2004. "Molecular markers in herbal drug technology." *Current Science* 87:159-165. http://www.jstor.org/stable/24108860.

[43] Karasev, A. V., Foulke, S., Wellens, C., Rich, A., Shon, K. J., Zwierzynski, I., Hone, D., Koprowski, H., and Reitz, M. 2005. "Plant based HIV-1 Vaccine Candidate: Tat Protein Produced in

Spinach." *Vaccine* 23(15):1875-80. doi: 10.1016/j.vaccine.2004.11.021.

[44] Khademi, M., Nazarian-Firouzabadi, F,, Ismaili, A, and Shirzadian Khorramabad, R. 2019. "Targeting microbial pathogens by expression of new recombinant dermaseptin peptides in tobacco." *Microbiologyopen.* 8(11):e837. doi: 10.1002/mbo3.837.

[45] Khan, S., Ullah, M. W., Siddique, R., Nabi, G., Manan, S., Yousaf, M., Hou, H. 2018. "Role of Recombinant DNA Technology to Improve Life." *International Journal of Genomics* 2016:2405954. doi: 10.1155/2016/2405954.

[46] Kim, D., Alptekin, B., and Budak, H. (2018). "CRISPR/Cas9 genome editing in wheat." *Funct. Integr. Genomics* 18:31-41. doi: 10.1007/s10142-017-0572-x.

[47] Kowalczyk, Tomasz, Przemyslaw Sitarek, Monika Toma, Laurent Picot, Marzena Wielanek, Ewa Skala, and Tomasz Sliwinski. 2020. "An Extract of Transgenic *Senna obtusifolia* L. Hairy Roots with Overexpression of PgSS1 Gene in Combination with Chemotherapeutic Agent Induces Apoptosis in the Leukemia Cell Line." *Biomolecules* 10:510. doi:10.3390/biom10040510.

[48] Kumar, A., Patil, D., Rajamohanan, P. R., and Ahmad, A. 2013. "Isolation, Purification and Characterization of Vinblastine and Vincristine from Endophytic Fungus *Fusarium oxysporum* Isolated from *Catharanthus roseus.*" *PLoS ONE* 8:e71805. doi: 10.1371/journal.pone.0071805.

[49] Kumar, K., Gambhir, G., Dass, A., Tripathi, A. K., Singh, A., Jha, A. K., Yadava, P., Choudhary, M., and Rakshit, S. 2020. "Genetically Modified Crops: Current Status and Future Prospects." *Planta* 251(4):91. doi: 10.1007/s00425-020-03372-8.

[50] Kung, S. H., Lund, S., Murarka, A., McPhee, D., and Paddon, C. J. 2018. "Approaches and Recent Developments for the Commercial Production of Semi-synthetic Artemisinin." *Front Plant Science* 9:87. doi: 10.3389/fpls.2018.00087.

[51] Kurup, V. M., Thomas, J. 2020. "Edible Vaccines: Promises and Challenges." *Molecular Biotechnology* 62(2):79-90. doi: 10.1007/s12033-019-00222-1.

[52] Kusari, S., Lamshöft, M., Zuhlke, S., and Spiteller, M. 2008. "An Endophytic Fungus from Hypericum Perforatum that Produces Hypericin." *Journal of Natural Products* 71:159-162. doi: 10.1021/np070669k.

[53] Kusari, S., Verma, V. C., Lamshoeft, M., and Spiteller, M. 2012. "An endophytic Fungus from *Azadirachta indica* A. Juss. that Produces Azadirachtin." *World Journal of Microbiology and Biotechnology* 28:1287-1294. doi: 10.1007/s11274-011-0876-2.

[54] Levisson, M., Patinios, C., Hein, S., de Groot, P. A., Daran, J. M., Hall, R. D., Martens, S., and Beekwilder, J. 2018. "Engineering de novo Anthocyanin Production in *Saccharomyces cerevisiae.*" *Microbial Cell Factories* 17(1):103. doi: 10.1186/s12934-018-0951-6.

[55] Li, H. Y., Ramalingam, S., and Chye, M. L. 2006. "Accumulation of Recombinant SARS-CoV Spike Protein in Plant Cytosol and Chloroplasts Indicate Potential for Development of Plant-Derived Oral Vaccines." *Society for Experimental Biology and Medicine (Maywood)* 231(8):1346-52. doi: 10.1177/153537020623100808.

[56] Li, Y., Li, S., Thodey, K., Trenchard, I., Cravens, A., and Smolke, C. D. 2018. "Complete Biosynthesis of Noscapine and Halogenated Alkaloids in Yeast." *Proceedings of the National Academy of Sciences of the United States of America* 115(17):E3922-E3931. doi: 10.1073/pnas.1721469115.

[57] Liu, K., Ding, X., Deng, B., and Chen, W., 2010. "10-Hydroxycamptothecin Produced by a New Endophytic *Xylaria* sp., M20, from *Camptotheca acuminate.*" *Biotechnology letters* 32:689-693. doi: 10.1007/s10529-010-0201-4.

[58] Lu, R. M., Hwang, Y. C., Liu, I. J., Lee, C. C., Tsai, H. Z., Li, H. J., and Wu, H. C. 2020. "Development of Therapeutic Antibodies

for the Treatment of Diseases." *Journal of Biomedical Science* 27(1):1. doi: 10.1186/s12929-019-0592-z.

[59] Lu, Suwen, Yin Zhang, Kaijie Zhu, Wei Yang, Junli Ye, Lijun Chai, Qiang Xu, and Xiuxin Deng. 2018. "The citrus transcription factor CsMADS6 Modulates Carotenoid Metabolism by Directly Regulating Carotenogenic Genes." *Plant Physiology* 176:2657-2676. doi:10.1104/pp.17.01830.

[60] Maehara, S., Simanjuntak, P., Maetani, Y., Kitamura, C., Ohashi, K., and Shibuya, H. 2013. "Ability of Endophytic Filamentous Fungi Associated with *Cinchona ledgeriana* to Produce *Cinchona alkaloids*." *Journal of Natural Medicines* 67:421-423. doi: 10.1007/s11418-012-0701-8.

[61] Malhotra, K., Subramaniyan, M., Rawat, K., Kalamuddin, M., Qureshi, M. I., Malhotra, P., Mohmmed, A., Katrina, C., Daniell, H., Kumar, S., 2016. "Compartmentalized Metabolic Engineering for Artemisinin Biosynthesis and Effective Malaria Treatment by Oral Delivery of Plant Cells." *Molecular Plant* 9:1464-1477. doi:10.1016/j.molp.2016.09.013.

[62] Marchev, A. S., Yordanova, Z. P., and Georgiev, M. I., 2020. "Green (cell) Factories for Advanced Production of Plant Secondary Metabolites." *Critical Reviews in Biotechnology* 40: 443-458. doi:10.1080/07388551.2020.1731414.

[63] Martineau, B., 2001. "First fruit: the Creation of the Flav-Savr Tomato and the Birth of Biotech Foods." McGraw-Hill 269.

[64] Mehrotra, S., Srivastava, V., Rahman, L. U., and Kukreja, A. K., 2013. "Overexpression of a *Catharanthus* Tryptophan decarboxylase (tdc) Gene Leads to Enhanced Terpenoid indole alkaloid (TIA) Production in Transgenic Hairy Root Lines of *Rauwolfia serpentine*." *Plant Cell, Tissue and Organ Culture* 115:377-384. doi:10.1007/s11240-013-0369-0.

[65] Meunier, B., de Visser, S. P., and Shaik, S. 2004. "Mechanism of Oxidation Reactions Catalyzed by Cytochrome p450 Enzymes." *Chemical Reviews* 104: 3947-3980, doi: 10.1021/cr020443g.

[66] Mirmazloum, Iman, Attila Kiss, Marta Ladanyi, and Zsuzsanna Gyorgy. 2019. "Production of Cinnamyl Alcohol Glycosides by Biotransformation in Roseroot Callus Cells." *Plant Cell, Tissue and Organ Culture* 139:29-37. doi:10.1007/s11240-019-01659-7.

[67] Moses, T., Papadopoulou, K. K., and Osbourn, A. 2014. "Metabolic and Functional Diversity of Saponins, Biosynthetic Intermediates and Semi-Synthetic Derivatives." *Critical Reviews in Biochemistry and Molecular Biology* 49(6):439-62. doi: 10.3109/10409238.2014.953628.

[68] Mukherjee, P. K., Harwansh, R. K., Bahadur, S., Banerjee, S., Kar, A., Chanda, J., Biswas, S., Ahmmed, S. M., and Katiyar, C. K. 2017. "Development of Ayurveda - Tradition to trend." *Journal of Ethnopharmacology.*197:10-24. doi: 10.1016/j.jep.2016.09.024.

[69] Nadeem, M., Mauji, R., Pravej, A., Ahmad, M. M., Mohammad, A., Qurainy, F. A., Khan, S., and Abdin, M. Z. 2012. "*Fusarium solani, P1,* a new endophytic podophyllotoxin-producing fungus from roots of *Podophyllum hexandrum.*" *African Journal of Microbiology Research.*6:2493–2499. https://doi.org/10.5897/AJMR11.1596.

[70] Nakagawa, A., Matsumura, E., Koyanagi, T., Katayama, T., Kawano, N., Yoshimatsu, K., Yamamoto, K., Kumagai, H., Sato, F., and Minami, H. 2016. "Total biosynthesis of opiates by stepwise fermentation using engineered *Escherichia coli.*" *Nature Communications* 5(7):10390. doi: 10.1038/ncomms10390.

[71] Newman, D. J. and Cragg, G. M. 2007. "Natural products as sources of new drugs over the last 25 years." *Journal of Natural Products* 70 (3):461-477, doi: 10.1021/np068054v.

[72] Ozber, N., Watkins, J. L., and Facchini, P. J. 2020. "Back to the Plant: Overcoming Roadblocks to the Microbial Production of Pharmaceutically Important Plant Natural Products." *Journal of Industrial Microbiology and Biotechnology* 47(9-10):815-828. doi: 10.1007/s10295-020-02300-9.

[73] Palem, P. P., Kuriakose, G. C., and Jayabaskaran, C. 2015. "An Endophytic Fungus, *Talaromyces radicus*, Isolated from

Catharanthus roseus, Produces vincristine and vinblastine, Which Induce Apoptotic Cell Death." *PLoS ONE* 10:12. doi.org/10.1371/journal.pone.0144476.

[74] Park, C. H., Xu, H., Yeo, H. J., Park, Y. E., Hwang, G. S., Park, N. I., and Park, S. U., 2021. "Enhancement of the Flavone Contents of *Scutellaria baicalensis* Hairy Roots via Metabolic Engineering Using Maize Lc and *Arabidopsis* PAP1 Transcription Factors." *Metabolic Engineering* 64:64-73. doi:10.1016/j.ymben.2021.01.003.

[75] Parmar, N., Singh, K. H., Sharma, D., Singh, L., Kumar, P., Nanjundan, J., Khan, Y. J., Chauhan, D. K., Thakur, A. K., 2017. "Genetic Engineering Strategies for Biotic and Abiotic Stress Tolerance and Quality Enhancement in Horticultural Crops: A Comprehensive Review." *Biotechnology* 7(4):239. doi: 10.1007/s13205-017-0870-y.

[76] Parthasarathy, R., Sathiyabama, M. 2014. "Gymnemagenin Producing Endophytic Fungus Isolated from a Medicinal Plant *Gymnema sylvestre* R Br." *Applied Biochemistry and Biotechnology* 172:3141-3152. doi: 10.1007/s12010-014-0754-0.

[77] Patil, M. P., Patil, R. H., and Maheshwari, V. L., 2015. "Biological activities and identification of bioactive metabolite from endophytic *Aspergillus flavus* L7 isolated from *Aegle marmelos.*" *Current Microbiology*71:39-48. doi: 10.1007/s00284-015-0805-y.

[78] Penney, C. A., Thomas, D. R., Deen, S. S., Walmsley, A. M., 2011. "Plant-made vaccines in support of the Millennium Development Goals." *Plant Cell Report* 30(5):789-98. doi: 10.1007/s00299-010-0995-5.

[79] Pillet, S., Couillard, J., Trepanier, S., Poulin, J. F., Yassine-Diab, B., Guy, B., Ward, B. J., and Landry, N., 2019. "Immunogenicity and Safety of a Quadrivalent Plant-Derived Virus Like Particle Influenza Vaccine Candidate-Two Randomized Phase II Clinical Trials in 18 to 49 and ≥50 Years Old Adults." *PLoS One* 14(6):e0216533. doi: 10.1371/journal.pone.0216533.

[80] Pyne, M. E., Narcross, L., and Martin, V. J. J., 2019. "Engineering Plant Secondary Metabolism in Microbial Systems." *Plant Physiology* 179(3):844-861. doi: 10.1104/pp.18.01291.

[81] Qiu, M., Xie, R., Shi, Y., Zhang, H., and Chen, H., 2010. "Isolation and Identification of two Flavonoid Producing Endophytic Fungi from *Ginkgo biloba* L." *Annals of Microbiology* 60:143-150. https://doi.org/10.1007/s13213-010-0016-5.

[82] Rodriguez, A., Strucko, T., Stahlhut, S. G., Kristensen, M., Svenssen, D. K., Forster, J., Nielsen, J., and Borodina, I., 2017. "Metabolic Engineering of Yeast for Fermentative Production of Flavonoids." *Bioresour Technology* 245:1645-1654. doi: 10.1016/j.biortech.2017.06.043.

[83] Saba, K., Sameeullah, M., Asghar, A., Gottschamel, J., Latif, S., Lössl, A. G., Mirza, B., Mirza, O., and Waheed, M. T. 2020. "Expression of ESAT-6 antigen from *Mycobacterium tuberculosis* in broccoli: An edible plant." *Biotechnology and Applied Biochemistry* 67(1):148-157. doi: 10.1002/bab.1867.

[84] Sarang, H., Rajani, P., Vasanthakumari, M. M., Kumara, P. M., Siva, R., Ravikanth, G., and Shaanker, R. U., 2017. "An endophytic fungus, *Gibberella moniliformis* from *Lawsonia inermis* L. Produces Laws one, an Orange-Red pigment." *Antonie Van Leeuwenhoek* 110:853-862. doi: 10.1007/s10482-017-0858-y.

[85] Satake, H., Koyama, T., Matsumoto, E., Morimoto, K., Ono, E., and Murata, J., 2017. Metabolic Engineering of Lignan Biosynthesis Pathways for the Production of Transgenic Plant-Based Foods. In: Jha S. (eds.) *Transgenesis and Secondary Metabolism*. Reference Series in Phytochemistry. Springer, Cham.

[86] Sathiyabama, M., and Parthasarathy, R., 2018. "Withanolide Production by Fungal Endophyte Isolated From *Withania somnifera*." *Natural Product Research* 32:1573-1577. doi: 10.1080/14786419.2017.1389934.

[87] Shakour, Z. T. A., Fayek, N. M., and Farag, M. A., 2020. "How do Biocatalysis and Biotransformation Affect *Citrus* Dietary Flavonoids Chemistry and Bioactivity? A review." *Critical*

Reviews in Biotechnology 40: 689-714. doi:10.1080/ 07388551.2020.1753648.

[88] Sharma, A., Shahzad, B., Rehman, A., Bhardwaj, R., Landi, M., and Zheng, B., 2019. "Response of Phenylpropanoid Pathway and the Role of Polyphenols in Plants Under Abiotic Stress." *Molecules* 24(13):2452. doi: 10.3390/molecules24132452.

[89] Sharma, P. K., Sangwan, N. S., Bose, S. K., Sangwan, R. S., 2013. "Biochemical Characteristics of a Novel Vegetative Tissue Geraniol Acetyltransferase From a Monoterpene Oil Grass (Palmarosa, *Cymbopogon martinii* var. Motia) leaf." *Plant Science* 203-204:(2013) 63-73. https://doi.org/10.1016/j.plantsci.2012.12. 013.

[90] Shelton, A. M., Olmstead, D. L., Burkness, E. C., Hutchison, W. D., Dively, G., Welty, C., 2013. "Multi State Trials of Bt Sweet Corn Varieties for Control of the Corn Earworm." *Journal of Economic Entomology* 106:2151-2159. doi: 10.1603/ec13146.

[91] Shimatani, Z., Kashojiya, S., Takayama, M., Terada, R., Arazoe, T., Ishii, H., et al. (2017). "Targeted base editing in rice and tomato using a CRISPR-Cas9 cytidine deaminase fusion." *Nature Biotechnol*ogy 35:441-443. doi: 10.1038/nbt.3833.

[92] Shweta, S., Zuehlke, S., Ramesha, B. T., Priti, V., Mohana, Kumar P., Ravikanth, G., Spiteller, M., Vasudeva, R., and Uma, Shaanker, R., 2010. "Endophytic fungal strains of *Fusarium solani,* from *Apodytes dimidiata* E. Mey.ex Arn (Icacinaceae) produce camptothecin, 10-hydroxycamptothecin and 9-methoxy-camptothecin." *Phytochemistry* 71(1):117-22. doi: 10.1016/j.phytochem.2009.09.030.

[93] Singh, A., Singh, D. K., Kharwar, R. N., White, J. F., and Gond, S. K., 2021. "Fungal Endophytes as Efficient Sources of Plant-Derived Bioactive Compounds and Their Prospective Applications in Natural Product Drug Discovery: Insights, Avenues, and Challenges." *Microorganisms* 9:197. https://doi.org/10.3390/ microorganisms9010197.

[94] Singh, B. P., Rateb, M. E., Rodriguez-Couto, S., Polizeli, M. L. T. M., and Li, W. J., 2019. "Microbial Secondary Metabolites: Recent Developments and Technological Challenges." *Frontier in Microbiology* 10:914. doi: 10.3389/fmicb.2019.00914.

[95] Singh, P., Khan, S., Pandey, S. K., Singh, M., Banerjee, S., Kitamura, Y., and Rahman, L. Ur., 2015. "Vanillin Production in Metabolically Engineered *Beta vulgaris* Hairy Roots through Heterologous Expression of *Pseudomonas fluorescens* HCHL gene." *Industrial Crops and Products* 74:839-848. doi:10.1016/j.indcrop.2015.05.037.

[96] Sliwoski, G., Kothiwale, S., Meiler, J., Edward, W. L. J., 2014. Computational Methods in Drug Discovery. *Pharmacological reviews* 66(1): 334-395. doi: 10.1124/pr.112.007336.

[97] Sohrab, S. S., Suhail, M., Kamal, M. A., Husen, A., and Azharm E. I., 2017. "Recent Development and Future Prospects of Plant-Based Vaccines." *Current Drug Metabolism* 18(9):831-841. doi: 10.2174/1389200218666170711121810.

[98] Soyk, S., Lemmon, Z. H., Oved, M., Fisher, J., Liberatore, K. L., Park, S. J., et al. (2017). "Bypassing Negative Epistasis On Yield In Tomato Imposed by a Domestication Gene." *Cell* 169:1142-1155. doi: 10.1016/j.cell.2017.04.032.

[99] Srinivasan, P., and Smolke, C. D., 2020. "Biosynthesis of medicinal tropane alkaloids in yeast." *Nature* 585(7826):614-619. doi: 10.1038/s41586-020-2650-9.

[100] Stierle, A. A., and Stierle, D. B., 2000. "Bioactive Compounds from Four Endophytic *Penicillium* sp. of a northwest pacific yew tree." *Studies in Natural Chemistry* 24:933-977. https://doi.org/10.1016/S1572-5995(00)80058-7.

[101] Stoger, E., Vaquero, C., Torres, E., Sack, M., Nicholson, L., Drossard, J., Williams, S., Keen, D., Perrin, Y., Christou, P., and Fischer, R., 2000. "Cereal Crops as Viable Production and Storage Systems for Pharmaceutical scFv Antibodies." *Plant Molecular Biology* 42:583-90. https://doi.org/10.1023/A:1006301519427.

[102] Sun, J., and Peebles, C. A. M., 2015. "Engineering over-expression of ORCA3 and strictosidine glucosidase in *Catharanthus roseus* Hairy Roots Increases Alkaloid Production." *Protoplasma* 253:1255-1264. Doi: 10.1007/s00709-015-0881-7.

[103] Sun, P., Huo, J., Kurtan, T., Mandi, A., Antus, S., Tang, H., Draeger, S., Schulz, B., Hussain, H., Krohn, K., et al. 2013. "Structural and Stereochemical Studies of Hydroxyanthraquinone Derivatives from the Endophytic Fungus *Coniothyrium sp.*" *Chirality* 25:141-148. https://doi.org/10.1002/chir.22128.

[104] Suprun A. R., Ogneva Z. V., Dubrovina, A. S., Kiselev, K. V., 2020. "Effect of spruce PjSTS1a, PjSTS2, or PjSTS3 Gene Over Expression on Stilbene Biosynthesis in Callus Cultures of *Vitis amurensis* Rupr." *Biotechnology and Applied Biochemistry* 67(2):234-9. https://doi.org/10.1002/bab.1839.

[105] Suwannarach, N., Kumla, J., Bussaban, B., Nuangmek, W., Matsui, K., and Lumyong, S., 2013. "Biofumigation with the endophytic fungus *Nodulisporium* spp. CMU-UPE34 to control postharvest decay of citrus fruit." *Crop Protection* 45:63-70. https://doi.org/10.1016/j.cropro.2012.11.015.

[106] Svitashev, S., Schwartz, C., Lenderts, B., Young, J. K., and Mark Cigan, A. (2016). "Genome editing in maize directed by CRISPR-Cas9 ribonucleoprotein complexes." *Nature Communications* 7:1-7. doi: 10.1038/ncomms13274.

[107] Tang, F., Yang, S., Liu, J., and Zhu, H. (2016). "Rj4, a Gene Controlling Nodulation Specificity in Soybeans, Encodes a Thaumatin Like Protein but Not the One Previously Reported." *Plant Physiology* 170:26-32. doi: 10.1104/pp.15.01661.

[108] Terryn, N., Van Montagu, M., Inze, D and Goosens, A., 2006. "Functional genomics approaches to study and engineer secondary natural metabolism in plant cell cultures," Chapter 21, in 17: 291-300 *Medicinal and Aromatic Plants: Agricultural, Commercial, Ecological, Legal, Pharmacological and Social Aspects* (eds. R. J. Bogers, L. E. Craker, and D. Lange), Springer.

[109] Van Drie, J. H., 2007. "Computer-aided drug design: the next 20 years." *Journal of Computer-Aided Molecular Design* 21:591-601. https://doi.org/10.1007/s10822-007-9142-y.

[110] Verma, V. C., Lobkovsky, E., Gange, A. C., Singh, S. K., and Prakash, S., 2011. "Piperine production by endophytic fungus *Periconia sp.* isolated from *Piper longum L.*" *Journal of Antibiotics* 64:427-431. https://doi.org/10.1038/ja.2011.27.

[111] Villar, D., Almagro, L., Palazon, J., Corchete, P., 2020. "Alterations in the Silymarin Metabolism in Transgenic *Silybum marianum* Cultured Cells by The Heterologous Expression of the *Arabidopsis thaliana* V-myb myeloblastosis Viral Oncogene Homolog Transcription Factor MYB12 and *Cicerarietinum* Chalcone synthase." *Industrial Crops and Products* 155:112794. doi:10.1016/j.indcrop.2020.112794.

[112] Vines, G., 2004. "Herbal harvests with a future: towards sustainable sources for medicinal plants." *Plantlife International*; Available at: http://www.plantlife.org.uk/uploads/documents/ Herbal Harvests with a Future.

[113] Wang, J., Yao, L. Y., and Lu, Y. H., 2013. "*Ceriporia lacerata* DMC1106, a New Endophytic Fungus: Isolation, Identification, and Optimal Medium for 20, 40 -dihydroxy-60 -Methoxy-30, 50-dimethylchalcone Production." *Biotechnology and Bioprocess Engineering* 18:669-678. https://doi.org/10.1007/s12257-012-0846-z.

[114] Wang, X. J., Min, C. L., Ge, M., and Zuo, R. H., 2014. "An Endophytic Sanguinarine-Producing Fungus From *Macleaya cordata, Fusarium proliferatum BLH51.*" *Current Microbiology* 68:336-341. https://doi.org/10.1007/s00284-013-0482-7.

[115] Wang, Y., Xu, L., Ren, W., Zhao, D., Zhu, Y., and Wu, X., 2012. "Bioactive metabolites From *Chaetomium globosum* L18, an Endophytic Fungus in The Medicinal Plant *Curcuma wenyujin.*" *Phytomedicine* 19:364-368. https://doi.org/10.1016/j.phymed.2011.10.011.

[116] Whaley, K. J., Hiatt, A., Zeitlin, L., (2011). "Emerging antibody products and *Nicotiana* manufacturing." *Human Vaccine* 7:349-356. https://doi.org/10.4161/hv.7.3.14266.

[117] Williams, D. C., McGarvey, D. J., Katahira, E. J., and Croteau, R., 1998. "Truncation of limonene synthase preprotein provides a fully active "pseudomature" form of this monoterpene cyclase and reveals the function of the aminoterminal arginine pair." *Biochemistry* 37:12213-12220, https://doi.org/10.1021/bi980854k.

[118] Wu, H., Yang, H. Y., You, X. L., and Li, Y. H. 2013. *Diversity of Endophytic Fungi from Roots of Panax ginseng and Their Saponin Yield Capacities.* Springer Plus 2:107. https://doi.org/10.1186/2193-1801-2-107.

[119] Xing, B., Liang, L., Liu, L., Hou, Z., Yang, D., Yan, K., Zhang, X., and Liang, Z., 2018. "Over Expression of SmbHLH148 Induced Biosynthesis of Tanshinones as well as Phenolic Acids in *Salvia miltiorrhiza* Hairy Roots." *Plant Cell Reports* 37:1681-1692. doi:10.1007/s00299-018-2339-9.

[120] Xu, L. L., Han, T., Wu, J. Z., Zhang, Q. Y., Zhang, H., Huang, B. K., Rahman, K., and Qin, L. P., 2009. "Comparative Research of Chemical Constituents, Antifungal and Antitumor Properties of Ether Extracts of *Panax ginseng* and Its Endophytic Fungus." *Phytomedicine* 16:609-616. https://doi.org/10.1016/j.phymed.2009.03.014.

[121] Yamazaki, T., and Tokiwa, T., 2010. "Isofraxidin, a coumarin component from *Acanthopanax senticosus,* inhibits matrix metalloproteinase-7 expression and cell invasion of human hepatoma cells." *Biological and Pharmaceutical Bulletin* 33:1716-1722. https://doi.org/10.1248/bpb.33.1716.

[122] Yang, C. Y., Chen, S. Y., and Duan, G. C., 2011. "Transgenic peanut (*Arachis hypogaea* L.) Expressing the Urease Subunit B Gene of *Helicobacter pylori.*" *Current Microbiology* 63(4):387-91. doi: 10.1007/s00284-011-9991-4.

[123] Yao, J., Weng, Y., Dickey, A., and Wang, K. Y., 2015. "Plants as Factories for Human Pharmaceuticals: Applications and

Challenges." *International Journal of Molecular Science* 16(12):28549-65. doi: 10.3390/ijms161226122.

[124] Zetsche, B., Heidenreich, M., Mohanraju, P., Fedorova, I., Kneppers, J., DeGennaro, E. M., et al. (2017). "Multiplex gene editing by CRISPR-Cpf1 using a single crRNA array. *Nature Biotechnology."* 35, 31-34. doi: 10.1038/nbt.3737.

[125] Zhang, Q., Wei, X., and Wang, J., 2012. "Phillyrin produced by *Colletotrichum gloeosporioides,* an endophytic fungus isolated from *Forsythia suspense.*" *Fitoterapia* 83:1500-1505. https://doi.org/10.1016/j.fitote.2012.08.017.

[126] Zhao, J. H., Zhang, Y. L., Wang, L. W., Wang, J. Y., and Zhang, C. L., 2012. "Bioactive secondary metabolites from *Nigrospora sp. L*LGLM003, an endophytic fungus of the medicinal plant *Moringa oleifera Lam." Journal of Microbiology and Biotechnology* 28:2107-2112. https://doi.org/10.1007/s11274-012-1015-4.

[127] Zhao, L., Chang, W. C., Xiao, Y., Liu, H. W., and Liu, P., 2013. "Methyl-erythritol phosphate pathway of isoprenoid biosynthesis." *Annual Review of Biochemistry.* 82:497-530. doi: 10.1146/annurev-biochem-052010-100934.

[128] Zhou, S. L., Zhou, S. L., Wang, M. X., and Chen, S. L., 2011. "Two Compounds From the Endophytic *Colletotrichum* sp. of *Ginkgo biloba." Natural Product Communications* 6:1131-113. https://doi.org/10.1177/1934578X1100600821.

In: The Future of Metabolic Engineering ISBN: 978-1-68507-362-6
Editors: A. Sharma and D. Amin

Chapter 11

PLANT METABOLIC ENGINEERING: PROMOTION OF HUMAN HEALTH

Rita Mahapatra*
Neotech Institute of Applied Science and Research,
Virod, Vadodara, Gujarat, India

ABSTRACT

Plant synthesizes various natural chemicals via biosynthetic pathways, which are interlinked by complex networking. Humans are dependent on plants to fulfill their needs since ancient times, but with population increase, we are forced to think about other sources for future generations. To accomplish this, the metabolic engineering approach in plants will help to produce higher amount of products in least time. Metabolic engineering is the term which gives a concept of alteration in metabolomics of any organism by regulating biosynthetic mechanism by changing its genetic level. Metabolic engineering is the enhancement of desired product through up-regulation of some gene or down regulation of other genes. The knowledge about molecular biology techniques such as isolation and cloning of gene(s), and its transformation in other suitable systems is necessary in plant metabolic engineering. In addition to basic techniques, Bioinformatics tools can

* Corresponding Author's Email: ritamahapatra28@gmail.com.

be used to refine the targets (genes/enzymes) to apply the metabolic engineering processes in plants. CRISPR/cas9 technique to create mutation could shed new light on plant metabolic engineering regulations. Metabolic engineering and synthetic biology combination can significantly facilitate the rapid development of a wild variety of plant products for a world need of sustainable solutions. Plant tissue culture technology has resulted in many pharmaceutical substances, including secondary metabolites such as alkaloids, terpenoids, steroids, saponins, phenolics, flavonoids and amino acids. The *Agrobacterium*-mediated transformation has also been used to produce transgenic plantlets showing an increase amount of secondary metabolites, which can be used for pharmaceutical substances for new therapeutics. By using the plant metabolic engineering approach, remarkable changes at the metabolic level in different plants have been achieved. This area will gradually result in to a miracle growth and generate a solution for all globally challenged problems regarding food, food supplements, medicines, and other required items from plants.

Keywords: secondary metabolites, metabolomics, biosynthetic pathways, therapeutics, tissue culture, biofortification

INTRODUCTION

Human health is directly or indirectly dependent on plants as they play vital roles as components of our diet and provides oxygen to breathe. Plants offer us macronutrients (carbohydrates, proteins and lipids) and micronutrients (vitamins A, B, C, some vitamin D, E and K) along with most of our essential minerals, as well as fibre. Phytonutrients include polyphenols (flavonoids, stilbenoids and hydroxycinnamates), polyunsaturated fatty acids (PUFAs), plant sterols, and carotenoids are compounds in plant-based foods that play beneficial roles in the prevention of disease. There is evidence that the consumption of plant-based food such as fresh fruit and vegetables helps in lowering the risk of major chronic diseases, particularly cardiovascular disease (Bradbury et al. 2014, 394), metabolic diseases such as obesity and Type 2 diabetes (Mozaffarian et al. 2011, 2392) and certain types of cancer (Doll & Peto, 1981,1192; Martin et al. 2011,1685 2013, 19; Parkin et al. 2011, 77;

Traka & Mithen, 2011, 2483; Liu, 2013a, 18, b, 384; Norat et al. 2014, 35; Wang et al. 2014a, 1, b, 349). Though plant provides all the necessary components of human diet, many of our staple food such as wheat, rice, maize, cassava, potatoes, and sorghum are white starchy foods and deficient in provitamin-A. Sorghum is the second most important cereal crop in Africa, with 300 million people dependent on it as a staple, and it is seriously deficient in provitamin A, iron and zinc (Che et al. 2016, 11040). To improve food qualities previously, cross or backcross breeding technologies were commonly used to generate multiple mutants for plant metabolic engineering research (Lu et al. 2016, 210). However, these simple and conventional methods are relatively time-consuming. Therefore, metabolic engineering is now more likely to be carried out at the molecular level.

PLANT METABOLIC ENGINEERING

Metabolic engineering or genetic engineering is a newly evolved technique in the early 1990s (Bourgaud et al. 2001, 839), using which a plant can be engineered to produce desired product in less time. Recent advances in metabolic engineering have open a new way for the production of secondary metabolites in higher quantities. This process can be further improved by using an appropriate genetic transformation procedure. So far, most of the biosynthetic pathway strategies developed for producing secondary metabolites were through various ways which include isolating and expressing of the respective genes in more efficient organisms, construction of promoters to enhance the expression of a target gene, or antisense and co-suppression techniques for knock-down of particular plants for the desired traits (Bourgaud et al. 2001, 839). Metabolic engineering encompasses the objectives and focuses modification of metabolic pathways occurs in an organism. It would deliver improved knowledge and usage of various cellular pathways for supramolecular assembly, transduction of energy, and alteration of chemical (Lessard 1996, 1654). This method implemented on plants will allow endogenous pathways (biochemical pathways) to be influenced and could result in the development of transgenic crops. Thus in the process, the synthesis of natural products by the plants is altered to deliver valuable biomolecules of therapeutic significance (Kinney 1998, 173). As in numerous studies, fabrication of primary and secondary metabolites is excessively low to be used commercially; therefore, metabolic engineering can offer several approaches as illustrated in the Figure 1:

- Categorize for a rate of transcription regulation of primary metabolites, enzyme or hindering feedback and competitive inhibition mechanism.

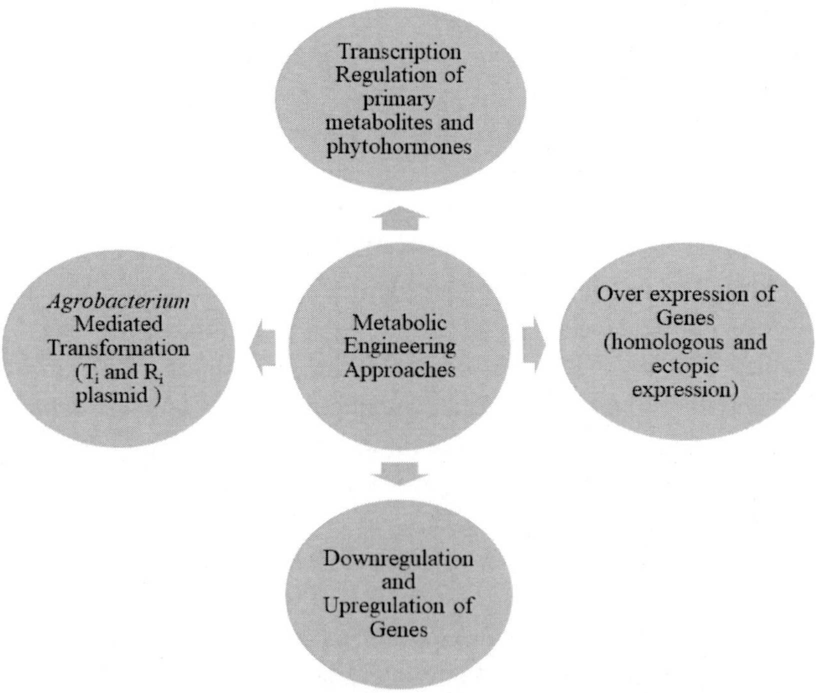

Figure 1. Plant metabolic engineering approaches employed for stimulating the production of various phytochemicals.

- Over expression of homologous and ectopic genes could increase the carbon flux by using a biosynthetic pathway.
- Better output, like up regulation and down regulation of genes and the number of cells employed for producing secondary metabolites.
- *Agrobacterium*-mediated transformation of genes through T_i and R_i Plasmid.

Metabolic engineering in plant involves the alteration of metabolic pathways to increase the flux toward desired secondary metabolites or to attain better understanding of metabolic pathways and use of cellular pathways for chemical transformation, energy transduction, and supramolecular assembly (Chandra and Chandra 2011, 371; Hussain et al. 2012, 10). In otherwords, metabolic engineering is the alteration or improvement of the cellular activities involving transport and enzymatic and regulatory functions of the cell by using rDNA-technology (Bourgaud et al. 2001, 839; Hussain et al. 2012, 10). Now, multi step metabolic engineering is possible, which overtakes single-step engineering, and it is the best way to produce secondary metabolites in transgenic plants (Capell and Christou 2004, 148). The main advantage of this method is that it is convenient and cost-effective and produces industrially important secondary metabolites continuously (Hussain et al. 2012, 10). Also, this technique is used as a tool for improving crop plants that are resistant to various diseases, plants producing allelopathic compounds to control the weeds, pest-resistant plants to improve the importance of ornamentals and fruits, and enhanced pollination by modifying scent profiles (Chandra and Chandra 2011, 371). Another advantage is the production of valuable secondary metabolites under controlled environment which is free from climate and soil conditions (Hussain et al. 2012, 10). Engineering or structural design of secondary metabolite pathways are quite difficult in plants, because it requires a detailed knowledge of the whole biosynthetic pathways and a detailed perception of its regulatory mechanisms. But, such information is not explored in many medicinal plants known to have vast variety of bioactive metabolites (Oksman-Caldentey and Inze 2004, 433). Up-regulation/overexpression of transcription factors (TFs) has led to the increased gene expression and finally the enhanced metabolite in the plant cell (Van der Fits and Memelin 2000, 295; Li et al. 2014, 1; Olivoto et al. 2017, 71). Gene expression can be manipulated at any step during transcriptional initiation, to RNA processing and finally to the post translational modification of protein products (Petrillo et al. 2014, 1215). A number of genes coding biosynthetic pathways of alkaloids such as nicotine, berberine, and scopolamine were engineered and executed. Cloned gene expression of two enzymes, viz., putrescine *N*-methyl transferase and (*S*)-scoulerine 9-*O*-methyltransferase, in *Atropa belladonna* and *Nicotiana sylvestris* (transgenic plants), respectively, in cell culture of *Coptis japonica* and *Eschscholzia californica* respectively, was performed. The results reveal that putrescine *N*-methyl transferase was overexpressed and amplified the content of nicotine in *N. sylvestris* (Sato et al. 2001, 367).Plants are safe and economical expression systems for the production of biopharmaceuticals, vaccines, enzymes, plasma proteins and antibodies. Plants offer attractive advantages in the production of recombinant proteins. Using molecular biology techniques, the production of improved food, drugs etc. from plants can be achieved which can ultimately promote human health. Metabolic engineering is the term that gives a concept of alteration in metabolomics of any organism by regulating biosynthetic mechanism by changing its genetic level.

METABOLOMICS

The term metabolomics involves comprehensive, non-biased, high-throughput analyses of complex metabolite mixtures from plant extracts (Hall et al. 2002, 1437). According to Moghe and Last (2015, 1512), a metabolome comprises of all the chains of responses, relating to enzymes and its substrates in the metabolic process and finishing up in the materialization of the primary and secondary metabolites. The main purpose of metabolomics is comprehensive and simultaneous analysis of the metabolites produced in cells and organisms. It is an effective approach for analysing various chemical compounds occurring in plant cells (Pichersky and Gang 2000, 439). Plant secretory trichomes are considered essential repositories of important metabolites. For this reason, an integrated omics database, TrichOME (http://www.plant trichome.org/), has been developed by using bioinformatics tools, to pool the data and increase the understanding of trichome metabolome analysis. The database also provides for mining trichome-specific genes and is a valuable source for plant trichome research (Dai et al. 2010, 44). The database is helpful to elucidate unique characteristic features of plant trichomes that produce a large number of important metabolites by linking metabolites with their respective genes. Wang et al. (2018, 757) have developed an important herbal medicine omics database (HMOD) (http://herbalplant.ynau.edu.cn/) intending to provide a reliable omics resource of herbal medicine plants for all researchers. This database is helpful for the genome sequence viewing and BLAST search option. Medicinal Plant Metabolomics Resource (MPM) (http://metnetdb. org/mpmr_public/), another important database representing 14 important medicinal plants, has been developed, and it provides a platform for identifying the genes involved in the regulation and synthesis of important metabolites. It enables understanding metabolic networks that lead to specialized metabolites. Barbosa et al. (2017, 141) attempted untargeted metabolite profiling to compare the metabolic composition of leaves and rhizomes. They demonstrated that different species show differential expression of few metabolites, whereas few of them could be used as potential markers among the members of the Zingiberaceae family. Establishing metabolite maps for individual plant species and families and their compilation into databases is an important and emerging tool for the pharmaceutical industry, providing plant-based solutions to treat diseases. The metabolite profiles are also increasingly used for the biosynthesis and quality control studies of many medicinal plants.

This chapter focuses on, what are the different metabolic engineering techniques used to modify plants for the production of several plant derived compounds which are essential for human health such as carotenoid, pro vitamin-A, lutin, zeaxanthine, lycopene, folate, ascorbate to improve quality of food, food supplements, medicines and other required items from plants are discussed as follows.

CAROTENOID AND PRO-VITAMIN-A

Carotenoid pigments are essential for photosynthesis and function as attractants in some higher plant organs (Demmig-Adams et al. 1996, 403). Being natural plant compounds, carotenoids are an integral part of the human diet (Fraser and Bramley 2004, 228). Evidence is rapidly accruing that different carotenoids have different beneficial effects, although the mechanisms of action remain unclear. For example, lycopene appears to have a protective effect against prostate cancer (Gann and Khachik, 2003, 1536; Miller et al. 2002, 1435). Maize kernels naturally accumulate lutein and zeaxanthin, and there is significant variation in their levels, suggesting that conventional breeding could be used to improve nutrition (Kurilich and Juvik 1999, 1948). The widespread occurrence of vitamin A deficiency is well documented. It emphasizes the requirement for staple foods enhanced in carotenoids exhibiting pro-vitamin A activity (Ye et al. 2000, 303 and references therein). In humans and animals, the health benefits of dietary carotenoids are becoming increasingly apparent, besides the health benefits, carotenoids still have commercial utilities in the food, pharmaceutical, and cosmetic industries. One strategy, which has already been commercialised to increase carotenoids, is genetic modification (Fraser et al. 2009, 196). Golden rice is a persuasive illustration in plants, which contains β- carotene due to transformation with carotenoid biosynthetic genes (Ye et al. 2000). Other workers have produced tomato fruit with elevated levels of lycopene and β-carotene (Fraser et al. 2002, 1092). Metabolically engineer carotenoid biosynthesis in plants and algae has grown exponentially (Fraser and Bramley 2004, 228). The common strategies are "push" and "block" strategies. The former strategies mean overexpression of the rate-limiting pathway or introduction of the non-organism genes to synthesize new compounds for example, the overexpression of the PSY gene in canola (Shewmaker et al. 1999, 401), and tomato (Fraser et al. 2002, 1092), or "mini-pathways" guiding the formation of β- carotene in rice endosperm (Ye et al. 2000, 303). On the contrary, the "block strategies" signify the downstream of biosynthesis. Examples of "block" strategies are the increase in lycopene in tomato fruits obtained through the silencing of lycopene β-cyclases (Rosati et al. 2000, 413) or the increase of zeaxanthin in potato tubers obtained through the silencing of zeaxanthin epoxidase (Shewmaker et al. 1999, 401). Maize is a valuable model for carotenoid research because of its diverse gene pool, its amenability for genetic analysis, and the tendency for carotenoid variants to display distinct, colour-based phenotypes. To increase carotenoid levels in maize, several scientists, e.g., Aluru et al. (2008, 3551) introduced the bacterial crtB and crtI genes under the control of an enhanced γ-zein promoter to provide strong endosperm-specific expression, increasing the total carotenoid content to 33.6 g/g dry weight. A significant advance was achieved by Zhu et al. (2008, 18232) with the development of a combinatorial nuclear transformation system designed to dissect and modify the carotenoid biosynthetic pathway in maize, using the white endosperm elite South African inbred M37W. The method involved transforming plants with multiple genes encoding the enzymes involved in carotenoid biosynthesis and then screening a library of random transformants for plants with appropriate metabolic profiles. The pilot study for this technique involved introducing five genes (maize psy1, *Gentiana lutea* cb and bch, and bacterial crtI and crtW) under the control of endosperm specific promoters. This recreates the entire pathway from geranyl geranyl pyrophosphate (GGPP) to zeaxanthin and also adds the enzyme CrtW (β-carotene ketolase), which converts β-carotene into downstream ketocarotenoids. The M37W genetic background provided a blank template because the endosperm in this variety has no PSY activity and lacks all carotenoids. The recovery of plants carrying random combinations of genes resulted in a metabolically diverse library comprising plants with a range of carotenoid profiles, revealed by easily identifiable endosperm colours ranging from yellow to scarlet. The kernels contained high levels of β-carotene, lycopene, zeaxanthin, and lutein and further commercially relevant ketocarotenoids such as astaxanthin and adonixanthin (Zhu et al. 2008, 18232). Carotenoid content of potato was enhanced through metabolic engineering using tuber-specific overexpression of a bacterial mini-pathway by Diretto et al. (2007, 350). Another recent breakthrough in this area was the development of transgenic maize plants transformed with multiple genes selected to simultaneously modulate three metabolic pathways in the endosperm, thereby increasing the levels of the three key nutrients β-carotene, ascorbate, and folate (Naqvi et al. 2009, 7762). In addition to the pioneering work in rice (see Beyer 2010, 478; Potrykus 2010, 446 for reviews), carotenoid enhancement has been attempted by conventional breeding and/or genetic engineering in three major staple cereals: maize, wheat, and sorghum. Plant genetic engineering of pro-vitamin A accumulation has successfully increased pro-vitamin A content in crop plants as discussed by Bai et al. (2011, 205) knowledge of the underlying mechanisms controlling the accumulation of pro-vitamin A in plants is relatively limited, and the subject is currently an area of active research.

ENHANCING THE ACTIVITY OF VITAMIN E IN FOOD CROPS

The Vitamin E family of antioxidants in plants consists of two groups of vitamin E vitamers i.etocopherols and tocotrienols and collectively termed as tocochromanols. Tocopherols occur more widely than tocotrienols in plants and are the principal vitamin E components of leaves as well as seeds of most dicot species (Kamal-Eldin et al., 1996, 671). Tocotrienols, however, are the major form of vitamin E in seeds of most monocots (Padley et al., 1994, 47) and a limited number of dicots (Aitzetmüller et al., 1997). Tocotrienols typically account for >50% of the total vitamin E antioxidants of the seed endosperm of monocots, including palm and agronomically important cereal grains such as rice, wheat and oats (Padley et al., 1994, 47; Barnes 1983, 33; Peterson and Qureshi, 1993, 157). Cahoon et al. (2003, 1082) reported that expression of cDNAs for divergent forms of homogentisic phytyltransferase (HPT) from tocotrienol-accumulating seeds of barley, wheat and rice cDNAs could be used to generate large increases in amounts of vitamin E antioxidants (that is, tocotrienols and tocopherols) in leaves and seeds of transgenic plants. They also show that the expression of these cDNAs is shown to confer tocotrienol biosynthetic ability to transgenic plant tissues in a manner consistent with that of homogentisic acid geranylgeranyl transferase (HGGT). Overexpression of homogentisateprenyl transferase (HPT; vte2) which is the first enzyme committed to tocopherol synthesis but this strategy had only a modest impact on tocopherol concentrations in seeds (Falk et al. 2003, 35; Zhang et al. 2013, 628,) compared to overexpression of HGGT which significantly increased the concentrations of tocotrienols in maize and sorghum seeds (Cahoon et al. 2003, 1082; Dolde & Wang, 2011, 1367; Che et al. 2016, 11040).

BIOFORTIFICATION OF FOLATE AND IRON

In biofortification, nutritional food crops are improved through agronomic practices, conventional plant breeding or metabolic engineering using modern biotechnology. Folates are a group of water-soluble vitamin B9, derived from tetrahydrofolate (THF). Folate is an essential coenzyme involved in one carbon metabolism. Folate deficiency is associated with a higher risk to newborns of neural tube defects, spina bifida, and anencephaly, and an increased risk of cardiovascular diseases, cancer, and impaired cognitive function in adults. Mandatory fortification of wheat flour with folic acid in the United States in 1998 was followed by a significant reduction in the prevalence of neural tube defects (Honein et al., 2001, 2981). Folate deficiency also causes widespread megaloblastic anaemia during pregnancy and often exacerbates already existing iron deficiency anaemia (Rush, 2000, 212; Barber et al., 2000,

45; Persad et. al., 2002, 241). As with pro-vitamin A, folate levels are very low in rice endosperm, and so is the genetic variability of the trait. Thus, folate biofortification is another case for which genetic modification is required and successful proof-of principle rice has recently been reported (Storozhenko et al. 2007, 1277). Here, the transformation of two pathway genes from *Arabidopsisthaliana* shifted folate production to levels in the grains, which can be expected to meet the requirements necessary to combat its deficiency. Other results demonstrate that recommendations for folic acid supplementation alone did not appear to succeed in reducing the incidence. In contrast, the fortification of grain products with folic acid did significantly reduce the incidence (Persad et al., 2002, 241). Enhancing folate content in staple crops by metabolic engineering is a promising, cost-effective strategy to eradicate folate malnutrition worldwide. Over the last decade, significant progress has been made in this field.

Plastid Transformation

Plastid transformation systems using somatic embryos or embryogenic calli show the way to engineer the plastid genome of several major crops in which regeneration is mediated through somatic embryogenesis. Many crops, including cereals (rice, wheat, barley, oats, sorghum, corn), legumes (alfalfa, lentils, peanut, pea, soybean), oil crops (palm, sunflowers, coconut, canola, olive), cash crops (cotton, sugarcane, cassava), vegetable crops (potato, tomato, carrot, sweet potato, sugarbeet, squash, cucumber, lettuce, broccoli, cauliflower, snap bean, cabbage, celery, onion, garlic), fruits/trees and nuts (banana, grape, cantaloupe, muskmelon, watermelon, strawberry, orange, apple, mango, avocado, peach, grapefruit, pineapple, maple almond), beverages (coffee, tea, coca), timber trees (oak, black walnut, sycamore), and so on, are regenerated *in vitro* by somatic embryogenesis. Multigene engineering through the chloroplast genome is possible in a single transformation event. For example, the introduction of the cry operon from *Bacillus thuringiensis* (BT), coding for the insecticidal protein delta-endotoxin, expressed up to 46% of the total leaf protein (De Cosa et al., 2001, 71). Two bacterial enzymes that confer resistance to different forms of mercury-mercuric ion reductase (merA) and organomercurial lyase (merB) – were expressed as an operon in transgenic chloroplasts and conferred resistance to very high levels of mercury and organomercurial compounds (Ruiz et. al., 2003, 1344). In addition, plastid transformation has become a powerful tool for the study of plastid biogenesis and function. This approach has been used to investigate plastid DNA replication origins, RNA editing elements, promoters, RNA stability determinants, processing of polycistrons, intron maturases, translation elements and proteolysis, import of proteins and several other processes (Daniell et al., 2004, 423). Thus, recent advancements predict well for the production of therapeutic proteins, vaccines and biomolecules in transgenic plastids, and the introduction of agronomic traits into crops using an environmentally friendly approach, and will increase our understanding of plastid biochemistry and molecular biology (Daniell et al., 2005, 238; Daniell 2006, 1071).

PLANTS AS A SOURCE OF THERAPEUTICS

Plants are having a rich source of naturally existing medicines in Indian literature having medicinal importance. These medicinal plants are responsible for producing various phytochemicals which providing immunity to combat plant disease. Out of around 18,000 flowering plants on our planet earth, 44% are of medicinal importance and are the rich source of phytochemicals (Handa et al. 2006, 305). Phytochemical study of medicinal plants is indispensable to identify the presence of active constituents in them. Phytochemicalsalso termed as secondary metabolites comprise of alkaloids, flavonoids, steroidal lactones, saponins, tannins, terpenoids, etc. which exhibit medicinal as well as, pharmacological activities. The applications of plant secondary metabolites are tremendous. They may be utilized as therapeutic compounds because of their antimicrobial, anti-inflammatory, and anticancer properties. For example, vincristine (an alkaloid obtained from *Catharanthus roseus*) is an anticancer compound, diosgenin (a saponin obtained from *Dioscorea* species) is used as contraceptive, and menthol (a monoterpene obtained from oil of peppermint) is used in toothpaste. Paclitaxel, an important anticancer compound, is originally obtained from *Taxus brevifolia* (Vongpaseuth et al. 2007, 1180). Matrine is another example of a plant-derived immunosuppressant that acts by inducing the expression of anti-tumour immune defence activating ligands (Lu et al. 2015, 4764). Plant-derived immunostimulants such as curcumin stimulate immune cell production involved in tumor defence, increasing immune system identification of tumor cells and the subsequent activation of apoptotic pathways (Bose et. al., 2015, 1). Multinutrient and multivitamin deficiencies represent a global issue compromising human health. Hence the creation of engineered crops aimed at the enhancement of different vitamins and nutrients levels is desirable, and one such successful example of a multivitamin approach using metabolic engineering was the creation of biofortified white corn with enhanced levels of ascorbate, β-carotene and folate (Naqvi et al. 2009, 7762). Glucosinolates are a class of specialized metabolites restricted to the Brassicaceae family and has anticancer properties (cabbage, broccoli, cauliflower, mustard etc.). The transfer of genes involved in one of the glucosinolates, glucoraphanin, into *Nicotiana benthamiana* plants scaled up the production of glucoraphanin upto considerable amounts (Mikkelsen et al. 2010, 751). Further improvements in the production of the glucoraphanin were carried out recently by the co-expression of two other genes (large subunit of the heterodimeric isopropylmalate isomerase and bile acid transporter 5) (Crocoll et al. 2016, 14). Due to emerging safety records of herbal medicines compared to chemical drugs, medicinal plants-based drugs have regained popularity for treating even the chronic human illness. Recent discoveries are stating that, with modern therapeutic usages of bioactive compounds from medicinal plants, researchers are giving immense attention to the metabolic engineering of medicinal plants.

GENETICALLY MODIFIED PLANTS

More than two decades after the creation of Golden Rice (GR) in 2000 (Ye et al. 2000, 303), and a further improvement in 2005 (Paine et al. 2005, 487), GR is still not available for consumption. GR and the second generation of Genetically Modified Organisms (GMOs) and other nutritionally improved crops may already have been publicly accepted and known for their target consumers. Once the road is cleared by GR rice, folate-biofortified crops could become available to target populations within some years. Plant cells are used to produce metabolites, which otherwise cannot be produced in the heterologous systems. Because of the low concentration of metabolites in heterologous systems, transgenic plants are generated for the specific metabolite synthesis. The use of engineering target genes in plants has significantly increased the production of metabolites. The entire pathway genes involved in synthesis ofcyanogenic glycosides, dhurrin, have been successfully transferred from *Sorghum bicolor* to *Arabidopsis thaliana*. The transgenic *A. thaliana* plants stored large amounts of dhurrin and showed resistance to herbivore *Phyllotreta nemorum* (Tattersall et al. 2001, 1826). Paclitaxel is produced through a complex biosynthetic pathway; recently, as many as 19 putative steps within its biosynthesis pathway have been fully characterized (Huang et al. 2001, 2237; Croteau et al. 2006, 75; Nims et al. 2006, 385; Vongpaseuth et al. 2007, 1180; Meng et al. 2011, 205; Lenka et al. 2012, 1). Plants are also being used as natural chemical factories to produce unnatural derivatives of metabolites with varying specificities or improved medicinal properties. Reengineering the plants with strictosidine synthase having altered substrate specificity and cocultivation with commercially available precursors produced modified monoterpene indole alkaloids (Ryan and Bradley 2009, 140; Runguphan and O'Connor 2009, 151; Runguphan et al. 2010, 461). There have been efforts to enhance production through plant cell cultures resulting in sustainable consumption-production patterns of medicinal plants and avoiding the over-harvesting of medicinal plants.

BENEFITS OF PLANT TISSUE CULTURE OVER TRADITIONAL AGRICULTURAL PRACTICES

Basic plant tissue culture (PTC) methods were proposed for *in vitro* synthesis of phytochemicals from various plant species since 1957. PTC method is an alternative to conventional and synthetic cell culture techniques. PTC provides an excellent medium for sustainable, easily expandable production of secondary metabolites to restrict the hurdles. To boost up the yield, noticeable approaches like manipulating the supplements and bettering the culture environment and elicitation are considered (Kumar and Sopory 2008, 83). Secondary metabolites obtained from callus, cell, and cell suspension cultures (Pépin et al. 1995, 131; Shibli et al. 1997, 201, 1998, 119) and plant parts like leaves and flowers. To exhibit accumulation of secondary metabolites in callus and cell suspension culture, various distinct determinants are practiced; the substantial ones are the chemical composition of the media compared to the growth regulators (Nawa et al. 1993, 770), concentration and source of carbon (Decendit and Merillon 1996, 762; Mori and Sakurai 1994, 588), and concentration and source of nitrogen (Mori and Sakurai 1994, 588; Sato et al. 1996, 91). Secondary metabolites produced via plant cell culture are much more favored over the conventional agricultural production because: (i) It is independent of geographical and seasonal variations and various environmental factors; (ii) It offers a defined production system, which ensures the continuous supply of products, uniform quality and yields; (iii) It is possible to produce novel compounds that are generally not found in the parent plant (Rao and Ravishankar 2002, 101). Secondary metabolite production from plant system includes screening of high-yielding cell line, media modification, precursor feeding, elicitation, large-scale cultivation in bioreactor system, hairy root culture, plant cell immobilization, biotransformation, and others (Rao and Ravishankar 2002, 101; Vanisree et al. 2004, 1). As the *in vitro* produced plants are independent of different external factors like geographical and seasonal variations, they provide a continuous and standardized supply of metabolites with homogenous quality and yield compared to traditional production. Unique compounds which cannot be easily obtained through parent plants can easily be created through plant tissue culture. The primary role of secondary metabolites is to protect plants and released them as a result of defense responses. Still, because of its therapeutic properties, researchers have focused their attention on it. Secondary metabolites such as alkaloids, steroids, tannins, glycosides, volatile oils, fixed oils, resins, phenols, and flavonoids are found in abundance in various plant parts like leaves, flowers, bark, seeds, fruits, and roots. Phytochemicals obtained from secondary metabolism have been refined for pharmaceuticals, food additives, flavours, and fragrance and products like latex and tannins. Almost 20% of recognized plants have been used in pharmaceutical studies; useful biological functions such as anticancer, antimicrobial, antioxidant, antidiarrheal, and analgesic and bound-healing properties were reported from secondary metabolites (Naczk and Shahidi 2006, 1523). Using different biotechnological approaches, metabolically engineered plant cells can be developed and biologically active metabolites can be produced using these cells and tissue from callus cultures, cell suspension cultures, and/or organ cultures. One of the best examples where biotechnology in conjugation with biochemistry led to the significant growth in production of secondary metabolites by *Agrobacterium* mediated gene transfer.

Agrobacterium Mediated Gene Transfer: A Promising Approach for Secondary Metabolite Production

Agrobacterium (gramnegative, soilbacterium)-mediated plant transformation is more acceptable, because of low cost and single-gene integration into plant genome (Zalewski et al. 2012, 1; Alok et al. 2016, 261; Alok et al, 2017, 91; Bhati et al. 2016, 4379). *A. tumefaciens contain* T_i plasmid which causes crown gall disease in plants (Krens et al. 1982, 72). The crowngall-derived cell suspension cultures trigger the wild-typeT_i plasmids which have been used for production of specific secondary mtabolites (Saito et al. 1992, 149). Similarly, R_i plasmid present in *A. rhizogenes* induce hairyroot disease in dicotplants, and the induced hairyroots will grow rapidly in culture medium without addition of phytohormones (Giri and Narasu 2000, 1). Hairyroots are the by-products from the *A. rhizogenes* infected sites, commonly known as hairyroot disease or syndrome. In this methodology, the hairy roots of the plant are firstly sterilized and then interact with the parts of the plant by infecting it with *A. rhizogenes*. For the period of the interaction course, *A. rhizogenes* transmits the portion of DNA (T-DNA) situated in the root-persuading plasmid to plant cells, and the confined genes in the region are expressed in the identical manner as the normal endogenous genes of the plant cells. Some strains of *A. rhizogenes* (like A4) possess T-DNA divided into two segments (i.e., TL-DNA and TR-DNA). Thus, both are assimilated independently into the plant's genome (Jouanin 1984, 91). This soil bacterium transfers its T-DNA segment from R_i (root-inducing) plasmid into the host plant genome. The T-DNA region contains a set of genes encoding for the specific enzymes, which control the biosynthesis of natural auxins and cytokinins. The new changes, i.e., insert ion of new genes, cause hormonal imbalance in the host plant and induce the formation of proliferating roots (hairyroots) from the wounded sites infected with *A. rhizogenes* (Guillon 2006, 341). The promising results obtained by implementing hairy root culture, now bioreactors, are incorporated to achieve much more bio-active compounds (Mehrotra et al. 2008, 69). To obtain various valuable alkaloids and alkannins, hairy root cultures of plants, namely, *Lithospermum erythrorhizon*, *Harpagophytum procumbens* (Ludwig-Muller et al. 2008, 15), and adventitious roots of *Panax ginseng* (Jeong et al. 2008, 288) and *Scopolia parviflora* (Min et al. 2007, 1748) were examined in different volumes of bubble column bioreactors. Ginsenoside, which is a class of natural product steroid, glycosides, and triterpene saponins, can also be synthesized by employing adventitious root culture in combination with stirred tank bioreactors (Jeong et al. 2008, 288).

In a study, the gene of interestfor enhancement of secondary metabolism was introduced into hairy roots. In the process, 6-

hydroxylase gene of *Hyoscyamus muticus*was incorporated in hyoscyamine-rich *Atropa belladonna*the help of *A. rhizogenes*. An amplified quantity of enzymatic activity with five times more concentration of scopolamine was observed in engineered roots (Hashimoto et al. 1993, 731). The plant part is selected to infect with *A.rhizogenes* favouring highergenetic constancy and growth. Therefore bio-active compounds released to the medium can conveniently be separated and purified to get higher yields (Anand 2010, 1). Hence the higher yield of these bio-active compounds can efficiently be used for various applications in the food and pharmaceutical industries. Inrecent past, genetic modification through *Agrobacterium*-mediated gene transfer technology attained significant progress in the regulated genetic enhancement of traits in demand for several other plants in the Iridaceae family. This technology has emerged to be the key approach in modern molecular breeding. Several research achievements have been reported on gene manipulations and modifications, for instance, genes responsible for abiotic stress tolerance and insect resistance, regulation of genes involved in the biosynthetic pathways of secondary metabolites, etc.

Transformed hariyroot cultures with anti-solamargine (As)-scFv gene controlsandenhances the solasodineglycosideconcentrationupto 2.3-fold more in the transgenic *S. khasianum* than wild-type hairyroots (Putalun et al. 2003, 344) Metabolic engineering of the hairyroots is also used to make the de novo synthesis of secondary metabolites by introducing the specific genes that encode related enzymatic process in other organisms. For example, engineering of the flavonoid pathway in *Saussurea involucrata* by a transgenic approach increased the production of apigenin. The gene responsible for apigenin production in *S. medusa* was found to be chalcone isomerase (chi) gene. A complete cDNA sequence of chi gene construct was prepared under the control of the cauliflower mosaic virus (CaMV) 35S promoter. The chigene had introduced into the *S. involucrata* genome by *A. rhizogenes*-mediated transformation which resulted in the establishment of transgenic hairyroot lines. The enzyme chalcone isomerase converts naringeninchalcone into naringenin, which is the precursor of apigenin. After 5 weeks of incubation, C46 hairyroot line accumulated 32.1 mg/l of apigenin with total flavonoids at 647.8 mg/l. The accumulation of apigenin and flavonoid content was found to be 12 and 4 times, respectively, which is superior when compared to the wild-type hairyroots. The enhanced enzyme productivity was obtained due to the superior activity of chalconeisomerase (Li et al. 2006, 553). In addition to that, hairyroot metabolic engineering has been widely used to enhance the production of pharmaceutically important secondary metabolites and also the production of certain recombinant proteins. The transfer of three genes from *Ralstonia eutropha* bacterium into the genome of sugar beet hairyroots directed the accumulation of poly(3-hydroxybutyrate) (Menzel et al. 2003, 571). Recently, Hidalgo et al. (2017, 3882) reported the metabolism of tobacco hairyroot for the production of stilbenes. There are several techniques available now a day to perform genetic engineering but recently gene manipulation using CRISPER technology becomes a versatile technique to up regulate or down regulate the gene according to human need.

GENOME EDITING BY CRISPER TECHNOLOGY

Gene editing using CRISPR/Cas9 there are three main genome editing tools include TALEN (transcription activator-like effector nuclease), ZFN (zinc finger nuclease), and the CRISPR (clustered regularly interspaced short palindromic repeats)/CRISPR/Cas9 (CRISPR-associated protein9) systems (Ma et al. 2015, 1274; Zhang et al. 2017, 291). But the implementation of TALEN and ZFN tools is very laborious because of the complications in protein design, their synthesis, and validation. However, the above problem was overcome with the discovery of a CRISPR/Cas9 system, the most emerging and adopted tool because of its simplicity, flexibility, design, high efficiency, and lowcost (Ma et al. 2015, 1274). This system has an endonuclease-type activity which breaks double strand at the specific genome sites. Such breaks are further restored by the error-pronenon homologous end joining (NHEJ) pathway, which often leads to loss of gene function due to bases deletion, insertion, or substitution event (Wang et al. 2017, 1007; Ma et al. 2015, 1274). Genome editing by type II CRISPR system has become a widely used and robust technique to implement site-directed mutagenesis in a variety of species including model and crop plants. CRISPR-Cas system can be grouped into two classes: Class 1 including types I, III, IV, and Class 2 including types II, V, VI (Makarova et al. 2020, 67). Those CRISPR-Cas systems that have been widely used for genome engineering, such as CRISPR-Cas9 and CRISPR-Cas12, mainly belong to class 2 (Makarova et al. 2020, 67). CRISPR-Cas9 systems rely upon creating DNA double-strand breaks (DSBs) at target sites. This is followed by an exogenous gene insert or a copy variant based on homology-directed repair (HDR). The resulting insertions and deletions, with or without a DNA template, are often used for either silencing genes or, more generally, to reprogram the genome. CRISPR Cas9 might be the most extensively used system, which originally acts as Cas9-crRNA-trarRNA pattern (Jinek et al. 2012, 816). In this way, the RNA guided endonuclease activity requires a crRNA sequence complementary to target DNA, and base-pairing between the crRNA and a tracrRNA (Jinek et al. 2012, 816). The maturation of crRNAs requires endogenous RNase III, CRISPR-associated Cas9 protein and tracrRNA (Deltcheva et al. 2011, 602). The CRISPR-Cas9 system from *Streptococcus pyogenes* via fusing crRNA and tracrRNA were reconstructed into a single guide RNA, resulting to a Cas9-gRNA two-component endonuclease system (Jinek et al. 2012, 816). This simpler CRISPR-Cas endonuclease system vastly facilitated the works on genome engineering. Different from CRISPR-Cas9 system (type II), CRISPR-Cas12 system (type V) can work without tracrRNA and the Cas12 protein mature pre-crRNA itself (Fonfara et al. 2016, 517), providing an alternative accessible CRISPR-Cas endonuclease system for genome engineering. The first challenge of pathway engineering using CRISPR-Cas nucleases system was to ensure that the engineered cells survive the double strand breaks (DSB) in genomic DNA generated by a natural CAS endonuclease. Two major repair pathways are employed by cells to recover from a DSB, an error-prone and non-homology-directed repair mechanism designated as non-homologous end-joining (NHEJ), and a high-fidelity homology-directed repair mechanism (HDR), which is also commonly referred to as homologous recombination (HR) (Sonoda et al. 2006, 1021). Target gene suppression, including RNAi and antisense gene silencing, are commonly used methods that enhance metabolite production, often resulting in the incomplete loss of gene function (Schaart et al. 2016, 438). However, few studies manipulated metabolic pathways in plants using the CRISPR system. Li et al., (2018, 415) introduced the pYLCRISPR/Cas9 system with one or two single-site guide RNAs to target the tomato phytoene desaturase gene. Fifty-three genome-edited plants were obtained following single plant transformation, and these samples represented single to quadruple mutants. A non protein amino acid i.e., Gamma amino butyric acid (GABA) accumulation in both the leaves and fruits of genomically edited lines was significantly enhanced, and the GABA content in the leaves of quadruple mutants was 19-fold higher than that in wild-type plants. Li et al. (2018, 415) demonstrated that the multiplex CRISPR/Cas9 system can be exploited to precisely edit tomato genomic sequences and effectively create multisite knockout mutations, which could shed new light on plant metabolic engineering regulations. Some of the genome editing approaches applied in plants are given in Table-1. In *Salvia miltiorrhiza*, which is known for its anti arrhythmic and vasorelaxation effects, CRISPR/Cas9 system has been employed to knockout the committed geneSmCPS1 coding for diterpenesynthase involved in the biosynthesis of tanshinone using *A. rhizogenes*-mediated transformation (Li et al. 2017, 1). The diterpene synthase is the key enzyme that utilizes geranylgeranyl pyrophosphate (GGPP) as its substrate for producing tanshinones (Arora and Narula 2017, 1932). Likewise, GGPP acts as a precursor during the biosynthesis of taxol by blocking the metabolic flux of forming GGPP to tanshinone and switching it to biosynthesize Taxol. CRISPR/Cas9 system generated 3 homozygous and 8 chimeric mutants from the 26 independent transgenic hairyroot lines of Salvia. Metabolomics analysis also proved the above results evidenced by the zero accumulation of tanshinone in homozygous mutants and a declined fraction in chimeric mutants. Similarly, a large number of studies have been carried out using CRISPR/Cas systems for various applications, such as improving the biotic and abiotic stress tolerance, enhanced yield, production of various bio-active metabolites, enhancement of plant quality, and biofortification (Ricroch et al. 2017, 169). Reports also suggest the successful applications of CRISPR-Cas9 genome editing tool for editing carotenoid biosynthetic pathway genes intomato (Pan et al. 2016, 1; Hayut et al. 2017, 1). Similarly, CRISPR/Cas9 tool has been utilized for targeted mutagenesis of the DFR-B genelocus (dihydroflavonol-4-reductase-B), a key gene of anthocyanin biosynthetic pathway in *Ipomoeanil* (Watanabe et al. 2017, 1). Čermák et al. 2015, 1, observed drastic changes in the stem color of the plant during the initial stages of transgenic plant development under *in vitro* conditions. About 75% of the transgenic plants exhibited anthocyanin-less white flowers with the biallelic mutations at Cas9 cleavagesite in DFR-B gene obtained by either a single-base insertion or by deletions of more than two bases. *Agrobacterium*-mediated homologous recombination was achieved in tomato by targeted insertion of strong cauliflowermosaic virus 35S promoter upstream of the endogenous anthocyanin mutant 1 (ANT1) coding sequence using CRISPR/Cas9 system (for detail see Čermák et al. 2015). ANT1 encodes a MYB transcription factor and its overexpression resulted in intense purple coloration in various tissues like flowers, fruit, and foliage in the gene-targeted plants as compared to the wild type due to anthocyanin accumulation.

Table 1. Genome editing approaches applied in plants using CRISPR/Cas9techniques

Plant	Target genes	References
Tomato	Regions in the viral genome	Ali et al. (2015)
Maize	ALS	Svitashev et al. (2015)
Rice	GW2, GW5, and TGW6	Xu et al. (2016)
Rice	SBEI and SBEIIb	Sun et al. (2017)
Maize	Waxy	Chilcoat et al. (2017)
Soya bean	*Bar*, *GmFEI1*, *GmFEI2*	Jacobs et al. (2015)

CONCLUSION

Metabolic engineering aims to change the cell's metabolic composition, and when applied to plants, food and feed quality improvements are the main goals.Plant systems are more economical in that they can be used as bioreactors to produce metabolites on a large scale than industrial methods. Genomic knowledge by metabolic engineering for fabrication of secondary metabolites derived from plants is presently well innovated. Thus, metabolic engineering andbiotechnologicalexercisescan be appliedas a substitute for the production of naturally active, economically valuable, and pharmaceutically important secondary metabolite. Developments in bio-techniques, mainly the technique of plant cellcultures, could deliver new worth for therapeutically and economicallybio-activecompounds. The main benefit of the *in vitro* cellcultures is elicitation and extraction of phytochemicals, cultivatedin controlled environmental conditions. As earlier reviewed, studies associated with human subjects have established the absorption of secondary metabolites derived from plants, but elucidating the consequence of thetechnique and the metabolic outcome still faces methodological challenges. The approaches like using bioinformatical tools (next-generation sequencing or meta-genomics) and animal/human model studies would improve our knowledge of the secondary metabolite interaction with tissue and its metabolism. It would elucidate the pathways of metabolism of secondary metabolite and the development of future nutritional food with enhancing agricultural output, nutritional value, disease resistance and other traits which could be obtained by metabolic engineering ultimately, it can improve human health and fulfil the demand of food.

ACKNOWLEDGMENTS

The author is grateful to the editors for the invitation to contribute to this chapter.

References

Aitztmüller, K. "Antioxidative effects of Carum seeds. 1997. "*Journal of the American Oil Chemists' Society* 74, no. 2: 185-185.

Ali, Zahir, AalaAbulfaraj, Ali Idris, Shakila Ali, Manal Tashkandi, and Magdy M. Mahfouz. 2015. "CRISPR/Cas9-mediated viral interference in plants." *Genome biology* 16, no. 1: 1-11.

Alok, Anshu, Shivam Sharma, Jitesh Kumar, Subodh Verma, and Hemant Sood. 2017. "Engineering in plant genome using agrobacterium: progress and future." In *Metabolic engineering for bioactive compounds*, Springer, Singapore: 91-111.

Alok, Anshu, Vishnu Shukla, Zarna Pala, Jitesh Kumar, Subhash Kudale, and Neetin Desai. 2016. "In vitro regeneration and optimization of factors affecting Agrobacterium mediated transformation in Artemisia Pallens, an important medicinal plant." *Physiology and Molecular Biology of Plants* 22, no. 2: 261-269.

Aluru, Maneesha, Yang Xu, Rong Guo, Zhenguo Wang, Shanshan Li, Wendy White, Kan Wang, and Steve Rodermel. 2008. "Generation of transgenic maize with enhanced provitamin A content." *Journal of experimental Botany* 59, no. 13: 3551-3562.

Anand, Sanjeev. 2010. "Various approaches for secondary metabolite production through plant tissue culture." *Pharmacia* 1, no. 1: 1-7.

Arora, Leena, and Alka Narula. 2017. "Gene editing and crop improvement using CRISPR-Cas9 system." *Frontiers in plant science* 8: 1932.

Bai, Chao, Richard M. Twyman, Gemma Farré, Georgina Sanahuja, Paul Christou, Teresa Capell, and Changfu Zhu. 2011. "A golden era—pro-vitamin A enhancement in diverse crops." *In Vitro Cellular & Developmental Biology-Plant* 47, no. 2: 205-221.

Barber, Robert, Stuart Shalat, Kate Hendricks, Brigitte Joggerst, Russell Larsen, Lucina Suarez, and Richard Finnell. 2000. "Investigation of folate pathway gene polymorphisms and the incidence of neural tube defects in a Texas Hispanic population." *Molecular Genetics and Metabolism* 70, no. 1: 45-52.

Barbosa, Gina B., Nirupama S. Jayasinghe, Siria H. A. Natera, Ellen D. Inutan, Nonita P. Peteros, and Ute Roessner. 2017. "From common to rare Zingiberaceae plants-A metabolomics study using GC-MS." *Phytochemistry* 140: 141-150.

Barnes, P. J. 1983. "Cereal tocopherols." *Developments in food science*. Lipids in cereal technology, New York, USA, Academic Press: 33-55.

Beyer, Peter. 2010. "Golden Rice and 'Golden'crops for human nutrition." *New Biotechnology* 27, no. 5: 478-481.

Bhati, Kaushal Kumar, Anshu Alok, Anil Kumar, Jagdeep Kaur, Siddharth Tiwari, and Ajay Kumar Pandey. 2016. "Silencing of ABCC13 transporter in wheat reveals its involvement in grain development, phytic acid accumulation and lateral root formation." *Journal of experimental botany* 67, no. 14: 4379-4389.

Bose, Sayantan, Abir Kumar Panda, Shravanti Mukherjee, and Gaurisankar S. A. 2015 "Curcumin and tumor immune-editing: resurrecting the immune system." *Cell division* 10, no. 1: 1-13.

Bourgaud, Frederic, Antoine Gravot, Sandrine Milesi, and E. Gontier. 2001. "Production of plant secondary metabolites: a historical perspective." *Plant science* 161, no. 5: 839-851.

Bradbury, Kathryn E., Paul N. Appleby, and Timothy J. Key. 2014."Fruit, vegetable, and fiber intake in relation to cancer risk: findings from the European Prospective Investigation into Cancer and Nutrition (EPIC)." *The American journal of clinical nutrition* 100, no. suppl_1: 394S-398S.

Cahoon, Edgar B., Sarah E. Hall, Kevin G. Ripp, Thaya S. Ganzke, William D. Hitz, and Sean J. Coughlan. 2003. "Metabolic redesign of vitamin E biosynthesis in plants for tocotrienol production and increased antioxidant content." *Nature biotechnology* 21, no. 9: 1082-1087.

Capell, Teresa, and Paul Christou. 2004. "Progress in plant metabolic engineering." *Current opinion in biotechnology* 15, no. 2: 148-154.

Čermák, Tomáš, Nicholas J. Baltes, RadimČegan, Yong Zhang, and Daniel F. Voytas. 2015. "High-frequency, precise modification of the tomato genome." *Genome biology* 16, no. 1: 1-15.

Chandra, Sheela, and Ramesh Chandra. 2011. "Engineering secondary metabolite production in hairy roots." *Phytochemistry reviews* 10, no. 3: 371.

Che, Ping, Zuo-Yu Zhao, Kimberly Glassman, David Dolde, Tiger X. Hu, Todd J. Jones, Darren Fred Gruis, Silas Obukosia, Florence Wambugu, and Marc C. Albertsen. 2016. "Elevated vitamin E content improves all-trans β-carotene accumulation and stability in biofortified sorghum." *Proceedings of the National Academy of Sciences* 113, no. 39: 11040-11045.

Chilcoat, Doane, Zhan-Bin Liu, and Jeffry Sander. 2017. "Use of CRISPR/Cas9 for crop improvement in maize and soybean." *Progress in molecular biology and translational science* 149: 27-46.

Crocoll, Christoph, Nadia Mirza, Michael Reichelt, Jonathan Gershenzon, and Barbara Ann Halkier. 2016. "Optimization of engineered production of the glucoraphanin precursor dihomomethionine in *Nicotiana benthamiana*." *Frontiers in bioengineering and biotechnology* 4: 14.

Croteau, Rodney, Raymond E. B. Ketchum, Robert M. Long, RüdigerKaspera, and Mark R. Wildung. 2006. "Taxol biosynthesis and molecular genetics." *Phytochemistry Reviews* 5, no. 1: 75-97.

Dai, Xinbin, Guodong Wang, Dong Sik Yang, Yuhong Tang, Pierre Broun, M. David Marks, Lloyd W. Sumner, Richard A. Dixon, and Patrick Xuechun Zhao. 2010. "TrichOME: a comparative omics database for plant trichomes." *Plant physiology* 152, no. 1: 44-54.

Daniell, Henry, Paul R. Cohill, Shashi Kumar, and Nathalie Dufourmantel. "Chloroplast genetic engineering. 2004." In *Molecular biology and biotechnology of plant organelles*, pp. 423-468. Springer, Dordrecht.

Daniell, Henry, Shashi Kumar, and Nathalie Dufourmantel. 2005 "Breakthrough in chloroplast genetic engineering of agronomically important crops." *TRENDS in Biotechnology* 23, no. 5: 238-245.

Daniell, Henry. "Production of biopharmaceuticals and vaccines in plants via the chloroplast genome. 2006." *Biotechnology Journal: Healthcare Nutrition Technology* 1, no. 10: 1071-1079.

Decendit, A., and J. M. Mérillon. 1996. "Condensed tannin and anthocyanin production in Vitis vinifera cell suspension cultures." *Plant cell reports* 15, no. 10: 762-765.

De Cosa, Brandy, William Moar, Seung-Bum Lee, Michael Miller, and Henry Daniell. 2001. "Overexpression of the Bt cry 2Aa2 operon in chloroplasts leads to formation of insecticidal crystals." *Nature biotechnology* 19, no. 1: 71-74.

Deltcheva, Elitza, Krzysztof Chylinski, Cynthia M. Sharma, Karine Gonzales, Yanjie Chao, Zaid A. Pirzada, Maria R. Eckert, Jörg Vogel, and Emmanuelle Charpentier. 2011. "CRISPR RNA maturation by trans-encoded small RNA and host factor RNase III." *Nature* 471, no. 7340: 602-607.

Demmig-Adams, Barbara, Adam M. Gilmore, and William W. Adams Iii. 1996. "In vivo functions of carotenoids in higher plants." *The FASEB journal* 10, no. 4: 403-412.

Diretto, Gianfranco, Salim Al-Babili, RaffaelaTavazza, Velia Papacchioli, Peter Beyer, and Giovanni Giuliano. 2007. "Metabolic engineering of potato carotenoid content through tuber-specific overexpression of a bacterial mini-pathway." *PLoS One* 2, no. 4: e350.

Dolde, David, and Tong Wang. 2011. "Oxidation of crude corn oil with and without elevated tocotrienols." *Journal of the American Oil Chemists' Society* 88, no. 9:1367-1372.

Doll, Richard, and Richard Peto. 1981. "The causes of cancer: quantitative estimates of avoidable risks of cancer in the United States today." *JNCI: Journal of the National Cancer Institute* 66, no. 6: 1192-1308.

Ducreux, Laurence J. M., Wayne L. Morris, Peter E. Hedley, Tom Shepherd, Howard V. Davies, Steve Millam, and Mark A. Taylor. 2005. "Metabolic engineering of high carotenoid potato tubers containing enhanced levels of β-carotene and lutein." *Journal of experimental botany* 56, no. 409: 81-89.

Estévez, Juan M., Araceli Cantero, Andreas Reindl, Stuart Reichler, and Patricia León. 2001. "1-Deoxy-D-xylulose-5-phosphate synthase, a limiting enzyme for plastidic isoprenoid biosynthesis in plants." *Journal of Biological Chemistry* 276, no. 25: 22901-22909.

Falk, Jon, Gaby Andersen, Birgit Kernebeck, and Karin Krupinska. 2003. "Constitutive overexpression of barley 4-hydroxyphenylpyruvate dioxygenase in tobacco results in elevation of the vitamin E content in seeds but not in leaves." *FEBS letters* 540, no. 1-3: 35-40.

Farré, Gemma, Chao Bai, Richard M. Twyman, Teresa Capell, Paul Christou, and Changfu Zhu. 2011. "Nutritious crops producing multiple carotenoids–a metabolic balancing act." *Trends in plant science* 16, no. 10: 532-540.

Fonfara, Ines, Hagen Richter, MajdaBratovič, Anaïs Le Rhun, and Emmanuelle Charpentier. 2016. "The CRISPR-associated DNA-cleaving enzyme Cpf1 also processes precursor CRISPR RNA." *Nature* 532, no. 7600: 517-521.

Fraser, Paul D., and Peter M. Bramley. 2004. "The biosynthesis and nutritional uses of carotenoids." *Progress in lipid research* 43, no. 3: 228-265.

Fraser, Paul D., Eugenia M. A. Enfissi, and Peter M. Bramley. 2009. "Genetic engineering of carotenoid formation in tomato fruit and the potential application of systems and synthetic biology approaches." *Archives of Biochemistry and Biophysics* 483, no. 2: 196-204.

Fraser, Paul D., Susanne Romer, Cathie A. Shipton, Philippa B. Mills, Joy W. Kiano, Norihiko Misawa, Rachel G. Drake, Wolfgang Schuch, and Peter M. Bramley. 2002. "Evaluation of transgenic tomato plants expressing an additional phytoene synthase in a fruit-specific manner." *Proceedings of the National Academy of sciences* 99, no. 2: 1092-1097.

Gann, Peter H., and Frederick Khachik. 2003. *Tomatoes or lycopene versus prostate cancer: is evolution anti-reductionist?.*: 1563-1565.

Giri, Archana, and M. Lakshmi Narasu. 2000. "Transgenic hairy roots: recent trends and applications." *Biotechnology advances* 18, no. 1: 1-22.

Guillon, Stephanie, Jocelyne Trémouillaux-Guiller, Pratap Kumar Pati, Marc Rideau, and Pascal Gantet. 2006. "Hairy root research: recent scenario and exciting prospects." *Current opinion in plant biology* 9, no. 3: 341-346.

Hall, Robert, Mike Beale, Oliver Fiehn, Nigel Hardy, Lloyd Sumner, and Raoul Bino. 2002. *Plant metabolomics: the missing link in functional genomics strategies*. 1437-1440.

Handa, S. S., D. D. Rakesh, and K. Vasisht. 2006. "Compendium of medicinal and aromatic plants Asia." *ICS UNIDO Asia* 2: 305.

Hashimoto, Takashi, Dae-Jin Yun, and Yasuyuki Yamada. 1993. "Production of tropane alkaloids in genetically engineered root cultures." *Phytochemistry* 32, no. 3: 713-718.

Hayut, Shdema Filler, Cathy Melamed Bessudo, and Avraham A. Levy. 2017 "Targeted recombination between homologous chromosomes for precise breeding in tomato." *Nature communications* 8, no. 1: 1-9.

Hidalgo, Maria, M. Jose Oruna-Concha, Sofia Kolida, Gemma E. Walton, StamatinaKallithraka, Jeremy PE Spencer, and Sonia de Pascual-Teresa. 2012. "Metabolism of anthocyanins by human gut microflora and their influence on gut bacterial growth." *Journal of agricultural and food chemistry* 60, no. 15: 3882-3890.

Honein, Margaret A., Leonard J. Paulozzi, T. J. Mathews, J. David Erickson, and Lee-Yang C. Wong. 2001. "Impact of folic acid fortification of the US food supply on the occurrence of neural tube defects." *Jama* 285, no. 23: 2981-2986.

Horvath, Philippe, and Rodolphe Barrangou. 2010. "CRISPR/Cas, the immune system of bacteria and archaea." *Science* 327, no. 5962: 167-170.

Huang, Qiulong, Charles A. Roessner, Rodney Croteau, and A. Ian Scott. 2001. "Engineering Escherichia coli for the synthesis of taxadiene, a key intermediate in the biosynthesis of taxol." *Bioorganic & medicinal chemistry* 9, no. 9: 2237-2242.

Hussain, Md Sarfaraj, Sheeba Fareed, Md Saba Ansari, Akhlaquer Rahman, Iffat Zareen Ahmad, and Mohd Saeed. 2012. "Current approaches toward production of secondary plant metabolites." *Journal of pharmacy & bioallied sciences* 4, no. 1: 10.

Jacobs, Thomas B., Peter R. LaFayette, Robert J. Schmitz, and Wayne A. Parrott. 2015. "Targeted genome modifications in soybean with CRISPR/Cas9." *BMC biotechnology* 15, no. 1: 1-10.

Jeong, Cheol-Seung, Hosakatte Niranjana Murthy, Eun-Joo Hahn, and Kee-YoeupPaek. 2008. "Improved production of ginsenosides in suspension cultures of ginseng by medium replenishment strategy." *Journal of bioscience and bioengineering* 105, no. 3: 288-291.

Jinek, Martin, Krzysztof Chylinski, Ines Fonfara, Michael Hauer, Jennifer A. Doudna, and Emmanuelle Charpentier. 2012. "A programmable dual-RNA–guided DNA endonuclease in adaptive bacterial immunity." *Science* 337, no. 6096: 816-821.

Jouanin, L. 1984. "Restriction map of an agropine-type Ri plasmid and its homologies with Ti plasmids." *Plasmid* 12, no. 2: 91-102.

Kamal-Eldin, Afaf, and Lars-ÅkeAppelqvist. 1996. "The chemistry and antioxidant properties of tocopherols and tocotrienols." *Lipids* 31, no. 7: 671-701.

Kinney, Anthony J. 1998."Manipulating flux through plant metabolic pathways." *Current opinion in plant biology* 1, no. 2: 173-178.

Kumar, Ashwani, and Sudhir K. Sopory. 2008. "Recent advances in plant biotechnology and its applications." *Phytomorphology*: 83.

Kurilich, Anne C., and John A. Juvik. 1999. "Quantification of Carotenoid and Tocopherol Antioxidants in Zea m ays." *Journal of Agricultural and Food Chemistry* 47, no. 5: 1948-1955.

Lenka, Sangram K., Nadia Boutaoui, Bibin Paulose, Kham Vongpaseuth, Jennifer Normanly, Susan C. Roberts, and Elsbeth L. Walker. 2012 "Identification and expression analysis of methyl jasmonate responsive ESTs in paclitaxel producing Taxus cuspidata suspension culture cells." *BMC genomics* 13, no. 1: 1-10.

Lessard, Philip. "Metabolic engineering: the concept coalesces. 1996." *Nature biotechnology* 14, no. 13: 1654-1655.

Li, Bin, Guanghong Cui, Guoan Shen, Zhilai Zhan, Luqi Huang, Jiachun Chen, and Xiaoquan Qi. 2017. "Targeted mutagenesis in the medicinal plant *Salvia miltiorrhiza*." *Scientific reports* 7, no. 1: 1-9.

Li, Feng-Xia, Zhi-Ping Jin, De-Xiu Zhao, Li-Qin Cheng, Chun-Xiang Fu, and Fengshan Ma. 2006. "Overexpression of the Saussurea medusa chalcone isomerase gene in *S. involucrata* hairy root cultures enhances their biosynthesis of apigenin." *Phytochemistry* 67, no. 6: 553-560.

Li, Meng-Yao, Feng Wang, Zhi-Sheng Xu, Qian Jiang, Jing Ma, Guo-Fei Tan, and Ai-Sheng Xiong. 2014. "High throughput sequencing of two celery varieties small RNAs identifies microRNAs involved in temperature stress response." *BMC genomics* 15, no. 1: 1-9.

Li, Rui, Ran Li, Xindi Li, Daqi Fu, Benzhong Zhu, Huiqin Tian, Yunbo Luo, and Hongliang Zhu.2018."Multiplexed CRISPR/Cas9-mediated metabolic engineering of γ-aminobutyric acid levels in *Solanum lycopersicum*." *Plant biotechnology journal* 16, no. 2: 415-427.

Liu, Rui Hai. 2013(a). "Dietary bioactive compounds and their health implications." *Journal of food science* 78, no. s1: A18-A25.

Liu, Rui Hai. 2013 (b). "Health-promoting components of fruits and vegetables in the diet." *Advances in nutrition* 4, no. 3: 384S-392S.

Lu, Baiyi, Maiquan Li, and Ran Yin. 2016. "Phytochemical content, health benefits, and toxicology of common edible flowers: a review (2000–2015)." *Critical Reviews in Food Science and Nutrition* 56, no. sup1: S130-S148.

Lu, Shijin, Xungang Xiao, and Minghua Cheng. 2015. "Matrine inhibits IL-1β-induced expression of matrix metalloproteinases by suppressing the activation of MAPK and NF-κB in human chondrocytes in vitro." *International journal of clinical and experimental pathology* 8, no. 5: 4764.

Ludwig-Müller, Jutta, MilenGeorgiev, and Thomas Bley. 2008. "Metabolite and hormonal status of hairy root cultures of Devil's claw (*Harpagophytum procumbens*) in flasks and in a bubble column bioreactor." *Process Biochemistry* 43, no. 1: 15-23.

Ma, Xingliang, Qunyu Zhang, Qinlong Zhu, Wei Liu, Yan Chen, Rong Qiu, Bin Wang et al. 2015. "A robust CRISPR/Cas9 system for convenient, high-efficiency multiplex genome editing in monocot and dicot plants." *Molecular plant* 8, no. 8: 1274-1284.

Makarova, Kira S., Yuri I. Wolf, Jaime Iranzo, Sergey A. Shmakov, Omer S. Alkhnbashi, Stan JJ Brouns, Emmanuelle Charpentier et al. 2020. "Evolutionary classification of CRISPR–Cas systems: a burst of class 2 and derived variants." *Nature Reviews Microbiology* 18, no. 2: 67-83.

Martin, Cathie, Eugenio Butelli, Katia Petroni, and Chiara Tonelli. 2011. "How can research on plants contribute to promoting human health?." *The Plant Cell* 23, no. 5: 1685-1699.

Martin, Cathie, Yang Zhang, Chiara Tonelli, and Katia Petroni. 2013. "Plants, diet, and health." *Annual review of plant biology* 64: 19-46.

Mehrotra, Shakti, Arun Kumar Kukreja, Suman Preet Singh Khanuja, and Bhartendu Nath Mishra. 2008. "Genetic transformation studies and scale up of hairy root culture of *Glycyrrhizaglabra* in bioreactor." *Electronic Journal of Biotechnology* 11, no. 2: 69-75.

Meng, Hailin, Yong Wang, Qiang Hua, Siliang Zhang, and Xiaoning Wang. 2011. "In silico analysis and experimental improvement of taxadiene heterologous biosynthesis in *Escherichia coli*." *Biotechnology and Bioprocess Engineering* 16, no. 2: 205-215.

Menzel, G., H. J. Harloff, and C. Jung. 2003. "Expression of bacterial poly (3-hydroxybutyrate) synthesis genes in hairy roots of sugar beet (*Beta vulgaris* L.)." *Applied microbiology and biotechnology* 60, no. 5: 571-576.

Mikkelsen, Michael Dalgaard, Carl Erik Olsen, and Barbara Ann Halkier. 2010. "Production of the cancer-preventive glucoraphanin in tobacco." *Molecular plant* 3, no. 4: 751-759.

Miller, Elizabeth C., Craig W. Hadley, Steven J. Schwartz, John W. Erdman, Thomas W-M. Boileau, and Steven K. Clinton. 2002. "Lycopene, tomato products, and prostate cancer prevention. Have we established causality?." *Pure and Applied Chemistry* 74, no. 8: 1435-1441.

Min, Ji Yun, Hee Young Jung, Seung Mi Kang, Yong Duck Kim, Young Min Kang, Dong Jin Park, DoddananjappaTheertha Prasad, and Myung Suk Choi. 2007. "Production of tropane alkaloids by small-scale bubble column bioreactor cultures of Scopolia parviflora adventitious roots." *Bioresource technology* 98, no. 9: 1748-1753.

Moghe, Gaurav D., and Robert L. Last. 2015. "Something old, something new: conserved enzymes and the evolution of novelty in plant specialized metabolism." *Plant Physiology* 169, no. 3: 1512-1523.

Mori, Tsukasa, and Mlel Sakurai. 1994. "Production of anthocyanin from strawberry cell suspension cultures; effects of sugar and nitrogen." *Journal of food science* 59, no. 3: 588-593.

Mozaffarian, Dariush, Tao Hao, Eric B. Rimm, Walter C. Willett, and Frank B. Hu. 2011. "Changes in diet and lifestyle and long-term weight gain in women and men." *New England Journal of Medicine* 364, no. 25: 2392-2404.

Naczk, Marian, and Fereidoon Shahidi. 2006. "Phenolics in cereals, fruits and vegetables: Occurrence, extraction and analysis." *Journal of pharmaceutical and biomedical analysis* 41, no. 5: 1523-1542.

Naqvi, Shaista, Changfu Zhu, Gemma Farre, KoreenRamessar, LudovicBassie, Jürgen Breitenbach, Dario Perez Conesa et al. "Transgenic multivitamin corn through biofortification of endosperm with three vitamins representing three distinct metabolic pathways." *Proceedings of the National Academy of Sciences* 106, no. 19 (2009): 7762-7767.

Nawa, Yoshihiko, Satoshi Asano, Satoko Motoori, and Toshio Ohtani. 1993. "Production of anthocyanins, carotenoids, and proanthocyanidins by cultured cells of rabbiteye blueberry (*Vaccinium ashei* Reade)." *Bioscience, biotechnology, and biochemistry* 57, no. 5: 770-774.

Nims, Ezekiel, Camille P. Dubois, Susan C. Roberts, and Elsbeth L. Walker. 2006. "Expression profiling of genes involved in paclitaxel biosynthesis for targeted metabolic engineering." *Metabolic engineering* 8, no. 5: 385-394.

Norat, Teresa, Dagfinn Aune, Doris Chan, and Dora Romaguera. 2014. "Fruits and vegetables: updating the epidemiologic evidence for the WCRF/AICR lifestyle recommendations for cancer prevention." *Advances in nutrition and cancer*: 35-50.

Oksman-Caldentey, Kirsi-Marja, and Dirk Inzé. 2004. "Plant cell factories in the post-genomic era: new ways to produce designer secondary metabolites." *Trends in plant science* 9, no. 9: 433-440.

Olivoto, Tiago, MaiconNardino, Ivan Ricardo Carvalho, Diego NicolauFollmann, Vin&iciusJardelSzareski, Mauricio Ferrari, Alan Junior de Pelegrin, and Velci Queir & oz de Souza. 2017. "Plant secondary metabolites and its dynamical systems of induction in response to environmental factors: A review." *African Journal of Agricultural Research* 12, no. 2: 71-84.

Padley, F. B. 1994. "Occurrence and characteristics of oils and fats." *The Lipid Handbook*: 47-223.

Paine J. A., Shipton C. A., Chaggar S., et al. 2005. Improving the nutritional value of Golden Rice through increased pro-vitamin A content. *Nature Biotechnology* 23, 482–487.

Pan, Changtian, Lei Ye, Li Qin, Xue Liu, Yanjun He, Jie Wang, Lifei Chen, and Gang Lu. 2016. "CRISPR/Cas9-mediated efficient and heritable targeted mutagenesis in tomato plants in the first and later generations." *Scientific reports* 6, no. 1: 1-9.

Parkin, D. Maxetal, L. Boyd, and L. C. Walker. 2011. "16. The fraction of cancer attributable to lifestyle and environmental factors in the UK in 2010." *British journal of cancer* 105, no. 2: S77-S81.

Pépin, Marie-France, Jean Archambault, Claude Chavarie, and François Cormier. 1995. "Growth kinetics of *Vitis vinifera* cell suspension cultures: I. Shake flask cultures." *Biotechnology and bioengineering* 47, no. 2: 131-138.

Persad, Vidia L., Michiel C. Van den Hof, Johanne M. Dubé, and Pamela Zimmer. 2002. "Incidence of open neural tube defects in Nova Scotia after folic acid fortification." *Cmaj* 167, no. 3: 241-245.

Peterson, David. M., and Qureshi, Asaf. A. 1993. “Genotype and environmental effects on tocols of barleand oats.” *Cereal Chemistry*70, no. 2: 157–162.

Petrillo, Ezequiel, Micaela A. Godoy Herz, Andrea Barta, Maria Kalyna, and Alberto R. Kornblihtt. 2014. "Let there be light: regulation of gene expression in plants." *RNA biology* 11, no. 10: 1215-1220.

Pichersky, Eran, and David R. Gang. 2000. "Genetics and biochemistry of secondary metabolites in plants: an evolutionary perspective." *Trends in plant science* 5, no. 10: 439-445.

Potrykus, Ingo. 2010. "Lessons from the ‘Humanitarian Golden Rice’project: regulation prevents development of public good genetically engineered crop products." *New Biotechnology* 27, no. 5: 466-472.

Putalun, W., F. Taura, W. Qing, H. Matsushita, H. Tanaka, and Y. Shoyama. 2003. "Anti-solasodine glycoside single-chain Fv antibody stimulates biosynthesis of solasodine glycoside in plants." *Plant cell reports* 22, no. 5: 344-349.

Rao, S. Ramachandra, and G. A. Ravishankar. 2002. "Plant cell cultures: chemical factories of secondary metabolites." *Biotechnology advances* 20, no. 2: 101-153.

Ricroch, Agnès, Pauline Clairand, and Wendy Harwood. 2017. "Use of CRISPR systems in plant genome editing: toward new opportunities in agriculture." *Emerging Topics in Life Sciences* 1, no. 2: 169-182.

Rosati, Carlo, Riccardo Aquilani, Sridhar Dharmapuri, PatriziaPallara, Carla Marusic, RaffaelaTavazza, Florence Bouvier, Bilal Camara, and Giovanni Giuliano. 2000. "Metabolic engineering of beta-carotene and lycopene content in tomato fruit." *The Plant Journal* 24, no. 3: 413-420.

Ruiz, Oscar N., Hussein S. Hussein, Norman Terry, and Henry Daniell. 2003 "Phytoremediation of organomercurial compounds via chloroplast genetic engineering." *Plant Physiology* 132, no. 3: 1344-1352.

Runguphan, Weerawat, and Sarah E. O'connor. 2009. "Metabolic reprogramming of periwinkle plant culture." *Nature chemical biology* 5, no. 3: 151-153.

Runguphan, Weerawat, Xudong Qu, and Sarah E. O'connor. 2010. "Integrating carbon–halogen bond formation into medicinal plant metabolism." *Nature* 468, no. 7322: 461-464.

Rush, David. "Nutrition and maternal mortality in the developing world. 2000. " *The American journal of clinical nutrition* 72, no. 1: 212S-240S.

Ryan, Katherine S., and Bradley S. Moore. 2009. "Alkaloid biosynthesis takes root." *Nature chemical biology* 5, no. 3: 140-141.

Saito, Kazuki, Mami Yamazaki, and Isamu Murakoshi. 1992. "Transgenic medicinal plants: Agrobacterium-mediated foreign gene transfer and production of secondary metabolites." *Journal of natural products* 55, no. 2: 149-162.

Sato, Fumihiko, Takashi Hashimoto, Akira Hachiya, Ken-ichi Tamura, Kum-Boo Choi, Takashi Morishige, Hideki Fujimoto, and Yasuyuki Yamada. 2001. "Metabolic engineering of plant alkaloid biosynthesis." *Proceedings of the National Academy of Sciences* 98, no. 1: 367-372.

Sato, Kenji, Mamoru Nakayama, and Jun-ichi Shigeta. 1996. "Culturing conditions affecting the production of anthocyanin in suspended cell cultures of strawberry." *Plant Science* 113, no. 1: 91-98.

Schaart, Jan G., Clemens C. M. van de Wiel, Lambertus A. P. Lotz, and Marinus J. M. Smulders. "Opportunities for products of new plant breeding techniques. 2016. " *Trends in Plant Science* 21, no. 5: 438-449.

Shewmaker, Christine K., Julie A. Sheehy, Maureen Daley, Susan Colburn, and Dang Yang Ke. 1999. "Seed-specific overexpression of phytoene synthase: increase in carotenoids and other metabolic effects." *The Plant Journal* 20, no. 4: 401-412.

Shibli, Rida A., M. A. L. Smith, and M. A. Shatnawi. 1998. "Pigment recovery from encapsulated-dehydrated *Vaccinium pahalae* (ohelo) cryopreserved cells." *Plant cell, tissue and organ culture* 55, no. 2: 119-123.

Shibli, Rida A., M. A. L. Smith, and M. Kushad. 1997. "Headspace ethylene accumulation effects on secondary metabolite production in *Vaccinium pahalae* cell culture." *Plant Growth Regulation* 23, no. 3: 201-205.

Sonoda, Eiichiro, HelfridHochegger, AlihosseinSaberi, Yoshihito Taniguchi, and Shunichi Takeda. 2006. "Differential usage of non-homologous end-joining and homologous recombination in double strand break repair." *DNA repair* 5, no. 9-10: 1021-1029.

Storozhenko, Sergei, Veerle De Brouwer, Maarten Volckaert, Oscar Navarrete, Dieter Blancquaert, Guo-Fang Zhang, Willy Lambert, and Dominique Van Der Straeten. 2007. "Folate fortification of rice by metabolic engineering." *Nature biotechnology* 25, no. 11: 1277-1279.

Sun, Yongwei, Guiai Jiao, Zupei Liu, Xin Zhang, Jingying Li, Xiuping Guo, Wenming Du et al. 2017. "Generation of high-amylose rice through CRISPR/Cas9-mediated targeted mutagenesis of starch branching enzymes." *Frontiers in plant science* 8: 298.

Svitashev, Sergei, Joshua K. Young, Christine Schwartz, Huirong Gao, S. Carl Falco, and A. Mark Cigan. 2015. "Targeted mutagenesis, precise gene editing, and site-specific gene insertion in maize using Cas9 and guide RNA." *Plant physiology* 169, no. 2: 931-945.

Tattersall, David B., SørenBak, Patrik R. Jones, Carl Erik Olsen, Jens K. Nielsen, Mads L. Hansen, Peter B. Høj, and Birger Lindberg Møller. 2001. "Resistance to an herbivore through engineered cyanogenic glucoside synthesis." *Science* 293, no. 5536: 1826-1828.

Traka, Maria H., and Richard F. Mithen. 2011."Plant science and human nutrition: challenges in assessing health-promoting properties of phytochemicals." *The Plant Cell* 23, no. 7: 2483-2497.

Van der Fits, Leslie, and Johan Memelink. 2000. "ORCA3, a jasmonate-responsive transcriptional regulator of plant primary and secondary metabolism." *Science* 289, no. 5477: 295-297.

Vanisree, Mulabagal, Chen-Yue Lee, Shu-Fung Lo, Satish Manohar Nalawade, ChienYih Lin, and Hsin-Sheng Tsay. 2004. "Studies on the production of some important secondary metabolites from medicinal plants by plant tissue cultures." *Bot. Bull. Acad. Sin* 45, no. 1: 1-22.

Vongpaseuth, Kham, Ezekiel Nims, Melissa St Amand, Elsbeth L. Walker, and Susan C. Roberts. 2007. "Development of a particle bombardment-mediated transient transformation system for Taxus spp. cells in culture." *Biotechnology progress* 23, no. 5: 1180-1185.

Wang, Shu, Naima Moustaid-Moussa, Lixia Chen, Huanbiao Mo, Anuradha Shastri, Rui Su, Priyanka Bapat, InSook Kwun, and Chwan-Li Shen. 2014 (a). "Novel insights of dietary polyphenols and obesity." *The Journal of nutritional biochemistry* 25, no. 1: 1-18.

Wang, Xia, Yingying Ouyang, Jun Liu, Minmin Zhu, Gang Zhao, Wei Bao, and Frank B. Hu. 2014 (b). "Fruit and vegetable consumption and mortality from all causes, cardiovascular disease, and cancer: systematic review and dose-response meta-analysis of prospective cohort studies." *Bmj* 349.

Wang, Xiao, Jiajin Zhang, Simei He, Yuanni Gao, Xiaoqin Ma, Yun Gao, Guanghui Zhang et al. 2018. "HMOD: an omics database for herbal medicine plants." *Molecular plant* 11, no. 5: 757-759.

Watanabe, Kenta, Anna Kobayashi, Masaki Endo, Kimiyo Sage-Ono, Seiichi Toki, and Michiyuki Ono. 2017. "CRISPR/Cas9-mediated mutagenesis of the dihydroflavonol-4-reductase-B (DFR-B) locus in the Japanese morning glory Ipomoea (Pharbitis) nil." *Scientific reports* 7, no. 1: 1-9.

Xu, Rongfang, Yachun Yang, Ruiying Qin, Hao Li, ChunhongQiu, Li Li, Pengcheng Wei, and Jianbo Yang. 2016. "Rapid improvement of grain weight via highly efficient CRISPR/Cas9-mediated multiplex genome editing in rice." *Journal of Genetics and Genomics= Yi chuanxue bao* 43, no. 8: 529-532.

Ye, Xudong, Salim Al-Babili, Andreas Klöti, Jing Zhang, Paola Lucca, Peter Beyer, and Ingo Potrykus. 2000. "Engineering the provitamin A (β-carotene) biosynthetic pathway into (carotenoid-free) rice endosperm." *Science* 287, no. 5451: 303-305.

Zalewski, Wojciech, WacławOrczyk, Sebastian Gasparis, and Anna Nadolska-Orczyk. 2012. "HvCKX2 gene silencing by biolistic or Agrobacterium-mediated transformation in barley leads to different phenotypes." *BMC Plant Biology* 12, no. 1: 1-12.

Zhang, Chunyu, Rebecca E. Cahoon, Sarah C. Hunter, Ming Chen, Jixiang Han, and Edgar B. Cahoon. 2013. "Genetic and biochemical basis for alternative routes of tocotrienol biosynthesis for enhanced vitamin E antioxidant production." *The Plant Journal* 73, no. 4: 628-639.

Zhu, Changfu, Shaista Naqvi, Jürgen Breitenbach, Gerhard Sandmann, Paul Christou, and Teresa Capell. 2008 "Combinatorial genetic transformation generates a library of metabolic phenotypes for the carotenoid pathway in maize." *Proceedings of the National Academy of Sciences* 105, no. 47: 18232-18237.

http://herbalplant.ynau.edu.cn/

http://metnetdb.org/mpmr_public/

http://www.planttrichome.org/

Further Readings

Abbasin, Z., S. Zamani, S. Movahedi, G. Khaksar, and BE Sayed Tabatabaei. 2010 "In vitro micropropagation of yew (*Taxus baccata*) and production of plantlets." *Biotechnology* 9, no. 1: 48-54.

Aharoni, Asaph, Maarten A. Jongsma, Tok-Yong Kim, Man-Bok Ri, Ashok P. Giri, Francel W. A. Verstappen, Wilfried Schwab, and Harro J. Bouwmeester. 2006. "Metabolic engineering of terpenoid biosynthesis in plants." *Phytochemistry Reviews* 5, no. 1: 49-58.

Besumbes, Óscar, Susanna Sauret-Güeto, Michael A. Phillips, Santiago Imperial, Manuel Rodríguez-Concepción, and Albert Boronat. 2004. "Metabolic engineering of isoprenoid biosynthesis in *Arabidopsis* for the production of taxadiene, the first committed precursor of Taxol." *Biotechnology and bioengineering* 88, no. 2: 168-175.

Bonfill, Mercedes, Sonia Malik, M. Hossein Mirjalili, Marta Goleniowski, Rosa Cusido, and Javier Palazon. 2013. "Production and genetic engineering of terpenoids production in plant cell and organ cultures." *Ramawat, KG, MJM (Eds.), Natural Products. Springer, Berlin*: 2761-2796.

Jaganath, Balusamy, Kondeti Subramanyam, Subramanian Mayavan, Sivabalan Karthik, Dhandapani Elayaraja, RajangamUdayakumar, MarkandanManickavasagam, and Andy Ganapathi. 2014. "An efficient in planta transformation of *Jatropha curcas* (L.) and multiplication of transformed plants through in vivo grafting." *Protoplasma* 251, no. 3: 591-601.

Kovacs, Katalin, Lina Zhang, Robert S. T. Linforth, Benjamin Whittaker, Christopher J. Hayes, and Rupert G. Fray. 2007. "Redirection of carotenoid metabolism for the efficient production of taxadiene [taxa-4 (5), 11 (12)-diene] in transgenic tomato fruit." *Transgenic research* 16, no. 1: 121-126.

Krens, F. A., L. Molendijk, G. J. Wullems, and R. A. Schilperoort. 1982. "In vitro transformation of plant protoplasts with Ti-plasmid DNA." *Nature* 296, no. 5852: 72-74.

Matsuura, Hélio Nitta, Sonia Malik, Fernanda de Costa, MortezaYousefzadi, Mohammad Hossein Mirjalili, Randolph Arroo, Avninder S. Bhambra, Miroslav Strnad, Mercedes Bonfill, and Arthur Germano Fett-Neto. 2018. "Specialized plant metabolism characteristics and impact on target molecule biotechnological production." *Molecular biotechnology* 60, no. 2: 169-183.

Nishikawa, Kazutaka, and Kanji Ishimaru. "Flavonoids in root cultures of Scutellariabaicalensis." *Journal of plant physiology* 151, no. 5 (1997): 633-636.

Wang, Mugui, Yuming Lu, José Ramón Botella, Yanfei Mao, Kai Hua, and Jian-kang Zhu. 2017. "Gene targeting by homology-directed repair in rice using a geminivirus-based CRISPR/Cas9 system." *Molecular plant* 10, no. 7: 1007-1010.

Zhang, Hui, Jinshan Zhang, Zhaobo Lang, José Ramón Botella, and Jian-Kang Zhu. 2017. "Genome editing—principles and applications for functional genomics research and crop improvement." *Critical Reviews in Plant Sciences* 36, no. 4: 291-309.

In: The Future of Metabolic Engineering ISBN: 978-1-68507-362-6
Editors: A. Sharma and D. Amin

Chapter 12

IN SILICO STUDIES DESCRIBING MIRNAS AND THEIR REGULARITY PROCESSES IN INDUSTRIAL IMPORTANT MEDICINAL AND AROMATIC PLANTS

Anand Mishra[1,*] and Pankhuri Gupta[2,3]
[1]Genetics and Plant Breeding Division, CSIR- Central Institute for Medicinal and Aromatic Plants, Lucknow, UP, India
[2]Biotechnology Division, CSIR- Central Institute for Medicinal and Aromatic Plants, Lucknow, UP, India
[3]CSIR- Human Resource Development Centre Campus, Academy of Scientific and Innovative Research (AcSIR), Kamla Nehru Nagar, Ghaziabad, Uttar Pradesh, India

ABSTRACT

MicroRNAs (miRNA) are small endogenous, non-coding RNAs and show a significant role in plant signaling and regulatory pathways, stimulus-response environmental stress tolerance, growth, and its development. Identification of miRNA provides many options by which

[*] Corresponding Author's E-mail: mishra.anand 653 @gmail.com.

secondary metabolism in medicinal and aromatic plants can help in introducing novel drugs, it can alter the biosynthesis pathway to produce the desired product in large amounts. Since, industrial important MAPs contain valuable medicinal properties, flavoring agents, fragrances, insecticides, drugs, and dyes that help to elaborate metabolic and regulatory pathways in MAPs like *Ocimum, Vetiver, Mentha arvensis, Rauwolfia serpentina, Pelargonium, Gymnema sylvestre, Withania somnifera, Bacopa monnieri, Curcuma longa,* etc to use them in pharmacological applications and biotechnological tools in an effective way. Here in this book chapter, we present an update on the changing trends. Here we summarise different studies describing miRNAs and their role as regulatory elements in industrial important medicinal and aromatic plants to understand and exploit the secondary metabolism.

Keywords: MicroRNAs, regulatory elements, MAPs, secondary metabolism

Introduction

MicroRNAs (miRNA) are small non-coding RNA molecules that regulate gene expression utilizing transcriptional cleavage activity or translational inhibition in plants and animals (Bartel 2004) (Carrington and Ambros 2003). miRNAs have 21–24 nucleotides long and are non-coding in nature. The first ever miRNA was discovered in *Caenorhabditis elegans* in heterochronic gene lin-4 (Lee, Feinbaum, and Ambros 1993). The discovery of miRNA in the plant is less than two decades old but they show an important role in the developmental processes such as morphogenesis of leaf, flower, vegetative phase change, flowering time, and various responses to environmental effects (Millar 2020). They also function in various abiotic and biotic stresses in plants. Many experiments and reports confirmed that stress conditions can increase the abundance of miRNAs in plants (Gupta et al., 2014).

Maximal miRNAs target 3 prime ends (3′ UTR) of mRNAs to control the expression of a gene with some exceptions (Ha and Kim 2014) (Broughton et al., 2016). Transport of miRNA to the different subcellular compartments ensures the rate of translation and to some extent of transcription (Makarova et al., 2016). The capability of certain groups of miRNA to target and regulate multiple gene messenger RNAs (mRNAs) is important in cellular machinery (Mishra et al., 2020). Approximately, a total of 40% of miRNA sequences are remaining in gene intron and initiation of transcription by RNA Pol II or Pol III exceptionally may stop transcription and splicing of the coding genes within an intron. Nevertheless, evidence shows the ability of numerous intronic originated miRNAs to co-express with host genes by experimentation and bioinformatics approaches. The processing of primary-miRNA occurs in a conserved manner. Despite the genomic location, miRNA transcription in the nucleus leads to the mature product in the cytosol. Inside the nucleus, transcribed product, primary miRNAs (pri-miRNAs) serve as precursor RNAs which is 70–100 nucleotide hairpin-like structure (Bartel 2004). Further, pri-miRNAs are processed by DICERLIKE 1 (DCL1) protein to produce the pre-miRNAs. Further, it cleaves pre-miRNAs complex release the miRNA/ miRNA* duplex in the cytosol. Hua Enhancer 1 (HEN1) is a methyltransferase that methylates miRNA/ miRNA* duplex at the 3'-end. Further, it is exported by the plant export in protein HASTY to the cytoplasm, (Axtell, Westholm, and Lai 2011) (Rogers and Chen 2012). miRNA/miRNA* duplex separation takes place in the cytoplasm and the guide strand is bind to the RNA-induced silencing complex (RISC) with help of Argonaute (AGO) proteins. This fully assembled RISC, contain other protein like HYPONASTIC LEAVES 1 (HYL1) and SERRATE (SE) bind to the target transcript sequence through sequence complementarily with its mature miRNA strand which is around 18-22 nucleotide, to further direct either mRNA cleavage or translational control or inhibition Ren and Yu 2012). The canonical and non-canonical pathways for the biogenesis of miRNA exist in cells (O'Brien et al., 2018).

MIRNA IDENTIFICATION TOOL

miRNAs are relatively stable however, some factors make them difficult analytical targets as small sequence lengths, large variability/transcript abundance within a cell, and high sequence homology within the miRNA family (Ye et al., 2019). Classic forward genetics play an important role in the discovery of miRNA like the discovery of lin-4 in *C. elegans* and directional cloning for an endogenous small RNAs cDNA library. Over the past decades, different detection methods are developed (Huang et al., 2011) (de Planell-Saguer and Rodicio 2013) (Y. Shen et al., 2015) (Tian, Wang, and Zhou 2015). Next-generation sequencing plays a significant impact in miRNA identification. Current miRNA identification techniques can have two categories as conventional methods and modern technological methods. The identification from different source miRNAs is the basis of miRNA research. Traditional methods for miRNA detection include northern blotting (Várallyay, Burgyán, and Havelda 2008) (Kim et al., 2010) (Akmal, Baig, and Khan 2017), microarray analysis (W. Li and Ruan 2009), and quantitative polymerase chain reaction(qPCR) (Niu et al., 2015). In situ hybridization detection, invader miRNA assay, RT-PCR based assay (Cloning miRNAs with stem-loop RT primers, Primer-extension, Cloning miRNAs with adaptor), radiolabeled splinted ligation detection. Integrating heterogeneous data sources are used by computational methods to infer miRNA functions including miRNA mode of action and gene regulatory networks, which is important in miRNA discovery and their target identification an important module of computational biology (Liu, Li, and Cairns 2014). miRNA discovery reaches more than 10,000 different type of miRNA in animals, plants and viruses by using different detection methods in miRNAs database (Kozomara, Birgaoanu, and Griffiths-Jones 2019). Several other computational base approaches are MiRscan, miRSeeker, Phylogenetic shadowing, miRank, MiRanalyzer, miRDeep etc (Figure 1).

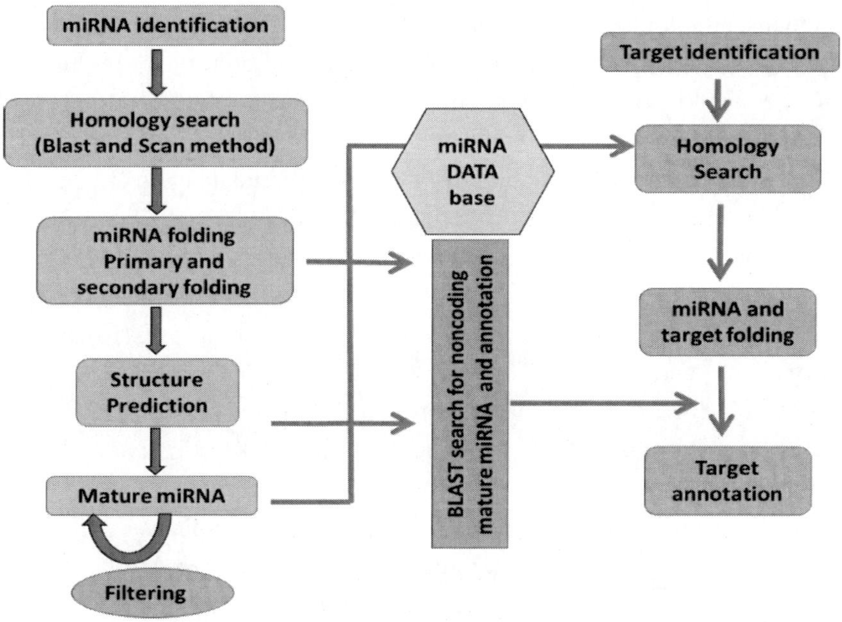

Figure 1. A putative homology base approach shows the process of miRNA Identification. (Mishra et al., 2020).

MIRNA IN MAPS PLANTS

Plants have been the main source of various medicines, to cure, heal and alleviate various diseases from a very early time and these plants are known as a medicinal plant. They are very important for pharmaceutical, cosmetic, and nutritional values and use for economic too. A group of compounds known as secondary metabolites produces from the plant for different types of stresses that are also useful for human health and commercially important. Plants synthesize a diverse array of aromatic substances in form of secondary metabolites of more than 12,000 isolates, which suggest only 10% of the total. Plant uses these compounds to defense against predation by microorganisms, herbivores, and insects. Different type of compounds as Glycosides, Flavonoids, and Proanthocyanidins, Tannins, Terpenoids and Phenylpropanoids, Resins,

Lignans, Alkaloids are synthesized from these plants (Ghorbanpour et al., 2017). Medicinal and aromatic plants will also uphold their significance in the exploration of novel important sources of drugs and lead compounds having diverse industrial uses such as flavoring, compounds, insecticides, nutraceuticals, and dyes useful for human health (Marcela et al., 2019). Recent reports showed that miRNAs in MAPs are not only key regulators in plant fundamental processes but act as a critical modulator in secondary metabolism and work cross-kingdom gene regulation. Some databases have information of miRNA report as miRbase (Kozomara, Birgaoanu, and Griffiths-Jones 2019), PMRD (Zhang et al., 2010), miRNEST (Szcześniak and Makałowska 2014), miREX (Zielezinski et al., 2015), PmiRKB (Meng et al., 2011) and PNRD (Yi et al., 2015). Some database exclusively gives information regarding miRNA present in MAP plants as MepmiRDB (Yu et al., 2019). Some plants families are very important in producing medicinal and aromatic plants like Lamiaceae, Solanaceae, Apiaceae. miRNA and their putative target are identified in *Swertia chirayita, Nicotiana tabacum*, plants for possible use in secondary metabolite synthesis. miRNA are reported to involve in Flavonoids producing plants in *Helianthus, Diospyros kaki, Lonicera japonica* and reported to interact with biosynthesis genes. miRNA in Alkaloids biosynthesis are reported in Opium poppy (*Papaver somniferum*), *Taxus baccata, Rauwolfia* (Marcela et al., 2019). *Ocimum* and *Mentha* are the important plants of the family Lamiaceae producing series of metabolites. Studies in the mint family revealed that miR156, miR414, and miR5021 regulate essential oil biosynthesis while miR156, miR5021, and miR5015b are involved in trichome development. (Singh et al., 2016) (Akter et al., 2014). In Ocimum, a putative miRNA was identified by nr ESTS and it was reported that seven miRNA families were identified. miRNA 5021 was suggested to regulate amino acid metabolism and cofactor metabolism (Singh and Sharma 2014). Similarly, miRNA in *Ocimum* was found to regulate target *O.basilicum* key human genes involved in diseases (Patel, Mangukia, et al., 2019). A putative *O.basilicum* miRNOME was reported by Patel and a co-worker for the cross-kingdom approach (Patel, et al., 2019). *Withania somnifera*,

(Ashwagandha) largely used in Ayurvedic and its metabolite withanolides have shown promising therapeutic properties (Pandey et al., 2017) (Sanchita et al., 2015). A comparative detail root and leaf miRNA revel endogenous, leaf-miR477, leaf-miR530 root-miR5140, and rootmiR159 were involved in withanolide metabolism (Srivastava et al., 2018). In Turmeric (*Curcuma longa*) miR156 and miR5015 regulate rhizome development. MiR5021 involve in isoquinoline alkaloid biosynthesis pathways and terpenoid backbone biosynthesis (Singh and Sharma 2017). In ginger, biosynthesis of gingerol was observed to regulate by miR5015 and inhibiting phenyl ammonia lyase (PAL), a precursor enzyme of gingerol biosynthesis. *Catharanthus roseus* is an important medicinal plant for its anticancerous property. Different miRNA are found to interact and regulate indole alkaloid biosynthetic pathways (Pani and Mahapatra 2013; E. M. Shen et al., 2017) (Gupta et al., 2015).

Role of miRNA in Secondary Metabolism

Secondary metabolite production can be regulated by miRNA as they control biological processes positively or negatively. Classical methods as plant cell culture are very difficult and tedious for stable production of plant secondary metabolite (Bulgakov and Avramenko 2015). Chemical structures are the basis of plant secondary metabolites are categorization (Hussein and Anssary 2018) (Teoh 2015). These secondary metabolites are broadly categorized in different categories (Verpoorte 1998). Different reports are published where the direct role of miRNA was found in alteration of secondary metabolism, which some gene silencing approaches are also followed to increase and decrease these metabolically active compounds. Different miRNAs are reported to involve in phenylpropanoid biosynthesis and flavonoid biosynthesis. Different miRNA are found involve in Anthocyanin biosynthesis like which target targeted SPL (Squamosa Promoter Binding Protein-like) which is negative regulator of anthocyanin whereas suppression of miR156 activity leads to high levels of flavonols. (Gou et al., 2011). Flavonoids biosynthetic pathways share shikimic acid and acetate biosynthetic pathways for the production of metabolites. Pathway regulates by R2R3-MYB transcription factors and miR858a, which putatively targets this transcription factor. This miRNA also regulates several factors that control the downstream pathway of flavonoid biosynthesis. (Sharma et al., 2016). miR156 interact with SPL9 protein and control metabolic flux in the flavonoid biosynthetic pathway. miR169a, miR169l, and miR319h are reportedly significant roles in control the biosynthesis of flavor compounds differentially expressed in tissues of the tea plant (Li et al., 2021) (Figure 2).

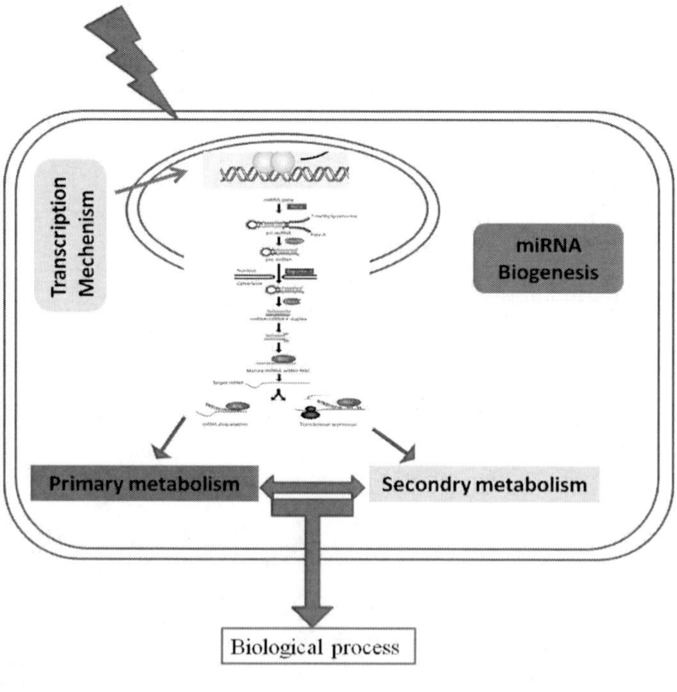

Figure 2. Role of miRNA in the biological process by regulation of primary and secondary metabolism.

HOW TO USE MIRNA FOR METABOLIC ENGINEERING IN PLANT

Metabolic engineering increases useful metabolites and reduces the extent of toxic metabolites, and synthesize novel metabolites in plants by modulating metabolic pathways (Sabzehzari and Naghavi 2019). Due to complexity in secondary metabolic pathway an adequate tool for mapping pathway and its regulating elements, miRNAs proved a better understanding and manipulation of biosynthetic pathways (Bulgakov and Avramenko, 2015). So, it follows different phases to production in its evaluation of plants for metabolite production. As plant secondary metabolites are widely used in medicine, industry, insecticides, drugs, dyes and agriculture toxins, and bio-warfare-terrorism, respectively. In natural conditions, production is too low. So we need to follow Metabolic Engineering in the plant. The use of RNAi is utilized to silence unwanted genes and reverse will over express genes of interest by using short interfering RNAs (siRNAs) and microRNAs (miRNAs (Hannon 2002). There is several information regarding the identification of miRNAs in plants. A few researchers hamper the miRNA for the regulation of secondary metabolites biosynthesis. Some examples are Alkaloids biosynthesis in opium poppy (Boke et al., 2015) Nicotine biosynthesis in Tobacco (F. Li et al., 2015) Taxol biosynthesis in Yew (Hao et al., 2012) Ginsenosides biosynthesis in ginseng (Jung et al., 2018; Wang et al., 2019). Picrosides biosynthesis in Indian gentian (Vashisht et al., 2015) Terpenoid indole alkaloids biosynthesis in *Catharanthus roseus* (Shen et al., 2017) Flavonoid biosynthesis in Clary Sage (Legrand et al., 2010) Tocopherols biosynthesis in Sunflower (Fritsche, Wang, and Jung 2017), (Pani et al., 2011), Terpenoid biosynthesis in cockleburr (Fan et al., 2015), PSMs biosynthesis in Gentianaceae, Terpenes biosynthesis in *Ferula gummosa*, Camalexin and Glucosinolate biosynthesis in Thale Cress (Om P. Gupta et al., 2017).

Conclusion

Since the discovery of miRNA, it increased the interest and concerns of a molecular biologist as it plays important role in developmental processes and cellular metabolism. The study of biosynthetic pathways suggests both plants and human are complementary to each other. Identification of and characterizing is important to delineate the functions of miRNAs. Computational approaches with the support of experimentation are contributing well to identification and target prediction. Next-generation sequencing makes a boost in the number of miRNAs in plants and animals. Deep sequencing technologies generate huge data for the discovery of miRNA and provide the earliest step with certain limitations. (Doran and Strauss 2007; Tay et al., 2008; Place et al., 2008). The ability to control multiple genes by a single miRNA can modulate the expression of multiple targets and lead to complexity in characterization. The ability of miRNA to regulate gene expression and control different development processes and cross-kingdom regulation can play an important role in therapeutic uses. The Discovery of miRNA in MAPs has opened a new door to hamper plant-based medicine in a good way. Most common miRNA are found as, miRNA159, miRNA156, miRNA157, miRNA160, miRNA161, miRNA171, and so on in diverse plant species(Pirrò et al., 2019). The miRNA study is a relatively recent field of study and regulation of secondary metabolism by miRNA in plants will assist to decipher the production in controlled systems and their use above discussed industries making these areas very attractive. miRNAs derived regulation of human health or targeting genes associated with diseases are another emerging area in herbal medicine. Such studies would help metabolic engineering of the complete biosynthetic pathway for generating novel phytochemicals or for producing desired combinations of such secondary metabolites.

References

Akmal, Mohd, Mirza S. Baig, and Jawaid A. Khan. 2017. "Suppression of Cotton Leaf Curl Disease Symptoms in Gossypium Hirsutum through over Expression of Host-Encoded MiRNAs." *Journal of Biotechnology* 263 (December): 21–29. https://doi.org/10.1016/j.jbiotec.2017.10.003.

Akter, Arzuba, Md Muzahidul Islam, Shakhinur Islam Mondal, Zabed Mahmud, Nurnabi Azad Jewel, Sabiha Ferdous, Md Ruhul Amin, and Md Mahfuzur Rahman. 2014. "Computational Identification of MiRNA and Targets from Expressed Sequence Tags of Coffee (Coffea Arabica)." *Saudi Journal of Biological Sciences* 21 (1): 3. https://doi.org/10.1016/j.sjbs.2013.04.007.

Axtell, Michael J., Jakub O. Westholm, and Eric C. Lai. 2011. "Vive La Différence: Biogenesis and Evolution of MicroRNAs in Plants and Animals." *Genome Biology* 12 (4): 221. https://doi.org/10.1186/gb-2011-12-4-221.

Bartel, David P. 2004. "MicroRNAs: Genomics, Biogenesis, Mechanism, and Function." *Cell* 116 (2): 281–97. https://doi.org/10.1016/S0092-8674(04)00045-5.

Boke, Hatice, Esma Ozhuner, Mine Turktas, Iskender Parmaksiz, Sebahattin Ozcan, and Turgay Unver. 2015. "Regulation of the Alkaloid Biosynthesis by MiRNA in Opium Poppy." *Plant Biotechnology Journal* 13 (3): 409–20. https://doi.org/10.1111/pbi.12346.

Broughton, James P., Michael T. Lovci, Jessica L. Huang, Gene W. Yeo, and Amy E. Pasquinelli. 2016. "Pairing beyond the Seed Supports MicroRNA Targeting Specificity." *Molecular Cell* 64 (2): 320–33. https://doi.org/10.1016/j.molcel.2016.09.004.

Bulgakov, Victor P., and Tatiana V. Avramenko. 2015. "New Opportunities for the Regulation of Secondary Metabolism in Plants: Focus on MicroRNAs." *Biotechnology Letters* 37 (9): 1719–27. https://doi.org/10.1007/s10529-015-1863-8.

Carrington, James C., and Victor Ambros. 2003. "Role of MicroRNAs in Plant and Animal Development." *Science (New York, N.Y.)* 301 (5631): 336–38. https://doi.org/10.1126/science.1085242.

Doran, Jonathon, and William M. Strauss. 2007. "Bio-Informatic Trends for the Determination of MiRNA-Target Interactions in Mammals." *DNA and Cell Biology* 26 (5): 353–60. https://doi.org/10.1089/dna.2006.0546.

Fan, Rongyan, Yuanjun Li, Changfu Li, and Yansheng Zhang. 2015. "Differential MicroRNA Analysis of Glandular Trichomes and Young Leaves in Xanthium Strumarium L. Reveals Their Putative Roles in Regulating Terpenoid Biosynthesis." *PLOS ONE* 10 (9): e0139002. https://doi.org/10.1371/journal.pone.0139002.

Fritsche, Steffi, Xingxing Wang, and Christian Jung. 2017. "Recent Advances in Our Understanding of Tocopherol Biosynthesis in Plants: An Overview of Key Genes, Functions, and Breeding of Vitamin E Improved Crops." *Antioxidants* 6 (4). https://doi.org/10.3390/antiox6040099.

Ghorbanpour, Mansour, Javad Hadian, Shahab Nikabadi, and Ajit Varma. 2017. "Importance of Medicinal and Aromatic Plants in Human Life." In *Medicinal Plants and Environmental Challenges*, edited by Mansour Ghorbanpour and Ajit Varma, 1–23. Cham: Springer International Publishing. https://doi.org/10.1007/978-3-319-68717-9_1.

Gou, Jin-Ying, Felipe F. Felippes, Chang-Jun Liu, Detlef Weigel, and Jia-Wei Wang. 2011. "Negative Regulation of Anthocyanin Biosynthesis in Arabidopsis by a MiR156-Targeted SPL Transcription Factor [W][OA]." *The Plant Cell* 23 (4): 1512–22. https://doi.org/10.1105/tpc.111.084525.

Gupta, Om P., Suhas G. Karkute, Sagar Banerjee, Nand L. Meena, and Anil Dahuja. 2017. "Contemporary Understanding of MiRNA-Based Regulation of Secondary Metabolites Biosynthesis in Plants." *Frontiers in Plant Science* 8 (March). https://doi.org/10.3389/fpls.2017.00374.

Gupta, Om Prakash, Nand Lal Meena, Indu Sharma, and Pradeep Sharma. 2014. "Differential Regulation of MicroRNAs in Response to Osmotic, Salt and Cold Stresses in Wheat." *Molecular Biology Reports* 41 (7): 4623–29. https://doi.org/10.1007/s11033-014-3333-0.

Ha, Minju, and V. Narry Kim. 2014. "Regulation of MicroRNA Biogenesis." *Nature Reviews Molecular Cell Biology* 15 (8): 509–24. https://doi.org/10.1038/nrm3838.

Hannon, Gregory J. 2002. "RNA Interference." *Nature* 418 (6894): 244–51. https://doi.org/10.1038/418244a.

Hao, Da-Cheng, Ling Yang, Pei-Gen Xiao, and Ming Liu. 2012. "Identification of Taxus MicroRNAs and Their Targets with High-Throughput Sequencing and Degradome Analysis." *Physiologia Plantarum* 146 (4): 388–403. https://doi.org/10.1111/j.1399-3054.2012.01668.x.

Huang, Yong, Quan Zou, Sheng Peng Wang, Shun Ming Tang, Guo Zheng Zhang, and Xing Jia Shen. 2011. "The Discovery Approaches and Detection Methods of MicroRNAs." *Molecular Biology Reports* 38 (6): 4125–35. https://doi.org/10.1007/s11033-010-0532-1.

Hussein, Rehab A., and Amira A. El-Anssary. 2018. *Plants Secondary Metabolites: The Key Drivers of the Pharmacological Actions of Medicinal Plants. Herbal Medicine*. IntechOpen. https://doi.org/ 10.5772/ intechopen.76139.

Jung, Inuk, Hyejin Kang, Jang Uk Kim, Hyeonsook Chang, Sun Kim, and Woosuk Jung. 2018. "The MRNA and MiRNA Transcriptomic Landscape of Panax Ginseng under the High Ambient Temperature." *BMC Systems Biology* 12 (2): 27. https://doi.org/10.1186/s12918-018-0548-z.

Kim, Sang Woo, Zhihua Li, Patrick S. Moore, A.Paula Monaghan, Yuan Chang, Mark Nichols, and Bino John. 2010. "A Sensitive Non-Radioactive Northern Blot Method to Detect Small RNAs." *Nucleic Acids Research* 38 (7): e98–e98. https://doi.org/10.1093/nar/gkp 1235.

Kozomara, Ana, Maria Birgaoanu, and Sam Griffiths-Jones. 2019. "MiRBase: From MicroRNA Sequences to Function." *Nucleic Acids Research* 47 (D1): D155–62. https://doi.org/10.1093/nar/gky1141.

Lee, Rosalind C., Rhonda L. Feinbaum, and Victor Ambros. 1993. "The C. Elegans Heterochronic Gene Lin-4 Encodes Small RNAs with Antisense Complementarity to Lin-14." *Cell* 75 (5): 843–54. https://doi.org/10.1016/0092-8674(93)90529-Y.

Legrand, Sylvain, Nadine Valot, Florence Nicolé, Sandrine Moja, Sylvie Baudino, Frédéric Jullien, Jean-Louis Magnard, Jean-Claude Caissard, and Laurent Legendre. 2010. "One-Step Identification of Conserved MiRNAs, Their Targets, Potential Transcription Factors and Effector Genes of Complete Secondary Metabolism Pathways after 454 Pyrosequencing of Calyx CDNAs from the *Labiate Salvia Sclarea* L." *Gene* 450 (1–2): 55–62. https://doi.org/10.1016/j.gene.2009.10.004.

Li, Fangfang, Weidi Wang, Nan Zhao, Bingguang Xiao, Peijian Cao, Xingfu Wu, Chuyu Ye, et al., 2015. "Regulation of Nicotine Biosynthesis by an Endogenous Target Mimicry of MicroRNA in Tobacco1 [OPEN]." *Plant Physiology* 169 (2): 1062–71. https://doi.org/10.1104/pp.15.00649.

Li, Hui, Qingqing Lin, Meilin Yan, Mingle Wang, Pu Wang, Hua Zhao, Yu Wang, Dejiang Ni, and Fei Guo. 2021. "Relationship between Secondary Metabolism and MiRNA for Important Flavor Compounds in Different Tissues of Tea Plant (*Camellia Sinensis*) As Revealed by Genome-Wide MiRNA Analysis." *Journal of Agricultural and Food Chemistry* 69 (6): 2001–12. https://doi.org/10.1021/acs.jafc.0c07440.

Li, Wei, and Kangcheng Ruan. 2009. "MicroRNA Detection by Microarray." *Analytical and Bioanalytical Chemistry* 394 (4): 1117–24. https://doi.org/10.1007/s00216-008-2570-2.

Liu, Bing, Jiuyong Li, and Murray J. Cairns. 2014. "Identifying MiRNAs, Targets and Functions." *Briefings in Bioinformatics* 15 (1): 1–19. https://doi.org/10.1093/bib/bbs075.

Makarova, Julia A., Maxim U. Shkurnikov, Daniel Wicklein, Tobias Lange, Timur R. Samatov, Andrey A. Turchinovich, and Alexander G. Tonevitsky. 2016. "Intracellular and Extracellular MicroRNA: An Update on Localization and Biological Role." *Progress in Histochemistry and Cytochemistry* 51 (3): 33–49. https://doi.org/10.1016/j.proghi.2016.06.001.

Marcela, Vargas-Hernández, Vázquez-Marrufo Gerardo, Aguilar-Ruiz Carlos Agustín, González-Márquez Marco Antonio, Rocha Oscar, Cerna-Pantoja Diego, and Andrés Cruz-Hernández. 2019. *MicroRNAs Associated with Secondary Metabolites Production. Plant Physiological Aspects of Phenolic Compounds*. IntechOpen. https://doi.org/10.5772/intechopen.83804.

Meng, Yijun, Lingfeng Gou, Dijun Chen, Chuanzao Mao, Yongfeng Jin, Ping Wu, and Ming Chen. 2011. "PmiRKB: A Plant MicroRNA Knowledge Base." *Nucleic Acids Research* 39 (Database issue): D181-187. https://doi.org/10.1093/nar/gkq721.

Millar, Anthony A. 2020. "The Function of MiRNAs in Plants." *Plants* 9 (2). https://doi.org/10.3390/plants9020198.

Mishra, Anand, Sanchita Gupta, Pankhuri Gupta, Sunita Singh Dhawan, and Raj Kishori Lal. 2020. "In Silico Identification of MiRNA and Targets from *Chrysopogon Zizanioides* (L.) Roberty with Functional Validation from Leaf and Root Tissues." *Applied Biochemistry and Biotechnology*, July. https://doi.org/10.1007/s12010-020-03381-z.

Niu, Yanqin, Limin Zhang, Huiling Qiu, Yike Wu, Zhiwei Wang, Yujia Zai, Lin Liu, Junle Qu, Kang Kang, and Deming Gou. 2015. "An Improved Method for Detecting Circulating MicroRNAs with S-Poly(T) Plus Real-Time PCR." *Scientific Reports* 5 (1): 1–10. https://doi.org/10.1038/srep15100.

O'Brien, Jacob, Heyam Hayder, Yara Zayed, and Chun Peng. 2018. "Overview of MicroRNA Biogenesis, Mechanisms of Actions, and Circulation." *Frontiers in Endocrinology* 9. https://doi.org/10.3389/fendo.2018.00402.

Pandey, Vibha, Waquar Akhter Ansari, Pratibha Misra, and Neelam Atri. 2017. "*Withania Somnifera*: Advances and Implementation of

Molecular and Tissue Culture Techniques to Enhance Its Application." *Frontiers in Plant Science* 8: 1390. https://doi.org/10.3389/fpls.2017.01390.

Pani, Alok, and Rajani Kanta Mahapatra. 2013. "Computational Identification of MicroRNAs and Their Targets in Catharanthus Roseus Expressed Sequence Tags." *Genomics Data* 1 (August): 2–6. https://doi.org/10.1016/j.gdata.2013.06.001.

Pani, Alok, Rajani Kanta Mahapatra, Niranjan Behera, and Pradeep Kumar Naik. 2011. "Computational Identification of Sweet Wormwood (*Artemisia Annua*) MicroRNA and Their MRNA Targets." *Genomics, Proteomics & Bioinformatics* 9 (6): 200. https://doi.org/10.1016/S1672-0229(11)60023-5.

Patel, Maulikkumar, Naman Mangukia, Neha Jha, Harshida Gadhavi, Kanisha Shah, Saumya Patel, Archana Mankad, Himanshu Pandya, and Rakesh Rawal. 2019. "Computational Identification of MiRNA and Their Cross Kingdom Targets from Expressed Sequence Tags of *Ocimum Basilicum*." *Molecular Biology Reports* 46 (3): 2979–95. https://doi.org/10.1007/s11033-019-04759-x.

Patel, Maulikkumar, Shanaya Patel, Naman Mangukia, Saumya Patel, Archana Mankad, Himanshu Pandya, and Rakesh Rawal. 2019. "*Ocimum Basilicum* MiRNOME Revisited: A Cross Kingdom Approach." *Genomics* 111 (4): 772–85. https://doi.org/10.1016/j.ygeno.2018.04.016.

Pirrò, Stefano, Ivana Matic, Arianna Guidi, Letizia Zanella, Angelo Gismondi, Rosella Cicconi, Roberta Bernardini, et al., 2019. "Identification of MicroRNAs and Relative Target Genes in *Moringa Oleifera* Leaf and Callus." *Scientific Reports* 9 (1): 15145. https://doi.org/10.1038/s41598-019-51100-4.

Place, Robert F., Long-Cheng Li, Deepa Pookot, Emily J. Noonan, and Rajvir Dahiya. 2008. "MicroRNA-373 Induces Expression of Genes with Complementary Promoter Sequences." *Proceedings of the National Academy of Sciences of the United States of America* 105 (5): 1608–13. https://doi.org/10.1073/pnas.0707594105.

Planell-Saguer, Mariàngels de, and Maria Celina Rodicio. 2013. "Detection Methods for MicroRNAs in Clinic Practice." *Clinical Biochemistry*, miRNAs in Clinical Practice, 46 (10): 869–78. https://doi.org/10.1016/j.clinbiochem.2013.02.017.

Prakash, Pravin, Dolly Ghosliya, and Vikrant Gupta. 2015. "Identification of Conserved and Novel MicroRNAs in Catharanthus Roseus by Deep Sequencing and Computational Prediction of Their Potential Targets." *Gene* 554 (2): 181–95. https://doi.org/10.1016/j.gene.2014.10.046.

Ren, Guodong, and Bin Yu. 2012. "Critical Roles of RNA-Binding Proteins in MiRNA Biogenesis in Arabidopsis." *RNA Biology* 9 (12): 1424–28. https://doi.org/10.4161/rna.22740.

Rogers, K., and X. Chen. 2012. "MicroRNA Biogenesis and Turnover in Plants." *Cold Spring Harbor Symposia on Quantitative Biology* 77 (January): 183–94. https://doi.org/10.1101/sqb.2013.77.014530.

Sabzehzari, M., and M. R. Naghavi. 2019. "Phyto-MiRNAs-Based Regulation of Metabolites Biosynthesis in Medicinal Plants." *Gene* 682 (January): 13–24. https://doi.org/10.1016/j.gene.2018.09.049.

Sanchita, null, Ruchi Singh, Anand Mishra, Sunita S. Dhawan, Pramod A. Shirke, Madan M. Gupta, and Ashok Sharma. 2015. "Physiological Performance, Secondary Metabolite and Expression Profiling of Genes Associated with Drought Tolerance in *Withania Somnifera*." *Protoplasma* 252 (6): 1439–50. https://doi.org/10.1007/s00709-015-0771-z.

Sharma, Deepika, Manish Tiwari, Ashutosh Pandey, Chitra Bhatia, Ashish Sharma, and Prabodh Kumar Trivedi. 2016. "MicroRNA858 Is a Potential Regulator of Phenylpropanoid Pathway and Plant Development." *Plant Physiology* 171 (2): 944–59. https://doi.org/10.1104/pp.15.01831.

Shen, Ethan M., Sanjay K. Singh, Jayadri S. Ghosh, Barunava Patra, Priyanka Paul, Ling Yuan, and Sitakanta Pattanaik. 2017. "The MiRNAome of Catharanthus Roseus : Identification, Expression Analysis, and Potential Roles of MicroRNAs in Regulation of

Terpenoid Indole Alkaloid Biosynthesis." *Scientific Reports* 7 (1): 43027. https://doi.org/10.1038/srep43027.

Shen, Yanting, Fei Tian, Zhenzhu Chen, Rui Li, Qinyu Ge, and Zuhong Lu. 2015. "Amplification-Based Method for MicroRNA Detection." *Biosensors and Bioelectronics* 71 (September): 322–31. https://doi.org/10.1016/j.bios.2015.04.057.

Singh, Noopur, and Ashok Sharma. 2014. "In-Silico Identification of MiRNAs and Their Regulating Target Functions in *Ocimum Basilicum*." *Gene* 552 (2): 277–82. https://doi.org/10.1016/j.gene.2014.09.040.

Singh, Noopur, and Ashok Sharma. 2017. "Turmeric (Curcuma Longa): MiRNAs and Their Regulating Targets Are Involved in Development and Secondary Metabolite Pathways." *Comptes Rendus Biologies* 340 (11–12): 481–91. https://doi.org/10.1016/j.crvi.2017.09.009.

Singh, Noopur, Swati Srivastava, Ajit K. Shasany, and Ashok Sharma. 2016. "Identification of MiRNAs and Their Targets Involved in the Secondary Metabolic Pathways of Mentha Spp." *Computational Biology and Chemistry* 64: 154–62. https://doi.org/10.1016/j.compbiolchem.2016.06.004.

Srivastava, Swati, null Sanchita, Ruchi Singh, Gaurava Srivastava, and Ashok Sharma. 2018. "Comparative Study of Withanolide Biosynthesis-Related MiRNAs in Root and Leaf Tissues of Withania Somnifera." *Applied Biochemistry and Biotechnology* 185 (4): 1145–59. https://doi.org/10.1007/s12010-018-2702-x.

Szcześniak, Michał W., and Izabela Makałowska. 2014. "MiRNEST 2.0: A Database of Plant and Animal MicroRNAs." *Nucleic Acids Research* 42 (Database issue): D74–77. https://doi.org/10.1093/nar/gkt1156.

Tay, Yvonne, Jinqiu Zhang, Andrew M. Thomson, Bing Lim, and Isidore Rigoutsos. 2008. "MicroRNAs to Nanog, Oct4 and Sox2 Coding Regions Modulate Embryonic Stem Cell Differentiation." *Nature* 455 (7216): 1124–28. https://doi.org/10.1038/nature07299.

Teoh, Eng Soon. 2015. "Secondary Metabolites of Plants." *Medicinal Orchids of Asia*, November, 59–73. https://doi.org/10.1007/978-3-319-24274-3_5.

Tian, Tian, Jiaqi Wang, and Xiang Zhou. 2015. "A Review: MicroRNA Detection Methods." *Organic & Biomolecular Chemistry* 13 (8): 2226–38. https://doi.org/10.1039/C4OB02104E.

Várallyay, Éva, József Burgyán, and Zoltán Havelda. 2008. "MicroRNA Detection by Northern Blotting Using Locked Nucleic Acid Probes." *Nature Protocols* 3 (2): 190–96. https://doi.org/10.1038/nprot.2007.528.

Vashisht, Ira, Prashant Mishra, Tarun Pal, Sreekrishna Chanumolu, Tiratha Raj Singh, and Rajinder Singh Chauhan. 2015. "Mining NGS Transcriptomes for MiRNAs and Dissecting Their Role in Regulating Growth, Development, and Secondary Metabolites Production in Different Organs of a Medicinal Herb, *Picrorhiza Kurroa*." *Planta* 241 (5): 1255–68. https://doi.org/10.1007/s00425-015-2255-y.

Verpoorte, Rob. 1998. "Exploration of Nature's Chemodiversity: The Role of Secondary Metabolites as Leads in Drug Development." *Drug Discovery Today* 3 (5): 232–38. https://doi.org/10.1016/S1359-6446(97)01167-7.

Wang, Yingfang, Mengyuan Peng, Wenjuan Wang, Yanlin Chen, Zhihua He, Jingjing Cao, Zhiyun Lin, Zemin Yang, Mengjuan Gong, and Yongqin Yin. 2019. "Verification of MiRNAs in Ginseng Decoction by High-Throughput Sequencing and Quantitative Real-Time PCR." *Heliyon* 5 (4): e01418. https://doi.org/10.1016/j.heliyon.2019.e01418.

Ye, Jiawei, Mingcheng Xu, Xueke Tian, Sheng Cai, and Su Zeng. 2019. "Research Advances in the Detection of MiRNA." *Journal of Pharmaceutical Analysis, Special Issue: Advances in Pharmaceutical Analysis* 2018, 9 (4): 217–26. https://doi.org/10.1016/j.jpha.2019.05.004.

Yi, Xin, Zhenhai Zhang, Yi Ling, Wenying Xu, and Zhen Su. 2015. "PNRD: A Plant Non-Coding RNA Database." *Nucleic Acids Research* 43 (D1). https://doi.org/10.1093/nar/gku1162.

Yu, Dongliang, Jiangjie Lu, Weishan Shao, Xiaoxia Ma, Tian Xie, Hidetaka Ito, Tingzhang Wang, Min Xu, Huizhong Wang, and Yijun Meng. 2019. "MepmiRDB: A Medicinal Plant MicroRNA Database." *Database* 2019 (baz070). https://doi.org/10.1093/database/baz070.

Zhang, Zhenhai, Jingyin Yu, Daofeng Li, Zuyong Zhang, Fengxia Liu, Xin Zhou, Tao Wang, Yi Ling, and Zhen Su. 2010. "PMRD: Plant MicroRNA Database." *Nucleic Acids Research* 38 (Database issue): D806-813. https://doi.org/10.1093/nar/gkp818.

Zielezinski, Andrzej, Jakub Dolata, Sylwia Alaba, Katarzyna Kruszka, Andrzej Pacak, Aleksandra Swida-Barteczka, Katarzyna Knop, et al., 2015. "MirEX 2.0 - an Integrated Environment for Expression Profiling of Plant MicroRNAs." *BMC Plant Biology* 15 (June). https://doi.org/10.1186/s12870-015-0533-2.

About the Editors

Abhishek Sharma is an Assistant Professor and Principal Investigator at C. G. Bhakta Institute of Biotechnology, Uka Tarsadia University, Gujarat, India. He is a plant metabolic engineering expert. He completed his MSc in Biotechnology as a University Topper and Gold Medallist from Jaipur National University, Jaipur, and PhD from Jawaharlal Nehru University, New Delhi - CSIR Central Institute of Medicinal and Aromatic Plants, Lucknow as DST-INSPIRE Fellow. Dr. Abhishek has more than 8 years of experience in teaching and research with a specialization in transgenic plant development and has developed genetically engineered transgenic plants of medicinal importance *Catharanthus roseus* and *Bacopa monnieri.* Dr. Abhishek's lab is working mainly on genomics, metabolomics, antifungal transgenic plant production, and biotic - abiotic stress studies in plants of high medicinal importance. Dr. Abhishek has published more than 15 research articles and 13 book chapters in peer-reviewed international journals and books of international publication houses viz. Elsevier, Springer and Taylor and Francis. Presently Dr. Abhishek has research grants of more than 56 Lakhs as PI and Co-PI from Centre and State Government Science Funding Agencies like Department of Science and Technology, Science and Engineering Board, Government of India (DST SERB Start-up Research Grant), and Gujarat State Biotechnology Mission, Government of Gujarat (Research Support Scheme).

Dhruti Amin is an Assistant Professor at C. G. Bhakta Institute of Biotechnology, Uka Tarsadia University, Gujarat, India, since 2017. She completed her education (BSc, MSc, PhD, Biotechnology) from Sardar Patel University, Gujarat. She has more than 10 years of research and academic experience and 2 research scholars are working under her guidance. She has experience in the field of plant biotechnology, plant-microbe interaction, and antimicrobial activity of medicinal plants. From

2011- 2016, she worked as Project Fellow for five years in BASF India Ltd., sponsored research project. She has published more than 20 research papers and book chapters and one book in peer-reviewed international publications. She has presented research work, delivered lectures, and participated in several academic events, technical seminars, and symposiums at various national and international conferences. She has received a research grant as PI and Co-PI from Uka Tarsadia University and GSBTM. She is actively involved as an editor and reviewer in various national and international journals.

INDEX

A

B

C

G

H

N

O

P

R

S

T